AF443451

MULTIHOP WIRELESS NETWORKS

Wiley Series on Wireless Communications and Mobile Computing

Series Editors: Dr Xuemin (Sherman) Shen, *University of Waterloo, Canada*
Dr Yi Pan, *Georgia State University, USA*

The "Wiley Series on Wireless Communications and Mobile Computing" is a series of comprehensive, practical and timely books on wireless communication and network systems. The series focuses on topics ranging from wireless communication and coding theory to wireless applications and pervasive computing. The books provide engineers and other technical professionals, researchers, educators, and advanced students in these fields with invaluable insight into the latest developments and cutting-edge research.

Other titles in the series:

Misic and Misic, *Wireless Personal Area Networks: Performance, Interconnection, and Security with IEEE 802.15.4*, January 2008, 978-0-470-51847-2

Takagi and Walke, *Spectrum Requirement Planning in Wireless Communications: Model and Methodology for IMT-Advanced*, April 2008, 978-0-470-98647-9

Pérez-Fontán and Espiñeira, *Modeling the Wireless Propagation Channel: A simulation approach with MATLAB®*, August 2008, 978-0-470-72785-0

Ippolito, *Satellite Communications Systems Engineering: Atmospheric Effects, Satellite Link Design and System Performance*, August 2008, 978-0-470-72527-6

Lin and Sou, *Charging for Mobile All-IP Telecommunications*, September 2008, 978-0-470-77565-3

Myung and Goodman, *Single Carrier FDMA: A New Air Interface for Long Term Evolution*, October 2008, 978-0-470-72449-1

Wang, Kondi, Luthra and Ci, *4G Wireless Video Communications*, April 2009, 978-0-470-77307-9

Cai, Shen and Mark, *Multimedia Services in Wireless Internet: Modeling and Analysis*, June 2009, 978-0-470-77065-8

Stojmenovic, *Wireless Sensor and Actuator Networks: Algorithms and Protocols for Scalable Coordination and Data Communication*, February 2010, 978-0-470-17082-3

Liu and Weiss, *Wideband Beamforming: Concepts and Techniques*, March 2010, 978-0-470-71392-1

Riccharia and Westbrook, *Satellite Systems for Personal Applications: Concepts and Technology*, July 2010, 978-0-470-71428-7

Qian, Muller and Chen, *Security in Wireless Networks and Systems*, February 2013, 978-0-470-512128

MULTIHOP WIRELESS NETWORKS

Opportunistic Routing

Kai Zeng
University of California, Davis, USA

Wenjing Lou
Worcester Polytechnic Institute, USA

Ming Li
Worcester Polytechnic Institute, USA

WILEY

A John Wiley & Sons, Ltd., Publication

Library of Congress Cataloging-in-Publication Data

Zeng, Kai.
  Multihop wireless networks : opportunistic routing / Kai Zeng, Wenjing Lou, Ming Li.
    p. cm.
  Includes bibliographical references and index.
  ISBN 978-0-470-66617-3 (hardback)
  1. Ad hoc networks (Computer networks) 2. Radio relay systems. I. Lou, Wenjing. II. Li, Ming, 1985- III. Title.
  TK5105.77.Z46 2011
  621.387'82–22
                                                          2011007718

A catalogue record for this book is available from the British Library.

Print ISBN: 978-0-470-66617-3
ePDF ISBN: 978-1-119-97361-4
oBook ISBN: 978-1-119-97360-7
ePub ISBN: 978-1-119-97429-1
eMobi ISBN: 978-1-119-97430-7

Typeset in 10/12 Times by Laserwords Private Limited, Chennai, India.

Contents

About the Series Editors   xi

Preface   xiii

List of Abbreviations   xvii

1 Introduction   **1**
1.1   Multihop wireless networks   1
1.2   Routing challenges in MWNs   3
1.3   Routing techniques in MWNs   4
     1.3.1   *Traditional routing*   4
     1.3.2   *Opportunistic routing*   7
1.4   Related work   9
     1.4.1   *Opportunistic techniques*   9
     1.4.2   *Network coding*   11
     1.4.3   *Opportunistic forwarding in opportunistic networks*   13
     1.4.4   *Geographic routing*   14
     1.4.5   *Multirate routing*   14
     1.4.6   *Energy-aware routing*   15
     1.4.7   *Capacity of MWNs*   15
     1.4.8   *Link-quality measurement*   16
1.5   Book contribution   16
1.6   System model and assumptions   20
   References   22

2 Taxonomy of opportunistic routing: principles and behaviors   **27**
2.1   EPA generalization   28
2.2   Principles of local behavior of GOR   28
     2.2.1   *EPA strictly increasing property*   28
     2.2.2   *Relay priority rule*   29
     2.2.3   *Containing property of feasible candidate set*   30
     2.2.4   *Concavity of maximum EPA*   32
     2.2.5   *Reliability increasing property*   32

2.3	Least cost opportunistic routing	33
	2.3.1 *Expected opportunistic transmission count (EOTX)*	33
	2.3.2 *End-to-end cost of opportunistic routing*	34
	2.3.3 *Properties of LCOR*	34
	2.3.4 *Dijkstra-based algorithm*	36
	2.3.5 *Bellman – Ford-based algorithm*	37
2.4	Conclusions	38
	References	38

3 Energy efficiency of geographic opportunistic routing **39**

3.1	EGOR problem formulation	40
	3.1.1 *Energy consumption model*	40
	3.1.2 *Tradeoff between EPA and energy consumption*	41
3.2	Efficient localized node-selection algorithms	42
	3.2.1 *Reformulate the node-selection optimization problem*	42
	3.2.2 *Efficient node-selection algorithms*	43
3.3	Energy-efficient geographic opportunistic routing	45
3.4	Performance evaluation	47
	3.4.1 *Simulation setup*	48
	3.4.2 *Simulation results and analysis*	49
3.5	Conclusion	59
	References	59

4 Capacity of multirate opportunistic routing **61**

4.1	Computing throughput bound of OR	62
	4.1.1 *Transmission interference and conflict*	62
	4.1.2 *Concurrent transmission sets*	63
	4.1.3 *Effective forwarding rate*	64
	4.1.4 *Lower bound of end-to-end throughput of OR*	65
	4.1.5 *Maximum end-to-end throughput of OR*	67
	4.1.6 *Multi-flow generalization*	69
4.2	Impact of transmission rate and forwarding strategy on throughput	69
4.3	Rate and candidate selection schemes	70
	4.3.1 *Least medium time opportunistic routing*	71
	4.3.2 *Per-hop greedy: most advancement per unit time*	75
4.4	Performance evaluation	76
	4.4.1 *Simulation setup*	77
	4.4.2 *Impact of source–destination distances*	77
	4.4.3 *Impact of forwarding candidate number*	82
	4.4.4 *Impact of node density*	84
4.5	Conclusion	85
	References	87

5 Multiradio multichannel opportunistic routing **89**

5.1	Introduction	90

	5.2	System model and opportunistic routing primer	91
		5.2.1 Opportunistic routing primer	92
	5.3	Problem formulation	93
		5.3.1 Concurrent transmission sets	93
		5.3.2 Effective forwarding rate	95
		5.3.3 Capacity region of an opportunistic module	96
		5.3.4 Maximum end-to-end throughput in multiradio, multichannel, multihop networks with OR capability	96
	5.4	Forwarding priority scheduling	98
		5.4.1 A scheduling based on LP	99
		5.4.2 A heuristic scheduling	99
	5.5	Performance evaluation	104
		5.5.1 Two scenarios with different link qualities	104
		5.5.2 Simulation of random networks	106
	5.6	Conclusions and future work	108
		References	108
6	**Medium access control for opportunistic routing – candidate coordination**		**111**
	6.1	Existing candidate coordination schemes	112
		6.1.1 GeRaF collision avoidance MAC	112
		6.1.2 ExOR batch-based MAC	114
		6.1.3 Contention-based forwarding (CBF)	115
		6.1.4 Slotted acknowledgment (SA)	117
		6.1.5 Compressed slotted acknowledgment (CSA)	118
	6.2	Design and analysis of FSA	119
		6.2.1 Design of FSA	119
		6.2.2 Analysis	120
		6.2.3 More on channel assessment techniques	122
	6.3	Simulation results and evaluation	122
		6.3.1 Simulation setup	123
		6.3.2 Simulation results and evaluation	125
	6.4	Conclusions	132
		References	132
7	**Integration of opportunistic routing and network coding**		**133**
	7.1	A brief review of MORE	134
	7.2	Mobile content distribution in VANETs	137
		7.2.1 Model and assumptions	138
	7.3	Related works on mobile content distribution in VANETs	140
		7.3.1 Cooperative downloading of general contents in VANETs	140
		7.3.2 Streaming of multimedia content in VANETs	142
	7.4	Background on symbol-level network coding	143
		7.4.1 A brief review of SLNC	143
		7.4.2 Motivation: why VANET content distribution benefits from SLNC	145

	7.5	CodeOn: a cooperative popular content broadcast scheme for VANETs based on SLNC	152
		7.5.1 Design objectives	152
		7.5.2 Design overview	152
		7.5.3 Network coding method	154
		7.5.4 Efficient exchange of content reception status	155
		7.5.5 Distributed relay selection in cooperative PCD	156
		7.5.6 Broadcast content scheduling	159
		7.5.7 Performance evaluation	160
		7.5.8 Simulation results	161
	7.6	CodePlay: a live multimedia streaming scheme for VANETs based on SLNC	171
		7.6.1 Design objectives	172
		7.6.2 Overview of codeplay	172
		7.6.3 LMS using symbol-level network coding	175
		7.6.4 Coordinated and distributed relay selection	176
		7.6.5 Transmission coordination of relays	179
		7.6.6 OLRR: opportunistic LRR scheduling for sparse VANETs	179
		7.6.7 Performance evaluation	181
	7.7	Conclusion	188
		References	190
8		**Multirate geographic opportunistic routing protocol design**	**193**
	8.1	System model	193
	8.2	Impact of transmission rate and forwarding strategy on OR performance	195
		8.2.1 One-hop packet forwarding time of opportunistic routing	196
		8.2.2 Impact of transmission rate	197
		8.2.3 Impact of forwarding strategy	198
		8.2.4 Impact of candidate coordination	198
	8.3	Opportunistic effective one-hop throughput (OEOT)	199
	8.4	Heuristic candidate selection algorithm	200
	8.5	Multirate link-quality measurement	202
	8.6	Performance evaluation	202
		8.6.1 Simulation setup	203
		8.6.2 Simulation results and analysis	204
	8.7	Conclusion	211
		References	211
9		**Opportunistic routing security**	**213**
	9.1	Attack on link quality measurement	213
		9.1.1 Existing link quality measurement mechanisms and vulnerabilities	215
		9.1.2 Performance demonstration	218
		9.1.3 Broadcast-based secure link quality measurement	219

9.2   Attacks on opportunistic coordination protocols   222
         9.2.1   *Attack on implicit-prioritized coordination protocol*   223
         9.2.2   *Attack on explicit-prioritized coordination protocol*   224
         9.2.3   *Attack on slotted ACK*   225
         9.2.4   *Attack on compressed slotted ACK*   226
         9.2.5   *Attack on fast slotted ACK*   226
9.3   Resilience to packet-dropping attack   226
9.4   Conclusion   227
References   228

10 Opportunistic broadcasts in vehicular networks   **231**
10.1   Related works on broadcasts in general MWNs   234
         10.1.1   *Stateful broadcast*   234
         10.1.2   *Stateless broadcast*   234
10.2   Related works on broadcasts in VANETs   235
         10.2.1   *Opportunistic forwarding in VANETs*   235
         10.2.2   *The reliability issue in VANET broadcast*   237
         10.2.3   *Broadcast in partitioned VANETs*   237
10.3   Problem statement   237
         10.3.1   *Model and assumptions*   237
         10.3.2   *Objectives*   238
10.4   Overview of OppCast   239
10.5   OppCast: main design   241
         10.5.1   *Fast-forward dissemination*   241
         10.5.2   *Makeup for reliability*   243
         10.5.3   *Broadcast coordination in OppCast*   246
         10.5.4   *Extension to disconnected VANET*   249
         10.5.5   *Implementation issues*   252
10.6   Parameter optimization   252
         10.6.1   *Optimize the forwarding range*   253
         10.6.2   *Optimal threshold density*   259
10.7   Performance evaluation   260
         10.7.1   *Simulation setup*   260
         10.7.2   *Results for OppCast without extension*   261
         10.7.3   *Results for OppCast with extension*   269
10.8   Conclusion   271
References   271

11 Conclusions and future research   **275**
11.1   Summary   275
11.2   Future research directions   279
References   281

Index   **283**

About the Series Editors

Xuemin (Sherman) Shen (M'97-SM'02) received his B.Sc degree in electrical engineering from Dalian Maritime University, China in 1982, and his M.Sc. and Ph.D. degrees (both in electrical engineering) from Rutgers University, New Jersey, USA, in 1987 and 1990 respectively. He is a Professor and University Research Chair, and the Associate Chair for Graduate Studies, Department of Electrical and Computer Engineering, University of Waterloo, Canada. His research focuses on mobility and resource management in interconnected wireless/wired networks, UWB wireless communications systems, wireless security, and ad hoc and sensor networks. He is a co-author of three books, and has published more than 300 papers and book chapters in wireless communications and networks, control and filtering. Dr. Shen serves as a founding area editor for *IEEE Transactions on Wireless Communications*; Editor-in-Chief for *Peer-to-Peer Networking and Application*; Associate Editor for *IEEE Transactions on Vehicular Technology*; *KICS/IEEE Journal of Communications and Networks, Computer Networks; ACM/Wireless Networks*; and *Wireless Communications and Mobile Computing* (Wiley), etc. He has also served as Guest Editor for *IEEE JSAC, IEEE Wireless Communications*, and *IEEE Communications Magazine*. Dr. Shen received the Excellent Graduate Supervision Award in 2006, and the Outstanding Performance Award in 2004 from the University of Waterloo, the Premier's Research Excellence Award (PREA) in 2003 from the Province of Ontario, Canada, and the Distinguished Performance Award in 2002 from the Faculty of Engineering, University of Waterloo. Dr. Shen is a registered Professional Engineer of Ontario, Canada.

Dr. Yi Pan is the Chair and a Professor in the Department of Computer Science at Georgia State University, USA. Dr. Pan received his B.Eng. and M.Eng. degrees in computer engineering from Tsinghua University, China, in 1982 and 1984, respectively, and his Ph.D. degree in computer science from the University of Pittsburgh, USA, in 1991. Dr. Pan's research interests include parallel and distributed computing, optical networks, wireless networks, and bioinformatics. Dr. Pan has published more than 100 journal papers with over 30 papers published in various IEEE journals. In addition, he has published over 130 papers in refereed conferences (including IPDPS, ICPP, ICDCS, INFOCOM, and GLOBECOM). He has also co-edited over 30 books. Dr. Pan has served as an editor-in-chief or an editorial board member for 15 journals including five IEEE Transactions and has organized many international conferences and workshops. Dr. Pan has delivered over 10 keynote speeches at many international conferences. Dr. Pan is an IEEE Distinguished Speaker (2000–2), a Yamacraw Distinguished Speaker (2002), and a Shell Oil Colloquium Speaker (2002). He is listed in *Men of Achievement, Who's Who in America, Who's Who in American Education, Who's Who in Computational Science and Engineering*, and *Who's Who of Asian Americans*.

Preface

Advances in communication and networking technologies are rapidly making ubiquitous network connectivity a reality. Wireless networks are indispensable for supporting such access anywhere and at any time. Among various types of wireless networks, multihop wireless networks (MWNs) have been attracting increasing attention for decades due to its broad civilian and military applications. Basically, a MWN is a network of nodes connected by wireless communication links. Due to the limited transmission range of the radio, many pairs of nodes in MWNs may not be able to communicate directly, hence they need other intermediate nodes to forward packets for them. Routing in such networks is an important issue and it poses great challenges.

On the one hand, due to its open-air nature, the wireless environment presents great challenges when attempting to ensure good routing performance. The wireless channel is unreliable due to fading and interference, which makes it hard to maintain a quality path between a source and a destination. A node's mobility also incurs frequent topology changes, which bring significant overheads on maintaining and recalculating paths. Furthermore, mobile devices and sensors are usually constrained by battery capacity and communication and computation capability, which imposes limitations on the functionality of routing protocols. On the other hand, the wireless medium possesses inherent unique characteristics, which can be exploited to enhance transmission reliability and routing performance. Opportunistic routing (OR) is one promising technique that takes advantages of the spacial diversity and broadcast nature of the wireless medium to improve the packet-forwarding reliability in multihop wireless networks. It combats the unreliable wireless links by involving multiple neighboring nodes (forwarding candidates) for packet relay. This book studies the properties, energy efficiency, capacity, throughput, protocol design and security issues related to OR in multihop wireless networks.

This book is intended for networking professionals working in wireless networks and communications, who are familiar with the fundamentals of networking and wireless communications. It may also be used as a supplement to graduate courses in wireless networking, mobile computing, and wireless communications.

The contents of each chapter are described as follows.

Chapter 1 presents the case for opportunistic routing and related work. We first introduce the background of multihop wireless networks (mesh networks, sensor networks, mobile ad hoc networks, vehicular networks, etc.). Next, we discuss general wireless multihop routing, including traditional routing (AODV, DSR, etc.), geographic routing, context-based routing and opportunistic routing. We will discuss the motivation of these routing techniques, how they evolved, and their advantages and disadvantages. We will then discuss related opportunistic and collaborative techniques, including cooperative communication, opportunistic scheduling, network coding and multiple access point (AP) collaboration. We will also introduce related issues about opportunistic routing, including capacity studies of multihop wireless networks, multirate routing, energy-efficient routing, and link quality measurement, etc.

Chapter 2 of this book presents the principles and properties of the local behaviors of opportunistic routing (including geographic and link state based opportunistic routing). We will demonstrate how the performance gain changes according to the selection, prioritization, and coordination of forwarding candidates in opportunistic routing. We will discuss in what scenario or situation, opportunistic routing will work or make sense. We will also present two polynomial algorithms to compute least cost opportunistic routing paths (anypath), and introduce properties of least cost anypath.

Chapter 3 of this book studies the energy efficiency of geographic opportunistic routing (GOR). First, we motivate the energy efficiency issues of opportunistic routing in the context of sensor networks. Next, we propose a metric, Expected Packet Advancement (EPA) per unit energy consumption, in order to balance the packet advancement, reliability and energy consumption of GOR. By leveraging the proved principles in Chapter 2, we then propose two efficient algorithms that select a feasible candidate set that maximizes this local metric. We validate our analysis results by simulations and justify the effectiveness of the new metric by comparing the performance of our GOR with those of the existing geographic and opportunistic routing schemes.

Chapter 4 of this book analyzes the throughput bound and capacity of opportunistic routing given the routing strategy, i.e. the forwarding candidates of each node and the corresponding relay priority. We will first give a brief introduction on computing end-to-end throughput of traditional routing and explain why the corresponding methodology cannot directly apply to opportunistic routing, which motivates the proposed framework and methodology. The maximum end-to-end throughput problem is formulated as a maximum-flow linear programming (LP) problem subject to the constraints of forwarding candidate set conflicts. The methodology establishes a theoretical foundation for the evaluation of the performance limits of variants of opportunistic routing protocols and strategies.

Chapter 5 extends the framework proposed in Chapter 4 to deal with dynamic opportunistic routing strategies and multi-radio, multi-channel scenario. An LP approach and a heuristic algorithm is proposed to obtain an opportunistic forwarding strategy scheduling that satisfies a traffic demand vector for a hyperlink, which contains all the outgoing links from a transmitter to all its forwarding candidates.

Chapter 6 of the book investigates the state-of-the-art of the candidate coordination schemes of opportunistic routing at the medium access control layer. These schemes include GeRaF collision avoidance MAC, contention-based forwarding,

ExOR batch-based MAC, slotted acknowledgment (ACK), and compressed slotted ACK. A new scheme, called "fast slotted acknowledgment (FSA)", is described in detail. The scheme adopts a single ACK to confirm the successful reception and suppress other candidates' attempts to forward the data packet with the help of a channel-sensing technique.

Chapter 7 shows how network coding can help ease the candidate coordination in opportunistic routing. It will include an introduction on network coding, how it can help ease the candidate coordination, and on integrating opportunistic routing/broadcast with network coding. A classical work integrating opportunistic routing and network coding, MORE, will be introduced. Recent advancements on integrating symbol-level network coding and opportunistic routing in wireless broadcast are introduced.

Chapter 8 of this book studies the impacts of multirate, candidate selection, prioritization, and coordination on the throughput of GOR under a contention-based medium-access scenario. It will also introduce distributed algorithms to compute the optimal path and transmission rate in multirate opportunistic routing.

Chapter 9 of the book discusses possible attacks on opportunistic routing and countermeasures. We analyze the security vulnerabilities of the existing link quality-measurement mechanisms, and their impacts on opportunistic routing and traditional routing. We present a broadcast-based secure link quality measurement mechanism that prevents a neighboring node from maliciously claiming a higher measurement result. The secure link quality measurement helps to secure the link-state-based opportunistic routing and traditional routing.

Chapter 10 studies the opportunistic broadcast in vehicular networks. Traditional connected dominant set-based broadcast or multi-point relay-based broadcast both suffer from unreliable wireless links in the similar way as that in traditional unicast routing. The broadcast performance can also be improved by introducing the concept of opportunistic forwarding.

Chapter 11 presents the conclusion of this book and discusses some future research topics related to opportunistic routing.

List of Abbreviations

ACK	ACKnowledgement
AODV	Ad hoc On-Demand Distance Vector Routing
AoI	Area of Interest
AP	Access Point
APRP	Accumulated Packet Reception Probability
ARQ	Automatic Repeat reQuest
BACK	Broadcast Acknowledgement
BBT	Broadcast Backoff Timer
CBD	Contention-Based Dissemination
CBF	Contention-Based Forwarding
CBR	Constant Bit Rate
CCA	Clear Channel Assessment
CCTS	Conservative Concurrent Transmission Set
CDS	Connected Dominant Set
CLP	Coordinated Local Push
CR	Communication Range
CSA	Compressed Slotted Acknowledgment
CSMA-CA	Carrier Sensing Multiple Access with Collision Avoidance
CTF	Clear-To-Forward
CTS	Concurrent Transmission Set/Clear-To-Send
DDB	Dynamic Delayed Broadcasting
DIFS	Distributed Inter-Frame Space
DSDV	Destination-Sequenced Distance Vector routing
DSR	Dynamic Source Routing
DSRC	Dedicated Short Range Communications
DTN	Delay/Disruption Tolerant Network
EAR	Expected Advancement Rate
ECC	Error Correction Coding
ED	Energy Detection
EGOR	Energy-efficient Geographic Opportunistic Routing
EMDV	Emergency Message Dissemination for Vehicular environments
EMT	Expected Medium Time

EOT	Expected One-hop Throughput
EOTX	Expected Opportunistic Transmission Count
EPA	Expected Packet Advancement
ETF	Expected number of Transmissions over Forward links
ETT	Expected Transmission Time
ETX	Expected Transmission count
EWMA	Exponentially Weighted Moving Average
ExOR	Extremely Opportunistic Routing
FFD	Fast-forward-Dissemination
FPSP	Forwarding Priority Scheduling Problem
FR	Forwarding Range
FSA	Fast Slotted Acknowledgement
FSR	Fisheye State Routing
GCTS	Greedy Concurrent Transmission Set
GeRaF	Geographic Random Forwarding
GOR	Geographic Opportunistic Routing
GPS	Global Positioning System
GR	Geographic Routing
GTR	Geographic Traditional Routing
IR	Interested Region
ITS	Intelligent Transportation Systems
LCOR	Least Cost Opportunistic Routing
LDPC	Low-Density Parity-Check
LMS	Live Multimedia Streaming
LMTOR	Least Medium Time Opportunistic Routing
LP	Linear Programming
LQM	Link Quality Measurement
LRR	Local Round-Robin Scheduling
MAC	Medium Assess Control
MANET	Mobile Ad hoc NETwork
MCD	Mobile Content Distribution
MCDS	Minimum Connected Dominant Set
MFR	Makeup-For-Reliability
MGOR	Multirate Geographic Opportunistic Routing
MIB	Management Information Base
MIMO	Multiple-Input Multiple-Output
MPR	Multipoint Relay
MRD	Multi-Radio Diversity
MTM	Medium Time Metric
MWN	Multihop Wireless Network
NADV	Normalized ADVance
NAV	Network Allocation Vector
NC	Network Coding
NCDD	Network Coding based Data Dissemination
OBCF	Opportunistic Broadcast Coordination Function
OBU	On-Board Unit

OB-VAN	Opportunistic Broadcast in VANETs
OEOT	Opportunistic Effective One-hop Throughput
OETT	Opportunistic Expected Transmission Time
OFDM	Orthogonal Frequency-Division Multiplexing
OLRR	Opportunistic Local Round-Robin scheduling
OLSR	Optimized Link State Routing
OppCast	Opportunistic Broadcast
OR	Opportunistic Routing
PCD	Popular Content Distribution
PD	Preamble Detection
PDR	Packet Delivery Ratio
PLNC	Packet Level Network Coding
PRP	Packet Reception Probability
PRR	Packet Reception Ratio
QoS	Quality of Service
RCR	Relay Candidate Region
RERR	Route Error
RLNC	Random Linear Network Coding
RREP	Route Reply
RREQ	Route Request
RSSI	Received Signal Strength Indicator
RSU	Road Side Unit
RTER	Reception to Transmission Energy Ratio
RTF	Ready-To-Forward
RTS	Ready-To-Send
SA	Slotted Acknowledgment
SAV	Self Allocation Vector
SB	Smart Broadcast
SIFS	Short Inter-Frame Space
SINR	Signal to Interference and Noise Ratio
SLNC	Symbol Level Network Coding
SLQM	Secure Link Quality Measurement
SNR	Signal to Noise Ratio
TC	Topology Control
TCP	Transport Control Protocol
TORA	Temporally-Ordered Routing Algorithm
TR	Traditional Routing
UDG	Unit Disk Graph
UMB	Urban Multihop Broadcast protocol
VANET	Vehicular Ad Hoc Network
WCETT	Weighted Cumulative Expected Transmission Time
WLAN	Wireless Local Area Network
WM	Warning Message
WMN	Wireless Mesh Network
WSN	Wireless Sensor Network
XOR	Exclusive OR

1

Introduction

This chapter presents the case for opportunistic routing and related work. We will first introduce the background of multihop wireless networks (mesh networks, sensor networks, mobile ad hoc networks, vehicular ad hoc networks, etc.). We then point out the routing challenges in multihop wireless networks. Secondly, we discuss general wireless multihop routing, including traditional routing (AODV, DSR, etc.), geographic routing, context-based routing and opportunistic routing. We will discuss the motivation of these routing techniques, how they evolved, and their advantages and disadvantages. We will then discuss related opportunistic and collaborative techniques, including cooperative communication, opportunistic scheduling, network coding, multiple AP collaboration, etc. This will help to put opportunistic routing in perspective. We will also introduce related issues about opportunistic routing, including capacity studies of multihop wireless networks, multirate routing, energy-efficient routing, and link-quality measurement, etc.

1.1 Multihop wireless networks

A multihop wireless network (MWN) is a network of nodes (e.g. computers) connected by wireless communication links. The links are usually implemented with digital packet radios. Due to the limited transmission range of the radio, many pairs of nodes in MWNs may not be able to communicate directly; hence they may need other intermediate nodes to forward packets for them. Multihop wireless networks have broad military and civilian applications in many critical situations. They have received increasing attention in the past decade due to their broad applications and easy deployment at low cost without relying on existing infrastructure (Akyildiz and Kasimoglu 2004; Akyildiz *et al.* 2002, 2005; Cerpa *et al.* 2001; Chong and

Multihop Wireless Networks: Opportunistic Routing, First Edition.
Kai Zeng, Wenjing Lou and Ming Li.
© 2011 John Wiley & Sons, Ltd. Published 2011 by John Wiley & Sons, Ltd.

Kumar 2003; Estrin *et al*. 2002; Lorincz *et al*. 2004). Different names are used to refer to them in different scenarios.

Mobile ad hoc networks (MANETs)   Generally speaking, a mobile ad hoc network (MANET) is a self-configuring network of mobile devices connected by wireless links. Each device in a MANET is free to move independently in any direction. So the node-to-node connection and network topology will change frequently. The primary challenge in MANETs is continuously to maintain the routing information at each node required to properly route traffic. The applications of MANETs include search-and-rescue operations. Such scenarios are characterized by a lack of installed communications infrastructure because all the equipment might already be destroyed or the region could be too remote. MANETs can also provide communications between autonomous vehicles, aircraft and ground troops in the battlefield where a fixed communication infrastructure is always unavailable and infeasible.

Wireless sensor networks (WSNs)   Wireless sensor networks (WSNs) (Akyildiz *et al*. 2002) are another variant of MWNs. They are normally used to monitor various physical or environmental conditions, such as temperature, sound, vibration, pressure, motion or pollutants. A large-scale WSN typically consists of hundreds or thousands of small and cheap sensor nodes with wireless communication capabilities. These sensor nodes may form local clusters, and reactively or periodically report the sensing results to one or multiple base stations via multihop routing. The sensors are usually powered by batteries with limited capacity. Energy efficiency is therefore the primary concern and key challenge in WSNs. The sensors are typically static but some more powerful sensor nodes may have mobile capability (Hu and Evans 2004).

Wireless-mesh networks (WMNs)   Wireless-mesh networks (WMN) (Akyildiz and Wang 2005) are another type of MWNs, and are usually used to provide the last mile wireless broadband Internet access for the civilian users. They can also support enterprise networking, healthcare and medical systems, and security surveillance systems. They consist of mesh routers and mesh clients, where mesh routers have minimal mobility and form the backbone of WMNs. The integration of WMNs with other networks such as the Internet, cellular networks, IEEE 802.11 WLAN, IEEE 802.15, IEEE 802.16, and sensor networks can be accomplished through the gateway and bridging functions in the mesh routers. Mesh clients can be either stationary or mobile, and can form a client mesh (multihop) network among themselves and with mesh routers. Network capacity in WMNs is an important issue. The capacity of WMNs is affected by many factors such as network topology, node density, traffic patterns, number of radios/channels used for each node, transmission power level, carrier sensing threshold, node mobility, and environment (indoor/outdoor), etc. A clear understanding of the relationship between network capacity and the above factors provides a guideline for protocol development, architecture design, deployment and operation of the network.

Vehicular ad hoc networks (VANETs)   In VANETs, every vehicle communicates with other vehicles (V2V) and with roadside infrastructures (V2I) by means of wireless communication equipment. The most important usage of these networks is to inform other vehicles in emergency situations such as car accidents, urgent braking or traffic jams. In such cases, a vehicle can inform other vehicles by broadcasting safety messages before facing the event. VANETs are a cornerstone of the envisioned Intelligent Transportation Systems (ITS). They will contribute to safer and more efficient roads in the future by providing timely information to drivers and concerned authorities. VANETs are similar to MANETs, but the key difference lies in that in VANETs, vehicles move in an organized fashion rather than randomly. The vehicles are restricted in their range of motion and their mobility can be predicted in the short term, because their movement should obey certain traffic rules.

Compared with traditional single-hop wireless networks, such as cellular networks and local area networks, MWNs have several advantages: 1. coverage extension and connectivity improvement; 2. reducing energy consumption; transmission over multiple short-range wireless links might require less transmission energy than that required over long-range single-hop links; 3. cost efficiency: they avoid wide deployment of cables and can be deployed in a cost efficient way; 4. robustness: in MWNs, multiple paths might exist between a pair of communication nodes, which can be used to increase robustness of the network.

1.2   Routing challenges in MWNs

The purpose of routing is generally to find a path or multiple paths from the source to the destination, maintain or update path(s) when the topology or link quality changes, and forward packets along the path(s). Routing protocol design in MWNs faces a great challenge mainly due to the following facts.

First, the wireless link is unreliable. The properties and quality of a wireless link may vary with the transmission power, transmission rate, distance and path loss between two nodes. Furthermore, channel fading (such as multipath fading and shadowing) results in fluctuations in the received signal strength and therefore intermittent link behavior. The difficulty in managing or controlling the link quality and reliability in wireless networks makes it very hard to find and maintain a good and stable path from the source to the destination.

Second, the wireless medium is broadcast in nature. Transmission on one link may interfere with the transmissions on the neighboring links. This broadcast medium contention brings fundamental constraints on the routing performance, such as throughput and delay. There is inevitable intra-path and inter-path contention due to the broadcast nature of the wireless medium. It is very challenging to achieve optimal routing performance even when there is a single flow, due to the complicate interdependence between the medium contention, route selection, and medium access control. When there are multiple flows (different source-destination pairs) in the network, optimizing overall routing performance becomes extremely hard.

Third, mobility is an inherent property and phenomenon in wireless networks. The node's mobility makes network topology change frequently, thus complicating the task of finding and maintaining a good path between the source and destination. Mobility also affects the link quality, which introduces further challenges in maintaining a timely good path.

Fourth, wireless-embedded devices, such as sensors and handheld devices, are typically battery powered. The lifetime of the battery imposes a limitation on the operation hours and connectivity of the network. Energy efficiency has been a critical concern in energy-constrained networks (e.g. wireless sensor networks). Finding paths that consume minimum energy to deliver the packets from the source to the destination is an important approach to save energy in wireless sensor networks, since the radio communication has been identified as the major source of energy consumption in such networks. However, finding minimum energy consumption path(s) is neither easy nor enough. It is hard mainly because of the unreliability of the wireless link, and it is not enough because we also have to achieve other performance goals, such as satisfying a delay constraint due to specific applications (e.g. surveillance). Choosing and maintaining path(s) that strike a good balance between energy consumption and performance is challenging, especially with the presence of unreliable wireless links.

A large body of research on routing protocols in MWNs has been motivated by the above challenges. Next, we will introduce the major routing protocols in the literature, and motivate the opportunistic routing.

1.3   Routing techniques in MWNs

The routing protocols in MWNs can be classified into different categories using different criteria. For example, they can be classified into link-state routing (e.g. OLSR – Jacquet *et al.* 2001) and distance-vector routing (e.g. AODV – Perkins and Royer 2001). They can also be categorized as proactive (e.g. DSDV – Perkins and Bhagwat 1994) and reactive (on-demand) (e.g. DSR – Johnson *et al.* 2001 – and TORA – Park and Corson 1997) routing. For a better understanding of the difference between opportunistic routing and other state-of-the-art routing protocols in MWNs, we would like to classify the routing protocols as traditional and opportunistic routing.

1.3.1   Traditional routing

Traditional routing protocols (Johnson *et al.* 2001; Perkins and Bhagwat 2001; Perkins and Royer 2001) for multihop wireless networks have followed the concept of routing in wired networks by abstracting the wireless links as wired links, and finding the shortest, least cost, or highest throughput path(s) between a source and destination. Most routing protocols rely on the consistent and stable behavior of individual links, so the intermittent behavior of wireless links can result in poor performance such as low packet delivery ratio and high control overhead. On the other hand, this abstraction ignores the unique broadcast nature and spacial diversity of the wireless medium. We introduce several well known traditional routing protocols in MWNs as follows.

AODV   The Ad hoc On Demand Distance Vector (AODV) routing protocol (Perkins and Royer 2001) is designed for routing in MANETs. It is called, on demand, because it builds routes between the source and destination nodes only when source nodes request it. It maintains and updates the route as long as it is needed by the source node. AODV uses sequence numbers to ensure the freshness of routes. It is loop-free, self-starting, and scales to large numbers of mobile nodes.

AODV establishes routes using a route request and route reply discovery cycle. When a source node requires a route to a destination for which it does not already have a route, it broadcasts a route request (RREQ) message throughout the network. The RREQ message contains the source node's IP address, current sequence number, a broadcast ID, and the most recent sequence number for the destination of which the source node is aware. Intermediate nodes that receive the RREQ update their information for the source node and set up backwards pointers to the source node in their routing tables. A node receiving the RREQ will unicast a route reply (RREP) message back to the source if it is either the destination or if it knows a route to the destination with corresponding sequence number no smaller than that contained in the RREQ. Otherwise, it rebroadcasts the RREQ. Nodes keep track of the RREQ's source IP address and broadcast ID, and discard the RREQ that they have received recently and do not forward it.

As the RREP message is relayed back to the source node along the reversing path, nodes on the path set up forward pointers to the destination. Once the source node receives the RREP, it may begin to forward data packets to the destination along the path. If the source later receives a RREP containing a greater sequence number or contains the same sequence number with a smaller hop count, it may update its routing information for that destination and begin using the new route. In this sense, AODV tries to find the shortest path with fewest hops.

The route is active and maintained as long as there are data packets routed through it. Once the source node stops sending the data packets, the links on the path will time out and eventually be deleted from the intermediate nodes' routing tables. If a link break occurs while the route is active, the upstream node of the break link propagates a route error (RERR) message to the source node to inform it of this broken link. After receiving the RERR, if the source node still desires a route, it will initiate a route discovery again.

The advantage of AODV lies in that due to its on-demand (reactive) nature, it can handle highly dynamic topology change in MANETs. However, it possesses several disadvantages. First, AODV lacks support for high throughput routing metrics. It is designed to support the shortest hop-count metric, thus favoring long and low-bandwidth links over short and high-bandwidth links. Second, it may incur long route discovery latency. AODV does not discover a route until a flow is initiated. This route-discovery latency can be high in large-scale networks. Third, it may introduce broadcast storm problems in bandwidth-limited MWNs. Each network node participates in the route discovery process by rebroadcasting RREQ messages, which leads to redundant rebroadcasts, contention and packet collision.

DSR   Dynamic Source Routing (DSR) (Johnson *et al.* 2001) is another well known on-demand routing protocol. It is similar to the AODV in that it forms a route on demand when a source node requests it. However, it uses source routing instead of relying on the routing table at each intermediate node.

In the route-discovery phase, like AODV, DSR uses RREQ and RREP messages. When the source node needs to send data packets to a destination and it does not know a route to it, it initiates a route discovery by broadcasting a RREQ message. The RREQ identifies the initiator (source) and target (destination) of the route discovery and contains a unique request ID. It also contains a record listing the address of each intermediate node through which this particular copy of RREQ has been forwarded. The route record is initialized to an empty list by the initiator.

When a node receives a RREQ, it checks if it is itself the target of this RREQ. If this is the case, the node returns a RREP, which contains the accumulated route record from the RREQ to the initiator. If it is not the case and it is the first time the node receives the RREQ, it appends its own address to the route record in the RREQ message and rebroadcast the RREQ. The node will drop the RREQ message if it is a duplicated one or when the node finds that its own address has already been in the route record.

When the source node receives a RREP message, it catches the route carried in the message and sends subsequent data packets to the destination along this route.

Dynamic source routing shares the same advantage as AODV: it is reactive, thus eliminating the need to flood the network periodically to update the routing table in a dynamic network. It actually does not maintain a routing table at each node but it maintains the whole path information to a destination at the source node. It is beaconless and hence does not require periodic Hello packet (beacon) transmissions, which saves bandwidth.

The disadvantage of DSR is that the route-maintenance mechanism does not repair a broken link locally. It may introduce a long connection setup delay due to its on-demand nature. Even though DSR performs well in static and low-mobility environments, the performance degrades rapidly with increasing mobility. Furthermore, a considerable routing overhead is involved due to the source-routing mechanism. This routing overhead is directly proportional to the path length. DSR also tends to find the shortest hop count path, which contains long and unreliable wireless links.

OLSR   Optimized Link State Routing (OLSR) (Jacquet *et al.* 2001) is a proactive link-state routing protocol, which uses Hello and Topology Control (TC) messages to discover and then disseminate link state information throughout the MWNs. Each node maintains the global topology information of the network, and computes the next hop for all the other nodes in the network using shortest hop forwarding paths.

The OLSR protocol discovers two-hop neighbor information for each node through one-hop Hello packet broadcasts. In order to minimize the topology information flooding overhead, OLSR performs a distributed election of a set of multipoint relays (MPRs). Each node selects its MPR set among its one-hop neighbors such that the set covers all the nodes that are two hops away. These MPR nodes generate and forward topology control (TC) messages that contain the MPR selectors.

The TC messages are generated periodically. The neighbors that are not in the MPR set read and process the TC message but do not rebroadcast the message. The information diffused in the network by the TC messages helps each node to build its topology table and the routing table is calculated based on this. This functioning of MPRs makes OLSR different from other link state routing protocols in several ways. Only a subset of the network nodes generates link-state information and not all the links of a node are advertised – only those that represent MPR selections.

The advantage of OLSR over reactive routing protocols (such as AODV and DRS) is that it does not introduce route-discovery delay for a flow because the route is computed in a proactive way. This favors situations where route requests for new destinations are very frequent. The OLSR protocol is adapted to the network, which is dense and where communication is assumed to occur frequently between a large number of nodes.

The drawback of OLSR is that it maintains the routing table for all the destinations at each node, which may not be necessary. When the number of the nodes increases, the overhead from the control messages also increases. By only using MPRs to flood topology information, OLSR removes some of the redundancy of the flooding process, which may be a problem in networks with weak wireless links.

1.3.2 Opportunistic routing

In a wireless network, when a packet is unicast to a specific next-hop node of the sender at the network layer, all the neighboring nodes in the effective communication range of the sender may be able to overhear the packet at the physical layer. It is possible that some of the neighbors may have received the packet correctly while the designated next-hop node did not. Based on this observation, a new routing paradigm, known as **opportunistic routing** (OR) (Ai *et al.* 2006; Biswas and Morris 2005; Bletsas *et al.* 2006; Fussler *et al.* 2003; Larsson 2001; Shah *et al.* 2004; Zhao and Valenti 2005; Zorzi and Rao 2003a,b) has recently been proposed. Opportunistic routing integrates the network and MAC layers. Instead of deterministically picking one node to forward a packet to, the network layer selects a set of candidate nodes to forward a packet to and at the MAC layer one node is selected dynamically as the actual forwarder based on the instantaneous wireless channel condition and node availability at the time of transmission. Opportunistic routing takes advantages of the spacial diversity and broadcast nature of wireless communications and is an efficient mechanism to combat the time-varying links. It improves the network throughput (Fussler *et al.* 2003; Biswas and Morris 2005; Zeng *et al.* 2008; 2007a,c) and energy efficiency (Zorzi and Rao 2003a; Zeng *et al.* 2007b) compared to traditional routing.

Some variants of opportunistic routing, such as ExOR (Biswas and Morris 2005) and opportunistic any-path forwarding (Zhong *et al.* 2006; Dubois-Ferriere *et al.* 2007; Zhong and Nelakuditi 2007), rely on the path-cost information or global knowledge of the network to select candidates and prioritize them. Another variant of OR is geographic opportunistic routing (GOR) (Fussler *et al.* 2003; Zorzi and Rao 2003a; Shah *et al.* 2004) which uses the location information of nodes to define the candidate set and relay priority. In GeRaF (Zorzi and Rao 2003a), the next-hop neighbors of the current forwarding node are divided into sets of priority regions with nodes closer to the destination having higher relay priorities. Like GeRaF, in (Shah *et al*. 2004), the network layer specifies a set of nodes by defining a forwarding region in space that consists of the candidate nodes. The data-link layer selects the first node available from that set to be the next hop node. (Fussler *et al*. 2003) discussed three suppression strategies of contention-based forwarding to avoid packet duplication in mobile ad hoc networks. We will discuss state-of-the-art opportunistic routing protocols in Chapter 6.

The performance of OR depends on several key issues. The first **key issue** is the selection of forwarding candidates. Although the most effective way seems to be to involve all the neighbors with smaller cost to the destination, the overhead is expected to grow with the increase of the number of forwarding candidates. In dense networks, this overhead might potentially be even higher than cost incurred due to repeated transmissions (Shah *et al*. 2005). The prioritization of the candidates is the second **key issue** that affects the performance. In general, we want to forward the packet along the "shortest" path. The lower priority forwarding candidates are essentially the backup to the node that is on the "shortest" path. However, due to the opportunistic nature of OR, the "distance" from a certain node to its multihop away destination will no longer be the same as that obtained by traditional shortest path routing. The path cost also depends on the spacial diversity opportunities along the path. How to quantify and incorporate the spacial diversity opportunities in OR has not been well understood. The third **key issue** is candidate coordination in the MAC layer, which ensures the multiple receivers of a packet to agree upon a next-hop forwarder in a distributed fashion (Choudhury and Vaidya 2004; Larsson 2001; Souryal and Moayeri 2005; Zhao and Valenti 2005; Zhong *et al*. 2005; Zorzi and Rao 2003a,b).

Although opportunistic routing has shown its effectiveness in achieving better energy efficiency (Zorzi and Rao 2003a,b) and higher throughput (Biswas and Morris 2005) than traditional routing, many important issues in OR remain unanswered or not well understood. First, none of the existing works provides a thorough understanding of how well the opportunistic routing can perform and how the selection of the forwarding candidate set will affect the routing efficiency. Questions, such as "a. how many and which neighbor nodes should be involved in the local forwarding?" and "b. What are the selection criteria and how do they affect the relay priority among the forwarding candidates?" remain unanswered. Second, there is a lack of theoretical analysis on the throughput bounds achievable by OR. Third, one of the current trends in wireless communication is to enable devices to operate using multiple transmission rates. For example, many existing wireless networking standards such as IEEE 802.11a/b/g include this multirate capability. The inherent rate–distance tradeoff of multirate transmissions has shown its impact on the throughput performance of traditional routing (Awerbuch *et al*. 2006; Zhai and Fang 2006a,b). Generally, low-rate communication covers a long transmission range, whereas high-rate communication must occur at short range. It is intuitive to expect that this rate–distance tradeoff will also affect the throughput of OR. Because different transmission ranges also imply different neighboring node sets, this results in different spacial diversity opportunities. The rate–distance–diversity tradeoffs in

OR are not well studied. Furthermore, existing OR coordination schemes have some inherent inefficiencies such as high time delay and potential duplicate forwarding. An improperly designed coordination scheme will aggravate these problems and even overwhelm the potential gain provided by OR. It is necessary to design more efficient candidate coordination schemes. Finally, most state-of-the-art OR protocols (Biswas and Morris 2005; Zeng *et al*. 2007c) rely on link quality (packet reception ratio) information to select and prioritize forwarding candidates. It is important to measure the link quality accurately in order to make OR operate optimally. However, the existing link quality measurement mechanisms are subject to malicious attacks. Thus they may not be able to provide accurate link quality information for OR and other link state-based traditional routing.

This book carries out a comprehensive study on the capacity, energy efficiency, throughput, and security issues in OR and the associated multirate, candidate selection, prioritization, and coordination problems. Our goal is to understand fully the principles, the tradeoffs, the gains of the node collaboration and its associated cost to provide insightful analysis and guidance for the design of more efficient routing protocols.

1.4   Related work

1.4.1   Opportunistic techniques

Cooperative communication   While opportunistic routing aims to harvest the diversity gain at the packet level, cooperative communication studies the diversity gain at the signal level. The idea of user cooperation diversity is usually attributed to Sendonaris, Erkip, and Aazhang in Sendonaris *et al*. (1998) but can also be traced back to the relay channel model first introduced in Meulen (1971). The relay channel generalizes the notion of a simple point-to-point channel with a single source and destination to include a relay whose sole purpose is to help transfer information from the source to the destination.

Cover and Gamal (1979) and Cover and Thomas (1991) are credited with developing most of the information theory results on relay channels. They analyzed the capacity of the relay channel under the assumption that all nodes operate in the same band. Under this assumption, the system can be decomposed into a broadcast channel from the viewpoint of the source and a multiple access channel from the viewpoint of the destination. The idea of user cooperation diversity first attracted the attention of the information theory after the paper by communit Sendonaris *et al*. (1998) was presented at the 1998 International Symposium on Information Theory. Many of the ideas and results that appeared in the literature shortly after Sendonaris *et al*. (1998) can be traced to Cover and Gamal (1979).

Sendonaris, Erkip, and Aazhang followed up on Sendonaris *et al*. (1998) with a more detailed information theory study of two-source transmission cooperation in a mobile uplink scenario in (Sendonaris *et al*. 2003a,b). This work was important in that it also exposed several practical implementation issues in cooperative transmission systems and attracted the interest of the communications and signal-processing communities. Also noteworthy are the contributions of Laneman and

Wornell (2002); Laneman (2002); and Laneman *et al.* (2004) for studying the performance of several practical cooperative transmission protocols in fading environments. Yet another important set of contributions came in the form of novel information theory results and new insights into information theory coding in Kramer *et al.* (2005). New information theory results and results on power control were also presented in Host-Madsen and Zhang (2005); Wang *et al.* (2005a). A variety of contributions to relaying including new bounds, cut-set theorems, power control strategies, LDPC relay code designs and some of the earliest results on half-duplex relaying were proposed in Khojastepour (2004). Researchers realized that relaying can mimic multiple-antenna systems even when the communicating entities were incapable of supporting multiple antennas. Prominent literature on the use of space-time codes with relays includes Laneman and Wornell (2002); Mitran *et al.* (2005); Nabar *et al.* (2004). Other interesting contributions are Boyer *et al.* (2004); Hasna and Alouini (2003); Liang and Veeravalli (2005); Toumpis and Goldsmith (2003).

The research in cooperative communication mainly focuses on the theoretical capacity under strict assumptions about user synchronization at the signal level. It is difficult to implement the proposed cooperative communication schemes in the real system. Opportunistic routing realizes the gains of cooperative diversity at the packet level, so it can be implemented in the standard radio hardware such as IEEE 802.11 devices.

Opportunistic scheduling Opportunistic scheduling aims to improve network utilization by taking advantage of the wireless channel fading across users and time. Knopp and Humblet (1995) showed that by scheduling transmission to the network user experiencing the best channel condition at the moment in cellular networks, significant system level gains can be realized. Thus fading essentially gives an opportunity for the network to ride on the peak channel condition at all times. However, opportunistic scheduling has its own costs and limitations (Liu *et al.* n.d.). In all the opportunistic scheduling schemes, signaling costs are unavoidable, because scheduling decisions inherently depend on channel conditions (and/or queuing status). Users need to estimate their channel conditions constantly and report to the base station. Hence, the actual scheduling gain should take into account the signaling costs. Furthermore, the timescale of channel fading plays an important role in opportunistic scheduling. The fluctuation of channels should be slow enough for user to estimate it and exploit it. On the other hand, the fluctuation should be fast enough, so that users do not experience extreme long delays.

In short, opportunistic routing is different from opportunistic scheduling in the following aspects. First, they target different problems. Opportunistic routing tries to improve packet forwarding reliability in multihop wireless networks by taking advantage of the spacial (user) diversity and the broadcast nature of the wireless medium whereas opportunistic scheduling aims to improve the system resource utilization by exploiting the channel fluctuations due to fading. Second, they are used in different networks. Opportunistic routing is used in multihop wireless networks whereas opportunistic scheduling is mainly used in single-hop cellular networks. Third, they are implemented at different layers. Opportunistic routing is a cross-layer design that is implemented at the network and MAC layers. In Chapter 7, we will see by integrating network coding, the opportunistic routing can be just implemented at network layer. While opportunistic scheduling is usually implemented at MAC layer with physical layer channel-state information.

AP collaborations   Protocols like MRD (Miu *et al.* 2005), SOFT (Woo *et al.* 2007) and Link-Alike (Jakubczak *et al.* 2009) all exploit different aspects of the same concept: multiple receiver diversity in WLANs. Consider, for example, a sender that has poor connectivity to multiple nearby APs. A transmitted packet is unlikely to reach any specified AP; any bit in the packet is likely to be received by at least one AP. All the above protocols exploit this receiver diversity by allowing APs to combine received bits or packets over the wired network and hence can increase uplink reliability without any retransmissions.

However, none of these schemes can similarly address a lossy downlink without expending medium time on retransmissions. SourceSync (Rahul *et al.* 2010) complements all these protocols by harnessing sender diversity to increase downlink reliability without any retransmissions, analogous to existing receiver diversity mechanisms on the uplink. Specifically, instead of requiring that a client receive packets from only one AP at a time, in SourceSync, multiple neighboring APs can transmit simultaneously to the client, and increase throughput.

In this category of research, the main focus is on how to improve the last-hop uplink or downlink transmission reliability, while routing is not a major concern. In opportunistic routing, the forwarding candidate selection and prioritization are not only dependent on the local one-hop instant link quality but also affected by the cost/distance from the candidates to the destination. Therefore, in comparison with AP collaborations, opportunistic routing faces more challenges.

1.4.2   Network coding

Candidate coordination is a challenging problem introduced by opportunistic routing. If the forwarding candidates are not well coordinated, multiple nodes may hear a packet and forward the same packet unnecessarily, which in turn degrades the network throughput. Fortunately, network coding can effectively alleviate this problem. Next we introduce the basic concepts of network coding, and then explain why it can be used to improve the routing performance in MWNs.

Network coding disposes of the traditional end-to-end packet forwarding paradigm, and enables intermediate nodes to mix the received packets and has the potential to increase network throughput. Intuitively, by combining received packets, a coded packet sent by an intermediate node could benefit multiple receivers simultaneously, thus improving the bandwidth efficiency. In the seminal paper by Ahlswede *et al.* (2000) it was shown that, for a butterfly network topology, the multicast capacity can be achieved by performing network coding at the routers. Later, Li *et al.* (2003) showed that linear codes are enough to achieve the maximum multicast capacity bounds under the same network topology, while Ho *et al.* (2006) extended their results to random linear codes. For unicast traffic, Li and Li (2004) showed that network coding results in higher throughput than pure forwarding for specific unicast topologies. The above results are for general networks; in the following, we give a brief review of how network coding can be applied to multicast/broadcast/unicast sessions with emphasis on multihop wireless networks and its advantages and limitations in both theory and practice.

From the theoretical viewpoint, Lun *et al.* (2008) proved that random linear network coding can be used to construct a capacity-approaching scheme for multicast over lossy wireless networks. Adjih *et al.* (2007) showed that by using a simple broadcast rate selection strategy, network coding can ensure that every transmission has a high probability of being useful. Fragouli *et al.* (2008) studied network coding-based efficient broadcast from both theoretical and practical points of view. They showed that network coding is able to increase the bandwidth/energy efficiency by a constant factor in fixed networks. They also proposed a probabilistic forwarding-based algorithm for random networks, which shows a significant reduction in the total transmission count compared with probabilistic flooding.

To bridge the gap between theory and practice, Chachulski *et al.* (2007) proposed MORE, which is the first practical network coding-based opportunistic routing protocol that achieves high throughput for both unicast and multicast sessions. The main motivation of MORE is to solve the candidate coordination challenge of opportunistic routing (Biswas and Morris 2005). Its idea is to combine opportunistic routing with network coding. By randomly mixing packets before forwarding them, MORE ensures that the routers hearing the same transmission do not forward the same packet. In this way, MORE eliminates the need of complicated coordination mechanism between multiple forwarders in pure opportunistic routing. However, MORE is inefficient when applied to multicast or broadcast, since almost every node in the network may become a forwarding node, which can cause heavy congestion (Koutsonikolas *et al.* 2009). Moreover, in mobile MWNs, traditional tree-based multicast schemes fall short in that they incur large overhead in maintenance of the tree structure as the topology changes very fast.

Recently, a special type of network coding, symbol-level network coding (SLNC) (Katti *et al.* 2008) is proposed. Taking a step further beyond the usual packet-level network coding (PLNC) method which processes information in the unit of packets, SLNC enables a node to combine information in the smaller granularity of "symbol", which may consist of several physical layer symbols. The immediate advantages brought by SLNC are enhanced error and interference tolerance, whereas the benefits of network coding are automatically inherited. In addition, SLNC also possesses the potential to enhance spacial reuse by encouraging concurrent transmissions, and it has been shown to be able to significantly improve the throughput of unicast in wireless mesh networks compared with PLNC (Katti *et al.* 2008). However, how and how much SLNC can improve the bandwidth efficiency in mobile WMNs is still not well understood.

In this book (Chapter 7), we exploit network coding and opportunistic listening in designing high-performance broadcast protocols in mobile MWNs, especially vehicular ad hoc network (VANET). In order to resolve the challenges posed by lossy links and fast-changing topology, SLNC is combined with a novel push-based broadcast method where the relay nodes (forwarders) are selected in a dynamic, opportunistic, and fully distributed manner, in contrast with the traditional tree-based multicast method in fixed MWNs. The coordination among relay nodes can be greatly simplified through the use of network coding. The gains of using SLNC and the opportunistic relay selection method are characterized, which again demonstrates the importance of being opportunistic (in the mobile setting).

1.4.3 Opportunistic forwarding in opportunistic networks

Opportunistic networks are an important class of Delay/Disruption Tolerant Networks (DTNs) in which contacts (time-windows when data can be exchanged) appear opportunistically without any prior information (Wang *et al.* 2005b). In opportunistic networks, source and destination nodes might never be fully connected at the same time. That is, there is no guarantee on the existence of a complete path between two nodes wishing to communicate. Examples of such networks are sparse mobile ad hoc networks, such as ZebraNet (Juang *et al.* 2002). Packet forwarding or routing in such networks is the most compelling issue. Nodes are not always connected to each other, so the forwarding algorithms in such networks follow a store-carry-forward paradigm. Typical algorithms differ based on their decisions as to who forwards the data, at what time the data is forwarded, and to whom the data is sent.

In general, the packet forwarding algorithms can be classified into three categories: flooding, simple replication, and history based.

In flooding algorithms, each node forwards any nonduplicated messages to any other node that it encounters. A representative protocol in this category is the epidemic routing protocol (Vahdat and Becker 2000), where messages diffuse in the network similarly to diseases or viruses, i.e., by means of pair wise contacts between individuals/nodes. The dissemination process is bounded because each message when generated is assigned a hop-count limit giving the maximum number of hops that the message is allowed to traverse before reaching the destination. The flooding algorithm delivers messages with the minimum delay but consumes a lot of bandwidth and node storage.

For simple replication algorithms, identical copies of the message are sent over the first k contacts. Only the source of the message sends multiple copies. The relay nodes are allowed to send only to the destination; they cannot forward it to another relay. This leads to small overhead as the message flooding is controlled to take place only near the source. This class of forwarding algorithms is also known as the two-hop relay algorithm (Chaintreau *et al.* 2005; Grossglauser and Tse 2002b). There is a natural tradeoff between overhead and data-delivery latency. A higher k leads to more storage/transmissions but has lower delays.

A history-based algorithm estimates the probability of delivery using the historical data forwarding record. Each node keeps track of the probability that a given node will deliver its messages. k highest ranked relays (based on delivery probability) are selected as forwarding nodes (Juang *et al.* 2002).

We should differentiate between the opportunistic forwarding in opportunistic networks and the opportunistic routing we study in this book. The former considers the node encounter probability as the opportunity, and store, carry, and forward the packet in a not completely connected network. It still abstracts each wireless link as a wired link, and does not consider exploiting spacial diversity to improve per-hop transmission reliability. The latter considers the broadcast nature of wireless medium as an opportunity and exploits the spacial node diversity in each transmission to improve the per-hop transmission reliability.

1.4.4 Geographic routing

Owing to its scalability, statelessness, and low maintenance overhead, geographical routing is considered as an efficient paradigm for data forwarding in multi-hop wireless ad hoc and sensor networks. Early work (Finn 1987; Karp and Kung 2000; Kuhn *et al.* 2003) on geographic routing exploited the concept of maximum advancement towards the destination to route packets in a greedy manner. However, recent empirical measurements (Couto *et al.* 2003; Zhao and Govindan 2003) have proved that the unit disk connectivity model, on which these solutions are based, often fails in real settings. More recent works on geographic routing are focused on lossy channel situations. Seada *et al.* (2004) articulated the distance–hop energy tradeoff for geographic routing. They concluded that packet advancement times packet reception ratio, the EPA, is an optimal metric for making localized geographic routing decisions in lossy wireless networks with ARQ (Automatic Repeat reQuest) mechanisms, and is also a good metric for Non-ARQ scenarios. Zorzi and Armaroli (2003) also independently proposed the same link metric. Lee *et al.* (2005) presented a more general framework called normalized advance (NADV) to normalize various types of link cost such as transmission times, delay and power consumption. Unfortunately, NADV only applies to geographic routing which involves a single forwarding candidate and cannot be directly used for geographic opportunistic routing.

1.4.5 Multirate routing

Multirate wireless networks have started attracting research attention recently. Draves *et al.* (2004) proposed using the weighted cumulative expected transmission time (WCETT) as a routing metric. Awerbuch *et al.* (2006) adopted the medium time metric (MTM). Zhai and Fang (2006b) studied the impact of multirate on carrier sensing range and spacial reuse ratio and demonstrated that the bandwidth distance product and the end-to-end transmission delay (the same as the medium time) are better routing metrics than the hop count. Zhai and Fang (2006a) also proposed the metric of interference clique transmission time to achieve a high path throughput. However, these metrics or protocols were proposed for routing on a fixed path following the concept of the traditional routing. In this book, we propose a framework to compute the end-to-end throughput bound of OR for different OR schemes in multiradio, multichannel and multirate networks.

The throughput bound derived in this book is the upper bound of the achievable throughput of the proposed and investigated OR schemes. We also study the impact of the protocol overhead and multirate capability on the performance of GOR under contention-based medium access protocols.

1.4.6 Energy-aware routing

Energy-aware routing has received significant attention over the past few years (Chang and Tassiulas 2000; Kar *et al.* 2003; Li *et al.* 2001; Singh *et al.* 1998). Singh *et al.* (1998) proposed five energy aware metrics such as *maximizing time to partition* and *minimizing maximum node cost*. These are important metrics for energy efficient routing. However, it is difficult to directly implement them in a local algorithm when even the global version of the same problem is NP-complete. Chang and Tassiulas (2000) proposed a class of flow-augmentation algorithms and a flow-redirection algorithm, which balances the energy consumption rates among the nodes in proportion to the energy reserves. The limitation of this approach is that it requires prior knowledge of the information generation rates at the origin nodes. Li *et al.* (2001) proposed an "online" power-aware routing and a zone-based routing, which maximizes the network lifetime without knowing the message generation rate. Following Li *et al.* (2001), another "online" routing algorithm was proposed in Kar *et al.* (2003), which aims to maximize the total number of successfully delivered messages. In this book, we study the energy efficiency of OR to tradeoff the routing performance and energy efficiency in terms of maximizing the bit advancement per unit energy consumption.

1.4.7 Capacity of MWNs

Theoretical work on the capacity of MWNs mainly focuses on two aspects. One is the asymptotic bounds of the network capacity (Grossglauser and Tse 2002a; Gupta and Kumar 2000). These works study the capacity trend with regard to the size of a wireless network under specific assumptions or scenarios. The other aspect of work on wireless network capacity is the computation of the exact performance bounds for a given network. Jain *et al.* (2003) proposed a framework to calculate the throughput bounds of traditional routing between a pair of nodes by adding wireless interference constraints into the maximum flow formulations. Zhai and Fang (2006a) studied the path capacity of traditional routing in a multirate scenario. Our work falls into this direction. However, distinct from the previous works, we propose a method to compute the end-to-end throughput bounds of opportunistic routing, which is different from the traditional routing in that we construct the transmitter (associated with multiple forwarding candidates) based conflict graph instead of a link conflict graph to capture the local broadcast nature of OR. Our framework can be used as a tool to calculate the end-to-end throughput bound of different OR variants, and is an important theoretical foundation for the performance study of OR. There has been recent work (Alicherry *et al.* 2005; Kodialam and Nandagopal 2005; Zhang *et al.* 2005) on capacity bound computation in multiradio multichannel networks. However, they are all based on the assumption that traditional routing is used at the network layer, where one transmitter can only deliver traffic to one receiver.

1.4.8   Link-quality measurement

The existing LQM mechanisms proposed in the literature (Couto *et al*. 2003; Kim and Shin 2006; Sang *et al*. 2007) can be generally classified into three types: active, passive and cooperative (Kim and Shin 2006). For broadcast-based active probing (Couto *et al*. 2003), each node periodically broadcasts Hello/probing packets and its neighbors record the number of received packets to calculate the packet reception ratios (PRRs) from the node to themselves. In passive probing (Kim and Shin 2006), the real traffic generated in the network is used as probing packets without introducing extra overheads. For cooperative probing (Kim and Shin 2006), a node overhears the transmissions of its neighbor to estimate the link quality from the neighbor to itself. However, for any of the existing LQM mechanisms, the inherent common fact is that a node's knowledge about the forward PRR from itself to its neighbor is informed by the neighbor. Since MWNs are generally deployed in an ad hoc style or in untrusted environments, nodes may be compromised and act maliciously. This receiver-dependent measurement opens up a door for malicious attackers to report a false measurement result and disturb the routing decision for all the PRR-based protocols.

1.5   Book contribution

The main contributions of this book are as follows:

- Chapter 2

    - We generalize the definition of EPA for an arbitrary number of forwarding candidates that follow a specific priority rule to relay the packet in OR.

    - Through theoretical analysis we prove that the maximum EPA can only be achieved by giving higher relay priorities to the forwarding candidates closer to the destination. This proof convinces us that given a forwarding candidate set, the relay priority among the candidates is only relevant to the advancement achieved by the candidate to the destination, but irrelevant to the packet delivery ratio between the transmitter and the forwarding candidate. The analysis result is the upper bound of the EPA that any GOR can achieve.

    - We find that given a set of M nodes that are available as next-hop neighbors, the candidate set achieving the maximum EPA with r $(r \leq M - 1)$ nodes is contained in at least one candidate set achieving the maximum EPA with $r + 1$ nodes.

    - We prove that the maximum EPA of selecting r $(r \leq M)$ nodes is a strictly increasing and concave function of r. This property indicates that although involving more forwarding candidates in GOR will increase the maximum EPA, the extra EPA gained by doing so becomes less significant.

- Chapter 3

  - We investigate the energy efficiency of GOR and propose two localized candidate selection algorithms with $\mathbf{O}(M^3)$ and $\mathbf{O}(M^2)$ running time in the worst case respectively and $\Omega(M)$ in the best case, where M is the number of available next-hop neighbors of the transmitter. The algorithms efficiently determine the optimal forwarding candidate set with respect to the EPA per unit of energy consumption.

  - We propose an energy-efficient geographic opportunistic routing (EGOR) framework applying the node selection algorithms to achieve the energy efficiency. Simulation results show that EGOR achieves better energy efficiency than geographic routing and blind opportunistic protocols in all the cases while maintaining very good routing performance. Our simulation results also show that the number of forwarding candidates necessary to achieve the maximum energy efficiency is mainly affected by the reception to transmission energy ratio but not by the node density under a uniform node distribution. Only a very small number of forwarding candidates (around 2) is needed on average. This is true even when the energy consumption of reception is far less than that of transmission.

- Chapter 4

  - We propose a new method of constructing transmission conflict graphs, and present a methodology for computing the end-to-end throughput bounds (capacity) of OR. We formulate the maximum end-to-end throughput problem of OR as a maximum-flow linear programming problem subject to the transmission conflict constraints and effective forwarding rate on each link. To the best of our knowledge, this is the first theoretical work on capacity problem of OR for multihop and multirate wireless networks.

  - We propose two metrics for OR under multirate scenario: one is *expected medium time* (EMT) and the other is *expected advancement rate* (EAR). Based on these metrics we propose two distributed and local rate and candidate selection schemes: least medium time OR (LMTOR) and multirate GOR (MGOR), respectively.

  - We show that OR has great potential to improve the end-to-end throughput under different settings, and our proposed multirate OR schemes achieve higher throughput bound than any single-rate GOR.

  - We make some observations about OR: 1. the end-to-end capacity gained decreases when the number of forwarding candidates is increased. When the number of forwarding candidates is larger than three, the throughput almost remains unchanged. 2. there exists a node-density threshold, higher than which 24 mbps GOR performs better than 12 mbps GOR, and lower than which, vice versa. The threshold is about 5.5 and 10.9 neighbors per node on 12 mbps for line and square topologies, respectively.

- Chapter 5

  - We propose a unified framework to compute the capacity of opportunistic routing between two end nodes in single/multi-radio/channel multihop wireless networks by allowing dynamic forwarding strategies.

  - We discuss the radio/channel and interference constraints when constructing concurrent transmission sets, and study the capacity region of an opportunistic module.

  - We propose an LP approach and a heuristic algorithm to obtain an opportunistic forwarding and scheduling strategy that satisfies a traffic demand vector.

  - Leveraging our analytical model, we find that OR can achieve comparable or even better performance than TR by using fewer radio resources.

- Chapter 6

  - We propose a new scheme "fast slotted acknowledgment" for candidate coordination in OR, which adopts single ACK to confirm successful reception and suppress other candidates' attempts to forward the data packet with the help of a channel-sensing technique.

  - Simulation shows that FSA can decrease the average end-to-end delay by up to 50% when the traffic is relatively light and can improve the throughput by up to 20% under heavy traffic load where other coordination schemes are already unable to delivery all the data packets.

  - The simulation results also validate that FSA can achieve performance similar to ideal coordination where relay priority can be ensured and duplicate packet forwarding is avoided.

- Chapter 7

  - We investigate the integration of network coding with opportunistic routing for easing the coordination in OR, and review MORE, a state-of-the-art MAC-layer independent OR protocol based on network coding.

  - We formulate the problem of mobile content distribution in a vehicular ad hoc network (VANET) and propose two mobile content broadcast schemes by leveraging symbol level network coding (SLNC) and combining it with opportunistic listening at the same time.

  - We propose two push-based broadcast protocols to exploit the benefits of SLNC fully, in which the content sources simply "pushes" information into the VANET actively, while a dynamic subset of temporary relay nodes from all the vehicles is determined in a fully distributed and localized way.

- We observe that, in addition to the advantage brought by network coding in simplifying the transmission scheduling in MORE, by using symbol-level network coding, another benefit is gained for the broadcast. That is, due to the higher error and interference tolerance from symbol-level diversity, the hidden terminal problem can be alleviated, which yields the possibility of using much simpler coordination mechanism in medium access, i.e., carrier sensing. In contrast, the traditional packet-level network coding does not achieve best performance under the same coordination method due to the interference from hidden terminals.

- Simulation shows that the proposed broadcast schemes achieve significant gains compared with state-of-the-art content distribution schemes in VANETs, where one important part of it comes from the use of SLNC and the other is attributed to the new push-based protocol design.

- **Chapter 8**

  - We investigate the impact of transmission rate and forwarding strategies (candidate selection, prioritization and coordination) on throughput of OR under a contention-based medium-access scenario.

  - We propose a local metric, *Opportunistic Effective One-hop Throughput* (OEOT), to characterize the tradeoff between the packet advancement and one-hop packet forwarding time under different data rates.

  - We propose a rate-adaptation and candidate-selection algorithm to approach the local optimum of this metric.

  - We propose a multirate link quality measurement mechanism.

  - We show that MGOR incorporating our algorithm achieves better throughput and delay performance than the corresponding opportunistic routing and geographic routing operating at any single rate, which indicates that OEOT is a good local metric to achieve high end-to-end throughput and low delay for MGOR.

- **Chapter 9**

  - We discuss possible attacks on opportunistic routing protocols and propose countermeasures.

  - We analyze the security vulnerabilities in the existing LQM mechanisms and propose an efficient broadcast-based secure LQM (SLQM) mechanism, which prevents the malicious receiver from reporting a higher PRR than the actual one.

  – We analyze the security strength, the cost and applicability of the proposed mechanism.

- Chapter 10

  – We study opportunistic broadcasting in vehicular networks, where we apply the concept of opportunistic routing to the design of a broadcast protocol.

  – In particular, we propose a multi hop *opportunistic broadcast* scheme, a fully distributed protocol that simultaneously achieves high reliability and fast message propagation while incurring low transmission overheads.

  – We propose a distributed *opportunistic broadcast coordination* mechanism to let the recipients of a single broadcast determine the "best" relay nodes in a localized manner. The proposed transmission coordination mechanism exploits the idea of opportunistic forwarding to enhance the reception reliability and reduce the hop delay in each single transmission.

  – Simulation results show that the proposed scheme achieves better performance than the state-of-the-art solutions and we characterize the tradeoff between broadcast reception reliability, end-to-end delay and transmission overhead.

1.6   System model and assumptions

We consider a multi hop wireless network with N nodes arbitrarily located on a plane. Each node n_i ($1 \leq i \leq N$) can transmit a packet at J different rates $R^1, R^2, \ldots, R^J$. We say there is a **usable** directed link l_{ij} from node n_i to n_j, when the **packet reception ratio** (PRR), denoted as p_{ij}, from n_i to n_j is larger than a non-negligible positive threshold p_{td}. The PRR we consider is an average value of the link quality in a long timescale (e.g. in tens of seconds). There exist several link-quality measurement mechanisms (Couto *et al.* 2003; Kim and Shin 2006) to obtain the PRR on each link. We assume that there is no power control scheme and the PRR on each link for each rate is given. We define the **effective transmission range** L_m at rate R^m ($1 \leq m \leq J$) as the sender-receiver distance at which the PRR equals p_{td}.

The basic module of opportunistic routing is illustrated in Figure 1.1. Assume node n_i is forwarding a packet to a remote sink/destination n_d. We denote the set of nodes within the effective transmission range of node n_i as the **neighboring node set** C_i (e.g., all the five nodes around n_i in Figure 1.1). Note that, for different transmission rates, the corresponding effective transmission ranges are different, then we have different neighboring node sets of node n_i, and the PRR on the same link l_{ij} may be different at different rates. We define the set $\mathcal{F}_i := \langle n_{i_1}, \ldots, n_{i_r} \rangle$ (e.g., $\langle n_{i_1}, n_{i_2}, n_{i_3} \rangle$ in Figure 1.1) as **forwarding candidate set**, which is a subset of C_i and includes r nodes selected to be involved in the local opportunistic forwarding based on a particular selection strategy. $\mathcal{F}_i$ is an ordered set, where the order of the elements corresponds to their priority in relaying a received packet.

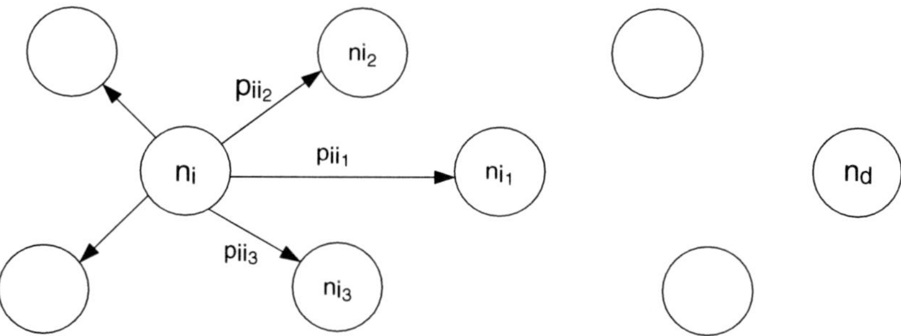

Figure 1.1   Node n_i is forwarding a packet to a remote destination n_d with a forwarding candidate set $\mathcal{F}_i = \langle n_{i_1}, n_{i_2}, n_{i_3} \rangle$ at some transmission rate. Reproduced by permission of © 2008 IEEE.

For GOR, we assume each node is aware of the location information[1] of itself, its one-hop neighbors and the destination. Given a transmitter n_i, one of its forwarding candidates n_{i_q}, and the destination n_d, we define the **packet advancement** d_{i_q} in Equation (1.1), which is the Euclidean distance between the transmitter and destination subtracting the Euclidean distance between the candidate n_{i_q} and the destination. This definition represents the advancement in distance made toward the destination when n_{i_q} forwards the packet sent by n_i.

$$d_{ii_q} = dist(n_i, n_d) - dist(n_{i_q}, n_d) \tag{1.1}$$

For GOR, because we are only interested in the neighbors that give positive advancement to the destination, we denote the set of those neighbors as C_i, the **available next-hop node set**.

Opportunistic routing works by the sender node n_s forwarding the packet to the nodes in its forwarding candidate set $\mathcal{F}_s$. One of the candidate nodes continues the forwarding based on their relay priorities—if the first node in the set has received the packet successfully it forwards the packet towards the destination while all other nodes suppress duplicate forwarding. Otherwise, the second node in the set is arranged to forward the packet if it has received the packet correctly. Otherwise the third node, the fourth node, and so forth. A forwarding candidate will forward the message only when all the nodes with higher priorities fail to do so. When no forwarding candidate has successfully received the packet, the sender will retransmit the packet if retransmission is enabled. The sender will drop the packet when the number of retransmissions exceeds the limit. The forwarding reiterates until the packet is delivered to the destination. Several MAC protocols have been proposed in Biswas and Morris (2005); Fussler *et al.* (2003); Zorzi and Rao (2003a); Zubow *et al.* (2007) to coordinate the forwarding candidates and ensure the relay priority among them. In this book, for all the analysis, we assume the relay priority can

---

[1] The node location information can be obtained by prior configuration, by the Global Positioning System (GPS) receiver, or through some sensor self-configuring localization mechanisms as in Bulusu *et al.* (2000); Savvides *et al.* (2001).

be perfectly realized. So there is no duplicate packet forwarding due to imperfect candidate coordination. We will show in Chapter 6 that it is a realistic assumption when our proposed candidate coordination scheme is used.

For capacity analysis in Chapters 4 and 5, we assume that packet transmissions at the individual nodes can be finely controlled and carefully scheduled by an omniscient and omnipotent central entity. So here we do not concern ourselves with issues such as MAC contention or coordination overhead that may be unavoidable in a distributed network. This is a very commonly used assumption for such theoretical studies (Jain *et al*. 2003; Zhai and Fang 2006a).

References

Adjih C, Cho SY and Jacquet P 2007 Near optimal broadcast with network coding in large sensor networks *First International Workshop on Information Theory for Sensor Networks, Sante Fe, USA*.

Ahlswede R, Cai N, Li SY and Yeung R 2000 Network information flow. *IEEE Transactions on Information Theory* **46**(4), 1204–1216.

Ai J, Abouzeid AA and Ye Z 2006 Cross-layer optimal decision policies for spatial diversity forwarding in wireless ad hoc networks. *Mobile Adhoc and Sensor Systems (MASS) IEEE International Conference on Mobile Ad-hoc and Sensor Systems (MASS '06)*, Vancouver, Canada.

Akyildiz IF and Kasimoglu IH 2004 Wireless sensor and actor networks: Research challenges. *Ad Hoc Networks Journal (Elsevier)* **2**(4), 351–367.

Akyildiz IF, Su W, Sankarasubramaniam Y and Cayirci E 2002 A survey on sensor networks. **40**(8), 102–116.

Akyildiz IF and Wang X 2005 A survey on wireless mesh networks. *Communications Magazine, IEEE* **43**(9), S23–S30.

Akyildiz IF, Wang X and Wang W 2005 Wireless mesh networks: a survey. *Computer Networks* **47**(4), 445–487.

Alicherry M, Bhatia R and Li LE 2005 Joint channel assignment and routing for throughput optimization in multi-radio wireless mesh networks *MobiCom '05*, pp. 58–72. ACM, New York, NY.

Awerbuch B, Holmer D and Rubens H 2006 The medium time metric: High throughput route selection in multi-rate ad hoc wireless networks. *MONET* **11**(2), 253–266.

Biswas S and Morris R 2005 Exor: Opportunistic multi-hop routing for wireless networks *SIGCOMM'05*, Philadelphia, Pennsylvania.

Bletsas A, Khisti A, Reed DP and Lippman A 2006 A simple cooperative diversity method based on network path selection. *IEEE Journal on Selected Areas in Communications* **24**(3), 659–672.

Boyer J, Falconer D and Yanikomeroglu H 2004 Multihop diversity in wireless relaying channels. *IEEE Transactions on Communications* **52**(10), 1820–1830.

Bulusu N, Heidemann J and Estrin D 2000 Gps-less low cost outdoor localization for very small devices. *IEEE Personal Communications Magazine* **7**(5), 28–34.

Cerpa A, Elson J, Estrin D, Girod L, Hamilton M and Zhao J 2001 Habitat monitoring: Application driver for wireless communications technology *ACM SIGCOMM Workshop Data Comm. Latin America and the Caribbean*, Costa Rica.

Chachulski S, Jennings M, Katti S and Katabi D 2007 Trading structure for randomness in wireless opportunistic routing *Proceedings of the 2007 Conference on Applications, Technologies, Architectures, and Protocols for Computer Communications*, pp. 169–180 SIGCOMM '07.

Chaintreau A, Hui P, Crowcroft J, Diot C, Gass R and Scott J 2005 Pocket Switched Networks: Real-world mobility and its consequences for opportunistic forwarding. Technical report, 2006 Computer Laboratory, University of Cambridge.

Chang J and Tassiulas L 2000 Energy conserving routing in wireless ad-hoc networks *IEEE INFOCOM'00*, Tel Aviv, Israel.

Chong CY and Kumar SP 2003 Sensor networks: evolution, opportunities, and challenges. *Proceedings of the IEEE* **91**(8), 1247–1256.

Choudhury R and Vaidya NH 2004 Mac layer anycasting in ad hoc networks. *SIGMOBILE Mobile Communication Review* **34**(1), 75–80.

Couto DD, Aguayo D, Bicket J and Morris R 2003 A high-throughput path metic for multi-hop wireless routing *ACM MobiCom'03*, San Diego, California.

Cover T and Gamal A 1979 Capacity theorems for the relay channel. *IEEE Transactions on Information Theory* **25**(5), 572–584.

Cover T and Thomas J 1991 *Elements of Information Theory*. Wiley, New York.

Draves R, Padhye J and Zill B 2004 Routing in multi-radio, multi-hop wireless mesh networks *MobiCom '04*.

Dubois-Ferriere H, Grossglauser M and Vetterli M 2007 Least-cost opportunistic routing. Technical Report LCAV-REPORT-2007-001, School of Computer and Communication Sciences, EPFL.

Estrin D, Culler D and Pister K 2002 Connecting the physical world with pervasive networks. *IEEE Pervasive Computing*.

Finn GG 1987 Routing and addressing problems in large metropolitan-scale internetworks. Technical Report ISI/RR-87-180, USC/ISI.

Fragouli C, Widmer J and Le Boudec JY 2008 Efficient broadcasting using network coding. *IEEE Transactions on Networking* **16**(2), 450–463.

Fussler H, Widmer J, Kasemann M, Mauve M and Hartenstein H 2003 Contention-based forwarding for mobile ad-hoc networks. *Elsevier's Ad Hoc Networks* **1**(4), 351–369.

Grossglauser M and Tse D 2002a Mobility increases the capacity of ad hoc wireless networks. *IEEE/ACM Transactions on Networking* **10**(4), 477–486.

Grossglauser M and Tse DNC 2002b Mobility increases the capacity of ad hoc wireless networks. *IEEE/ACM Transactions on Networking* **10**, 477–486.

Gupta P and Kumar PR 2000 The capacity of wireless networks. *Transactions on Information Theory* **46**(2), 388–404.

Hasna M and Alouini MS 2003 End-to-end performance of transmission systems with relays over rayleigh-fading channels. *IEEE Transactions on Wireless Communications* **2**(6), 1126–1131.

Ho T, Medard M, Koetter R, Karger D, Effros M, Shi J and Leong B 2006 A random linear network coding approach to multicast. *IEEE Transactions on Information Theory* **52**(10), 4413–4430.

Host-Madsen A and Zhang J 2005 Capacity bounds and power allocation for wireless relay channel *IEEE Transactions on Information Theory*, vol. **51**, pp. 2020–2040.

Hu L and Evans D 2004 Localization for mobile sensor networks *MobiCom '04: Proceedings of the 10th Annual International Conference on Mobile Computing and Networking*, pp. 45–57. ACM, New York, NY, USA.

Jacquet P, Muhlethaler P, Clausen T, Laouiti A, Qayyum A and Viennot L 2001 Optimized link state routing protocol for ad hoc networks *IEEE International Multi-Topic Conference (IEEE INMIC)*.

Jain K, Padhye J, Padmanabhan VN and Qiu L 2003 Impact of interference on multi-hop wireless network performance *MobiCom '03: Proceedings of the 9th Annual International Conference on Mobile Computing and Networking*, pp. 66–80. ACM Press, New York, NY, USA.

Jakubczak S, Andersen DG, Kaminsky M, Papagiannaki K and Seshan S 2009 Link-alike: using wireless to share network resources in a neighborhood. *SIGMOBILE Mobile Computing and Communication Review* **12**, 1–14.

Johnson DB, Maltz DA and Broch J 2001 Dsr: The dynamic source routing protocol for multi-hop wireless ad hoc networks In *Ad Hoc Networking* (ed. Perkins CE) Addison-Wesley chapter 5, pp. 139–172.

Juang P, Oki H, Wang Y, Martonosi M, Peh LS and Rubenstein D 2002 Energy-efficient computing for wildlife tracking: design tradeoffs and early experiences with zebranet. *SIGPLAN Not.* **37**, 96–107.

Kar K, Kodialam M, Lakshman TV and Tassiulas L 2003 Routing for network capacity maximization in energy-constrained ad-hoc networks *IEEE INFOCOM*, San Francisco.

Karp B and Kung H 2000 Gpsr: Greedy perimeter stateless routing for wireless networks *ACM MOBICOM*, Boston.

Katti S, Katabi D, Balakrishnan H and Medard M 2008 Symbol-level network coding for wireless mesh networks. *ACM SIGCOMM Computer Communication Review* **38**(4), 401–412.

Khojastepour, M. A. 2004 *Distributed Cooperative Communications in Wireless Networks* PhD thesis Rice University.

Kim KH and Shin KG 2006 On accurate measurement of link quality in multi-hop wireless mesh networks *MobiCom '06: Proceedings of the 12th Annual International Conference on Mobile Computing and Networking*, pp. 38–49. ACM, New York, NY, USA.

Knopp R and Humblet P 1995 Information capacity and power control in single-cell multiuser communications *IEEE International Conference on Communications, 1995. ICC '95 Seattle, "Gateway to Globalization"*, vol. **1**, pp. 331–335 vol.1.

Kodialam M and Nandagopal T 2005 Characterizing the capacity region in multi-radio multi-channel wireless mesh networks *MobiCom '05*, pp. 73–87. ACM, New York, NY.

Koutsonikolas D, Hu Y and Wang CC 2009 Pacifier: High-throughput, reliable multicast without "crying babies" in wireless mesh networks *INFOCOM 2009, IEEE*, pp. 2473–2481.

Kramer G, Gastpar M and Gupta P 2005 Cooperative strategies and capacity theorems for relay networks. *IEEE Transactions on Information Theory* **51**(9), 3037–3063.

Kuhn F, Wattenhofer R, Zhang Y and Zollinger A 2003 Geometric ad-hoc routing: Of theory and practice *22nd ACM Symposium on the Principles of Distributed Computing (PODC)*, Boston.

Laneman J and Wornell G 2002 Distributed space–time-coded protocols for exploiting cooperative diversity in wireless networks *Proceedings of the IEEE Global Communications Conference (GLOBECOMM)*, pp. 77–81, Taipei, Taiwan.

Laneman, J. N. 2002 *Cooperative Diversity in Wireless Networks: Algorithms and Architectures* PhD thesis Massachusetts Institute of Technology.

Laneman JN, Tse DNC and Wornell GW 2004 Cooperative diversity in wireless networks: Efficient protocols and outage behavior. *IEEE Transactions on Information Theory* **50**(12), 3062–3080.

Larsson P 2001 Selection diversity forwarding in a multihop packet radio network with fading channel and capture. *SIGMOBILE Mobile Communication Review* **5**(4), 47–54.

Lee S, Bhattacharjee B and Banerjee S 2005 Efficient geographic routing in multihop wireless networks *MobiHoc*.

Li Q, Aslam JA and Rus D 2001 Online power-aware routing in wireless ad-hoc networks *Mobicom'01*, Rome, Italy.

Li SY, Yeung R and Cai N 2003 Linear network coding. *IEEE Transactions on Information Theory* **49**(2), 371–381.

Li Z and Li B 2004 Network coding: The case of multiple unicast sessions *in Proceedings of the 42nd Allerton Annual Conference on Communication, Control, and Computing*.

Liang Y and Veeravalli V 2005 Gaussian orthogonal relay channels: Optimal resource allocation and capacity. *IEEE Transactions on Information Theory* **51**(9), 3284–3289.

Liu X, Chong EKP and Shroff NB n.d. Optimal opportunistic scheduling in wireless networks *IEEE Infocom*.

Lorincz K, Malan D, Fulford-Jones T, Nawoj A, Clavel A, Shnayder V, Mainland G, Moulton S and Welsh M 2004 Sensor networks for emergency response: Challenges and opportunities. *IEEE Pervasive Computing*.

Lun DS, Medard M, Koetter R and Effros M 2008 On coding for reliable communication over packet networks. *Physical Communication* **1**(1), 3–20.

Meulen ECVD 1971 Three-terminal communication channels. *Advances in Applied Probability* **3**(1), 120–154.

Mitran P, Ochiai H and Tarokh V 2005 Space-time diversity enhancements using collaborative communications. *IEEE Transactions on Information Theory* **51**(6), 2041–2057.

Miu A, Balakrishnan H and Koksal CE 2005 Improving loss resilience with multi-radio diversity in wireless networks *Proceedings of the 11th Annual International Conference on Mobile Computing and Networking*, pp. 16–30 MobiCom '05. ACM, New York, NY, USA.

Nabar R, Bolcskei H and Kneubuhler F 2004 Fading relay channels: performance limits and space-time signal design. *IEEE Journal on Selected Areas in Communications* **22**(6), 1099–1109.

Park VD and Corson MS 1997 A highly adaptive distributed routing algorithm for mobile wireless networks *IEEE INFOCOM '97*, pp. 1405–1413.

Perkins CE and Bhagwat P 1994 Highly dynamic destination-sequenced distance-vector routing (dsdv) for mobile computers *ACM SIGCOMM 1994*, pp. 234–244.

Perkins CE and Bhagwat P 2001 Dsdv: Routing over a multihop wireless network of mobile computers In *Ad Hoc Networking* (ed. Perkins CE) Addison-Wesley chapter 3, pp. 53–74.

Perkins CE and Royer EM 2001 The ad hoc on-demand distance-vector protocol In *Ad Hoc Networking* (ed. Perkins CE) Addison-Wesley chapter 6, pp. 173–219.

Rahul H, Hassanieh H and Katabi D 2010 Sourcesync: a distributed wireless architecture for exploiting sender diversity *Proceedings of the ACM SIGCOMM 2010 Conference on SIGCOMM*, pp. 171–182 SIGCOMM '10. ACM, New York, NY, USA.

Sang L, Arora A and Zhang H 2007 On exploiting asymmetric wireless links via one-way estimation *MobiHoc '07: Proceedings of the 8th ACM International Symposium on Mobile Ad hoc Networking and Computing*, pp. 11–21. ACM, New York, NY, USA.

Savvides A, Han C and Strivastava MB 2001 Dynamic finegrained localization in ad-hoc networks of sensors *IEEE/ACM MobiCom*.

Seada K, Zuniga M, Helmy A and Krishnamachari B 2004 Energy efficient forwarding strategies for geographic routing in wireless sensor networks *ACM Sensys'04*, Baltimore, MD.

Sendonaris A, Erkip E and Aazhang B 1998 Increasing uplink capacity via user cooperation diversity *Proc. 1998 IEEE International Symp. on Information Theory*, p. 156.

Sendonaris A, Erkip E and Aazhang B 2003a User cooperation diversity – Part I: System description. *IEEE Transactions on Communications* **51**(11), 1927–1938.

Sendonaris A, Erkip E and Aazhang B 2003b User cooperation diversity – part ii: Implementation aspects and performance analysis. *IEEE Transactions on Communications* **51**(11), 1939–1948.

Shah RC, Bonivento A, Petrovic D, Lin E, van Greunen J and Rabaey J 2004 Joint optimization of a protocol stack for sensor networks *IEEE Milcom*.

Shah RC, Wietholter S, Wolisz A and Rabaey JM 2005 When does opportunistic routing make sense? *IEEE PerSens*.

Singh S, Woo M and Raghavendra CS 1998 Power-aware routing in mobile ad hoc networks *ACM/IEEE MOBICOM'98*, Dallas, Texas.

Souryal MR and Moayeri N 2005 Channel-adaptive relaying in mobile ad hoc networks with fading *IEEE SECON*.

Toumpis S and Goldsmith A 2003 Capacity regions for wireless ad hoc networks. *IEEE Transactions on Wireless Communications* **2**(4), 736–748.

Vahdat A and Becker D 2000 Epidemic Routing for Partially-Connected Ad Hoc Networks. Technical report, Duke University.

Wang B, Zhang J and Host-Madsen A 2005a On the capacity of mimo relay channels. *IEEE Transactions on Information Theory* **51**(1), 29–43.

Wang Y, Jain S, Martonosi M and Fall K 2005b Erasure-coding based routing for opportunistic networks *Proceedings of the 2005 ACM SIGCOMM Workshop on Delay-tolerant Networking*, pp. 229–236 WDTN '05. ACM, New York, NY, USA.

Woo GR, Kheradpour P, Shen D and Katabi D 2007 Beyond the bits: cooperative packet recovery using physical layer information *Proceedings of the 13th Annual ACM International Conference on Mobile Computing and Networking*, pp. 147–158 MobiCom '07. ACM, New York, NY, USA.

Zeng K, Lou W and Zhai H 2008 On end-to-end throughput of opportunistic routing in multirate and multihop wireless networks *Infocom*, Phoenix, AZ.

Zeng K, Lou W and Zhang Y 2007a Multi-rate geographic opportunistic routing in wireless ad hoc networks *IEEE Milcom*, Orlando, FL.

Zeng K, Lou W, Yang J and Brown DR 2007b On geographic collaborative forwarding in wireless ad hoc and sensor networks *WASA'07*, Chicago, IL.

Zeng K, Lou W, Yang J and Brown DR 2007c On throughput efficiency of geographic opportunistic routing in multihop wireless networks *QShine'07*, Vancouver, British Columbia, Canada.

Zhai H and Fang Y 2006a Impact of routing metrics on path capacity in multirate and multihop wireless ad hoc networks *IEEE ICNP*.

Zhai H and Fang Y 2006b Physical carrier sensing and spatial reuse in multirate and multihop wireless ad hoc networks *IEEE Infocom*.

Zhang J, Wu H, Zhang Q and Li B 2005 Joint routing and scheduling in multi-radio multi-channel multi-hop wireless networks *IEEE Broadnets*.

Zhao B and Valenti M 2005 Practical relay networks: A generalization of hybrid-arq. *IEEE Journal of Selected Areas in Communications* **23**(1), 7–18.

Zhao J and Govindan R 2003 Understanding packet delivery performance in dense wireless sensor networks *ACM Sensys'03*, LA, CA.

Zhong Z and Nelakuditi S 2007 On the efficacy of opportunistic routing *IEEE SECON*.

Zhong Z, Wang J and Nelakuditi S 2006 Opportunistic any-path forwarding in multi-hop wireless mesh networks. Technical Report TR-2006-015, USC-CSE.

Zhong Z, Wang J, Lu GH and Nelakuditi S 2005 On selection of candidates for opportunistic anypath forwardning *ACM MOBICOM (Poster Session)*.

Zorzi M and Armaroli A 2003 Advancement optimization in multihop wireless networks. *Proceedings of VTC*.

Zorzi M and Rao RR 2003a Geographic random forwarding (geraf) for ad hoc and sensor networks: energy and latency performance. *IEEE Transactions on Mobile Computing* **2**(4), 349–365.

Zorzi M and Rao RR 2003b Geographic random forwarding (geraf) for ad hoc and sensor networks: multihop performance. *IEEE Transactions on Mobile Computing* **2**(4) 337–348.

Zubow A, Kurth M and Redlich JP 2007 Multi-channel opportunistic routing *IEEE European Wireless Conference*, Paris, France.

2

Taxonomy of opportunistic routing: principles and behaviors

This chapter analyzes the principles of the local behavior of GOR. We then introduce the least cost opportunistic/anypath routing, and present some of its important properties. Finally, we describe two polynomial-time algorithms that select the optimal forwarding candidates at each hop to achieve the shortest anypath cost from a source to a destination.

For GOR, we first generalize the definition of expected packet advancement (EPA) for an arbitrary number of forwarding candidates that follow a specific priority rule to relay the packet in OR. Through theoretical analysis, we prove that the maximum EPA can only be achieved by giving the forwarding candidates closer to the destination higher relay priorities. This **relay priority rule** convinces us that given a forwarding candidate set, the relay priority among the candidates is only relevant to the advancement achieved by the candidate to the destination, but is irrelevant to the packet delivery ratio between the transmitter and the forwarding candidate. The analysis result is the upper bound of the EPA that any GOR can achieve. We further prove that, given a set of M nodes that are available as next-hop neighbors, a subset of the available next-hop neighbors with r $(r < M)$ nodes achieving the maximum EPA is contained in a subset with more nodes achieving the maximum EPA. Leveraging the **containing property**, we demonstrate that the maximum EPA of selecting $r (r \leq M)$ nodes is a strictly **increasing** and **concave** function of r. This property indicates that although getting more forwarding

Multihop Wireless Networks: Opportunistic Routing, First Edition.
Kai Zeng, Wenjing Lou and Ming Li.
© 2011 John Wiley & Sons, Ltd. Published 2011 by John Wiley & Sons, Ltd.

candidates involved in GOR will increase the maximum EPA, the extra EPA gained by doing so becomes less significant. It also implies consistency between EPA and reliability. These principles of GOR will help us analyze the capacity of OR in Chapters 4 and 5, and design efficient local candidate selection and prioritization algorithms for achieving energy and throughput efficiency in Chapters 3 and 8, respectively.

2.1 EPA generalization

In this section, as we mainly focus on the local behavior of GOR, for a given transmitter n_i, we abbreviate its **forwarding candidate set** $\mathcal{F}_i$ as $\mathcal{F}$, and its **available next-hop node set** C_i as C. Note that, $\mathcal{F}$ is an ordered subset of C, which is a set of all the neighbors that are geographically closer to the destination than the transmitter n_i. n_i's neighbor n_{i_q}, its advancement to the destination d_{ii_q}, and the PRR p_{ii_q} on link l_{ii_q} are simplified as i_q, d_q, and p_q, respectively. We assume the packet reception ratios (p_q) are independent of each other. The independence has been validated in practice (Laufer and Kleinrock 2008a; Reis *et al.* 2006). We denote the number of nodes in $\mathcal{F}$ as r, and the number of nodes in C as M. Redefine $\mathcal{F} := \langle i_1, \ldots, i_r \rangle$, and $C := \{i_1, \ldots, i_M\}$. Note that, the subscript of i only represents the sequence number of each node in set $\mathcal{F}$ and C, and two nodes having the same subscript in $\mathcal{F}$ and C are not necessarily the same node. For example, i_1 in $\mathcal{F}$ does not necessarily indicate the same node as i_1 in C. Without loss of generality, we assume all the nodes in C and $\mathcal{F}$ are descending ordered according to the advancement s.t. given nodes i_m and i_n, we have $d_m > d_n$, $\forall m < n$.

Let $\pi(\mathcal{F}) = \langle i_{\pi_1}, i_{\pi_2}, \ldots, i_{\pi_r} \rangle$ be one permutation of nodes in $\mathcal{F}$, and the order indicates that nodes will attempt to forward the packet with priority $i_{\pi_1} > i_{\pi_2} > \ldots > i_{\pi_r}$. We define the EPA for the ordered forwarding candidate set $\pi(\mathcal{F})$ in Equation (2.1)

$$\text{EPA}(\pi(\mathcal{F})) = \sum_{k=1}^{r} d_{\pi_k} p_{\pi_k} \cdot \prod_{n=0}^{k-1} \overline{p}_{\pi_n} \tag{2.1}$$

where $\overline{p}_{\pi_n} = 1 - p_{\pi_n}$ and $\overline{p}_{\pi_0} := 1$. The physical meaning of Equation (2.1) is the expected packet advancement achieved by GOR in one transmission using the ordered forwarding candidate set $\pi(\mathcal{F})$. The EPA metric accurately indicates the relationship between the packet advancement and candidate selection and prioritization. Note that when $r = 1$, Equation (2.1) degenerates to the "distance $\times$ PRR" proposed in geographic routing (Lee *et al.* 2005; Seada *et al.* 2004).

2.2 Principles of local behavior of GOR

2.2.1 EPA strictly increasing property

Intuitively, increasing the number of forwarding candidates would result in a larger EPA. We present Lemma 2.1 to confirm this intuition.

Definition 2.1 *Define $EM(\mathcal{C}, r)$ be the maximum EPA (defined in Equation (2.1)) achieved by selecting r forwarding candidates from $\mathcal{C}$.*

Lemma 2.1 *(**Strictly increasing property**) $EM(\mathcal{C}, r)$ is a strictly increasing function of r.*

Proof. Assume $1 \leq m < n \leq M$, and without loss of generality, let $\mathcal{A} = \langle i_1, i_2, \ldots, i_m \rangle$ be the ordered node set achieving $EM(\mathcal{C}, m)$ with forwarding priority $i_1 > \ldots > i_m$. We then select a subset with $n - m$ nodes from the remaining node set $\{i_{m+1}, i_{m+2}, \ldots, i_M\}$, say $\mathcal{B} = \langle i_{m+1}, \ldots, i_n \rangle$. Assume we retain the relay priority of the m nodes in $\mathcal{A}$ unchanged and give the nodes in $\mathcal{B}$ lower priorities than those in $\mathcal{A}$. Then in $\mathcal{B}$, we give the nodes with smaller subscripts higher relay priorities. So we have

$$EM(\mathcal{C}, n) \geq EPA(\langle i_1, \ldots, i_n \rangle) = EM(\mathcal{C}, m) + EPA(\langle i_{m+1}, \ldots, i_n \rangle)$$

$$\times \prod_{k=1}^{m} \overline{p}_k > EM(\mathcal{C}, m)$$

Lemma 2.1 basically indicates that the more nodes get involved in GOR, the larger the EPA can be. The maximum EPA can be obtained by involving all the nodes in $\mathcal{C}$. Then, how can the candidates be prioritized to maximize the EPA? We answer this question in the following section.

2.2.2   Relay priority rule

Theorem 2.1 identifies the upper bound of EPA and the corresponding relay priority rule.

Theorem 2.1 *(**Relay priority rule**) $EM(\mathcal{F}, |\mathcal{F}|)$ can only be obtained by giving the node closer to the destination higher relay priority. That is*

$$EM(\mathcal{F}, |\mathcal{F}|) = \sum_{k=1}^{r} d_k p_k \cdot \prod_{n=0}^{k-1} \overline{p}_n \tag{2.2}$$

where $\overline{p}_0 := 1$.

Proof. We prove Theorem 2.1 by induction on r, the size of $\mathcal{F}$.
First, when r = 1, obviously Equation (2.2) holds.
Next, we assume Equation (2.2) holds for r = N (N $\geq$ 1). When $|\mathcal{F}| = N + 1$, $\mathcal{F}$ can be divided into $\mathcal{F}_1 = \mathcal{F} - \{i_m\}$ with N nodes and $\mathcal{F}_2 = \{i_m\}$ with 1 node. Then

$$EM(\mathcal{F}, |\mathcal{F}|) = \max_{1 \leq m \leq N+1} \left\{ \sum_{k=1}^{m-1} d_k p_k \prod_{w=0}^{k-1} \overline{p}_w \sum_{k=m+1}^{N+1} d_k p_k \frac{\prod_{w=0}^{k-1} \overline{p}_w}{\overline{p}_m} \right.$$

$$\left. + d_m p_m \frac{\prod_{w=0}^{N+1} \overline{p}_w}{\overline{p}_m} \right\}$$

Thus we only need to prove for any integer m ($1 \leq m \leq N$),

$$A := \sum_{k=1}^{m-1} d_k p_k \prod_{w=0}^{k-1} \overline{p}_w + \sum_{k=m+1}^{N+1} d_k p_k \frac{\prod_{w=0}^{k-1} \overline{p}_n}{\overline{p}_m} + d_m p_m \frac{\prod_{w=0}^{N+1} \overline{p}_w}{\overline{p}_m}$$

$$< B := \sum_{k=1}^{N+1} d_k p_k \prod_{w=0}^{k-1} \overline{p}_w$$

Subtracting A from B, we have

$$B - A = \frac{1}{\overline{p}_m} \sum_{k=m+1}^{N+1} (d_m - d_k) p_m p_k \prod_{w=0}^{k-1} \overline{p}_w > 0$$

Then Equation (2.2) holds for r = N + 1. So it holds for any r ($r \geq 1$).

Theorem 2.1 indicates that when a forwarding candidate set is chosen, the maximum EPA can only be achieved by assigning the relay priority to each node based on its distance from the destination. That is, the furthest node should try to forward the packet first; if it fails (i.e., does not receive the packet correctly), the second furthest node should try next, and so on. The analysis result is the upper bound of the EPA that any GOR can achieve.

Based on the **relay priority rule**, we will next identify and prove two important principles about the maximum EPA. First, we look at the characteristics of the forwarding candidates that are selected to achieve $EM(C, r)$ with various sizes r. We prove the **containing property** for those node sets. Following that, the **concavity** of the function $EM(C, r)$ is proved.

2.2.3 Containing property of feasible candidate set

Let $\mathcal{F}_r^*$ be a feasible ordered node set that achieves the $EM(C, r)$. We have the following containing property of $\mathcal{F}_r^*$'s.

Lemma 2.2 *(Containing property) Given the available next-hop node set C with M nodes, $\forall \, \mathcal{F}_{r-1}^*, \, \exists \, \mathcal{F}_r^*$, s.t.*

$$\mathcal{F}_{r-1}^* \subset \mathcal{F}_r^* \; \forall \, 1 \leq r \leq M \tag{2.3}$$

Proof. Let $\mathcal{A} = \langle a_1, \ldots, a_M \rangle^1$ be an ordered node set with M nodes, and $\mathcal{B} = \langle b_1, \ldots, b_N \rangle$ with N nodes. $\mathcal{B} \subset \mathcal{A}$ and $b_N = a_M$. For any node $q \notin \mathcal{A}$ with $d_q < d_{a_M}$, we have

$$EPA(\langle \mathcal{A}, q \rangle) - EPA(\langle q, \mathcal{A} \rangle) > EPA(\langle \mathcal{B}, q \rangle) - EPA(\langle q, \mathcal{B} \rangle) \tag{2.4}$$

We then prove Lemma 2.2 by induction on r.

First, for an arbitrary N, when $r = 1$, as $\mathcal{F}_0^* = \emptyset$, and $\mathcal{F}_1^* \neq \emptyset$, it is obvious that the containing property holds.

---

[1] For simplicity, we denote a node using its subscript in this proof.

Then, we assume $\forall \mathcal{F}_{m-1}^*$, $\exists$ an $\mathcal{F}_m^*$, s.t. $\mathcal{F}_{m-1}^* \subset \mathcal{F}_m^*$, when $r = m$ ($m \geq 1$). We first prove for any feasible $\mathcal{F}_m^*$ and $\mathcal{F}_{m+1}^*$, the first node in $\mathcal{F}_{m+1}^*$ cannot be the nodes from the second place to the last place in $\mathcal{F}_m^*$, that is $(m+1)_1 \neq m_i$, $\forall\, 2 \leq i \leq m$.

We prove this by contradiction. Assume $(m+1)_1 = m_i$. Let node $(m+1)_j$ be the first node in $\mathcal{F}_{m+1}^*$ but not in $\mathcal{F}_m^*$. We have

$$\text{EPA}(\mathcal{F}_m^*) \geq \text{EPA}(\mathcal{F}_{m+1}^* - \{(m+1)_j\}) \tag{2.5}$$

then,

$$\text{EPA}(\langle (m+1)_j, \mathcal{F}_m^* \rangle) \geq \text{EPA}(\langle (m+1)_j, \mathcal{F}_{m+1}^* - \{(m+1)_j\} \rangle) \tag{2.6}$$

Assume $(m+1)_{j-1} = m_l$, and according to inequality (2.4), we have

$$\Delta 1 > \Delta 2 \tag{2.7}$$

where

$$\Delta 1 := \text{EPA}(\langle m_1, \ldots, m_l, (m+1)_j, m_{l+1}, \ldots, m_m \rangle) - \text{EPA}(\langle (m+1)_j, \mathcal{F}_m^* \rangle)$$
$$= \text{EPA}(\langle m_1, \ldots, m_l, (m+1)_j \rangle) - \text{EPA}(\langle (m+1)_j, m_1, \ldots, m_l \rangle) \tag{2.8}$$
$$\Delta 2 := \text{EPA}(\mathcal{F}_{m+1}^*) - \text{EPA}(\langle (m+1)_j, \mathcal{F}_{m+1}^* - \{(m+1)_j\} \rangle)$$
$$= \text{EPA}(\langle (m+1)_1, \ldots, (m+1)_{j-1}, (m+1)_j \rangle) \tag{2.9}$$
$$- \text{EPA}(\langle (m+1)_j, (m+1)_1, \ldots, (m+1)_{j-1} \rangle)$$

Then combining this with inequality (2.6), we get

$$\text{EPA}(\langle m_1 \ldots m_l, (m+1)_j, m_{l+1} \ldots m_m \rangle) > \text{EPA}(\mathcal{F}_{m+1}^*) \tag{2.10}$$

The inequality (2.10) contradicts the fact that $\text{EPA}(\mathcal{F}_{m+1}^*)$ is the largest EPA achieved by selecting $m+1$ nodes. So the assumption $(m+1)_1 = m_i$ is wrong. Therefore $(m+1)_1$ cannot be m_i, $\forall\, 2 \leq i \leq m$. So there are two cases for $(m+1)_1$:

1. $(m+1)_1 \neq m_1$. Then $\langle (m+1)_1, \mathcal{F}_m^* \rangle$ should be one $\mathcal{F}_{m+1}^*$.

2. $(m+1)_1 = m_1$. By the inductive hypothesis, we have $\mathcal{F}_m^* - \{m_1\} \subset \langle (m+1)_2, \ldots, (m+1)_{m+1} \rangle$, then $\mathcal{F}_m^* \subset \mathcal{F}_{m+1}^*$.

From the induction above, we know that for an arbitrary N, we have $\forall\, \mathcal{F}_{r-1}^*$, $\exists\, \mathcal{F}_r^*$ s.t. $\mathcal{F}_{r-1}^* \subset \mathcal{F}_r^*$, $\forall\, 1 \leq r \leq M$.

Lemma 2.2 indicates that an $r-1$-node set that achieves $\text{EM}(\mathcal{C}, r-1)$ is a subset of at least one of the feasible r-node sets that achieve $\text{EM}(\mathcal{C}, r)$. It also implies that when more forwarding candidates are selected to increase the maximum EPA, the transmission reliability also increases.

2.2.4 Concavity of maximum EPA

Following Lemma 2.2, we have the concave property of $EM(\mathcal{C}, r)$ as in Theorem 2.2.

Theorem 2.2 (Concavity of maximum EPA) $EM(\mathcal{C}, r + 1) - EM(\mathcal{C}, r) < EM(\mathcal{C}, r) - EM(\mathcal{C}, r - 1)$, $\forall r$, s.t. $1 \leq r < N$.

Proof. According to Lemma 2.2, assume $\mathcal{F}^*_{r+1} - \mathcal{F}^*_r = \{i_k\}$, and $\mathcal{F}^*_r - \mathcal{F}^*_{r-1} = \{i_j\}$. There are two cases for d_k and d_j.

1. $d_k > d_j$. Then $\mathcal{F}^*_{r+1}$, $\mathcal{F}^*_r$ and $\mathcal{F}^*_{r-1}$ can be represented as

$$\mathcal{F}^*_{r+1} = \langle \mathcal{A}_1, i_k, \mathcal{A}_2, i_j, \mathcal{A}_3 \rangle, \mathcal{F}^*_r = \langle \mathcal{A}_1, \mathcal{A}_2, i_j, \mathcal{A}_3 \rangle, \mathcal{F}^*_{r-1} = \langle \mathcal{A}_1, \mathcal{A}_2, \mathcal{A}_3 \rangle$$

where $\mathcal{A}_i$ ($1 \leq i \leq 3$) is ordered node set and can be $\varnothing$.
We have

$$B := EPA(\mathcal{F}^*_r) - EPA(\langle \mathcal{A}_1, i_k, \mathcal{A}_2, \mathcal{A}_3 \rangle) \geq 0 \qquad (2.11)$$

Then,

$$[EPA(\mathcal{F}^*_r) - EPA(\mathcal{F}^*_{r-1})] - [EPA(\mathcal{F}^*_{r+1}) - EPA(\mathcal{F}^*_r)]$$
$$= B + \overline{p}_{\mathcal{A}_1} \overline{p}_{\mathcal{A}_2} p_k p_j (d_j - EPA(\mathcal{A}_3)) > 0$$

where $\overline{p}_{\mathcal{A}_i}$ is the probability of none of nodes in $\mathcal{A}_i$ receiving the packet correctly.

2. $d_k < d_j$. Similarly, with

$$B := EPA(\mathcal{F}^*_r) - EPA(\langle \mathcal{A}_1, \mathcal{A}_2, i_k, \mathcal{A}_3 \rangle) \geq 0 \qquad (2.12)$$

we can derive

$$[EPA(\mathcal{F}^*_r) - EPA(\mathcal{F}^*_{r-1})] - [EPA(\mathcal{F}^*_{r+1}) - EPA(\mathcal{F}^*_r)]$$
$$= B + \overline{p}_{\mathcal{A}_1} \overline{p}_{\mathcal{A}_2} p_k p_j (d_k - EPA(\mathcal{A}_3)) > 0$$

From the analysis above, we know $EM(\mathcal{C}, r)$ is a concave function of r.

Combining Lemma 2.1 and Theorem 2.2, we know that giving an available next-hop node set $\mathcal{C}$ with N nodes, the maximum EPA of selecting r ($1 \leq r \leq N$) nodes is a ***strictly increasing and concave function*** of r. This means that although the maximum EPA keeps increasing when more nodes get involved, the speed of the increase slows down. When many nodes are involved, the extra EPA gained becomes marginal.

2.2.5 Reliability increasing property

Following the **containing property** in Lemma 2.2, we have the **reliability increasing property** in Corollary 2.1.

Denote $\mathcal{F}_r^* = \langle i_{r_1}, i_{r_2}, \ldots, i_{r_r} \rangle$. Define the one-hop reliability $P_{\mathcal{F}_r^*}$ in Equation (2.13) which is the probability of at least one node in $\mathcal{F}_r^*$ correctly receiving the packet sent by node i for one transmission.

$$P_{\mathcal{F}_r^*} = 1 - \prod_{n=0}^{r} (1 - p_{r_n}) \tag{2.13}$$

where $p_{r_0} = 0$.

$P^*(r)$ is defined in Equation (2.14). This is the maximum one-hop reliability achieved by one of the feasible $\mathcal{F}_r^*$'s.

$$P^*(r) = \max_{\forall \mathcal{F}_r^*} \{P_{\mathcal{F}_r^*}\} \tag{2.14}$$

Corollary 2.1 $P^*(r)$ *defined in Equation (2.14) is an increasing function of r.*

Proof. The proof is straightforward following Lemma 2.2. Assume one $\mathcal{F}_r^*$ achieves $P^*(r)$, then $\exists \, \mathcal{F}_{r+1}^*$ s.t. $\mathcal{F}_{r+1}^* \supset \mathcal{F}_r^*$. According to the definitions of $P^*(r)$ in Equation (2.14) and the one-hop reliability in Equation (2.13), we have

$$P^*(r+1) \geq P_{\mathcal{F}_{r+1}^*} > P_{\mathcal{F}_r^*} = P^*(r)$$

So $P^*(r)$ is an increasing function of r.

Corollary 2.1 indicates that the maximum one-hop reliability corresponding to the forwarding candidate set that maximizes the EPA also increases when more forwarding candidates are involved. The increasing of the maximum EPA implies increasing the reliability. Therefore, the EPA is a good metric for balancing the packet advancement and reliability.

2.3 Least cost opportunistic routing

In the above analysis, we assume the node location information is known and make opportunistic routing decision locally. Next, we introduce least cost opportunistic routing and two polynomial time distributed algorithms to compute the end-to-end opportunistic paths that achieve the least cost. First, we introduce the metric of Expected Opportunistic Transmission Count (EOTX).

2.3.1 Expected opportunistic transmission count (EOTX)

The expected opportunistic transmission count is proposed in Dubois-Ferriere et al. (2007). It is a generalization of the ETX metric in traditional routing (Couto et al. 2003). For a node n_i and its forwarding candidate set $\mathcal{F}$, the EOTX is defined in Equation (2.15).

$$d_i^{\mathcal{F}} = \frac{1}{P_{\mathcal{F}}} \tag{2.15}$$

where $P_{\mathcal{F}}$ has the same definition in Equation (2.13). The physical meaning of $P_{\mathcal{F}}$ is the probability of at least one forwarding candidate receiving the packet correctly sent by n_i per transmission. So the EOTX is the expected number of transmissions necessary for at least one candidate in $\mathcal{F}$ correctly receiving the packet transmitted by n_i.

2.3.2 End-to-end cost of opportunistic routing

The end-to-end cost of opportunistic routing from a node n_i to the destination n_d is defined in Equation (2.16).

$$D_i = d_i^{\mathcal{F}} + D_{\mathcal{F}} \tag{2.16}$$

where $D_{\mathcal{F}}$ is the remaining opportunistic path cost intuitively defined below.

$$D_{\mathcal{F}} = \sum_{n_j \in \mathcal{F}} w_j D_j \tag{2.17}$$

where

$$w_j = \frac{p_j \prod_{k=1}^{j-1}(1 - p_k)}{P_{\mathcal{F}}}, \text{ with } \sum_{n_j \in \mathcal{F}} w_j = 1 \tag{2.18}$$

2.3.3 Properties of LCOR

The goal of least cost opportunistic routing (Dubois-Ferriere *et al.* 2007) or shortest anypath routing (Laufer and Kleinrock 2008b) is to find the forwarding candidates at each hop to achieve a minimum end-to-end cost D_s in Equation (2.16) from the source node n_s to the destination node n_d. This problem looks complicated because the least cost or shortest anypath from a source to a destination depends on the least cost from the forwarding candidates of the source node, and those costs are further dependent on the forwarding candidates of those forwarding candidates, etc. It seems that we need to enumerate all the possible forwarding candidates and their forwarding priorities to find the shortest anypath. However, as proved in (Laufer and Kleinrock 2008b), polynomial algorithms exist to solve this combinatorial problem. Now, we present five lemmas proved in (Laufer and Kleinrock 2008b), and they will help us to understand the polynomial algorithms that will be presented in Section 2.3.4 and Section 2.3.5.

Let δ_i be the distance of the shortest anypath from a node n_i to the destination n_d, and Φ_i be the corresponding optimal forwarding candidate set.

Lemma 2.3 *Let D_i be the distance of a node n_i via forwarding set $\mathcal{F}$ and let D_i' be the distance via forwarding set $\mathcal{F}' = \mathcal{F} \cup n_k$, where $D_k \geq D_j$ for every node $n_j \in \mathcal{F}$. We have $D_i' \leq D_i$ if and only if $D_i \geq D_k$.*

This lemma indicates that for a node n_i, it is always beneficial to involve its neighbor node n_k into the forwarding candidate set in order to obtain a shorter distance to the destination if the distance from n_k to the destination is larger than the distance from any current candidate to the destination but is smaller than the distance from n_i to the destination.

Lemma 2.4 *The shortest distance δ_i of a node n_i is always no shorter than the shortest distance δ_j of any node n_j in the optimal forwarding set Φ_i. That is, we have $\delta_i \geq \delta_j \forall n_j \in \Phi_i$.*

Lemma 2.4 guarantees that if a node n_i uses its neighbor node n_j in its optimal forwarding set Φ_i, the shortest distance δ_i can never be smaller than δ_j. This is equivalent to the restriction that all weights in the graph must be non-negative in Dijkstra's algorithm.

Lemma 2.5 *If a node n_i uses a node n_k in its optimal forwarding set Φ_i and $\delta_i = \delta_k$, node n_k can be safely removed from Φ_i without changing δ_i. The link $n_i \rightarrow n_k$ is said to be "redundant".*

Lemma 2.5 says if the shortest distances from node n_i and n_k to the destination are the same, the shortest distance δ_i via forwarding candidate set Φ_i (where $n_k \in \Phi_i$) is the same as the shortest distance via forwarding set $\Phi_i - n_k$. That is, the shortest distance from node n_i to the destination does not change no matter whether it uses n_k in its forwarding candidate set or not.

Lemma 2.6 *If the shortest distances from the neighbors of a node n_i to a destination are $\delta_1 \leq \delta_2 \leq \ldots \leq \delta_k$, Φ_i is always of the form $\Phi_i = \langle n_1, n_2, \ldots, n_r \rangle$, for some $r \in \{1, 2, \ldots, k\}$.*

By Lemma 2.6, for a node n_i, the optimal forwarding candidate set Φ_i is a subset of n_i's neighbors with the shortest distances to the destination. That is, given a set of neighbors with distances $\delta_1 \leq \delta_2 \leq \ldots \leq \delta_k$, the best forwarding candidate set Φ_i is always one of $\langle n_1 \rangle$, $\langle n_1, n_2 \rangle$, $\langle n_1, n_2, n_3 \rangle$, $\ldots$, $\langle n_1, n_2, \ldots, n_r \rangle$. As a result, any forwarding candidate set with gaps between the neighbors, such as $\langle n_2, n_3 \rangle$ or $\langle n_1, n_4 \rangle$, can never yield the shortest distance to the destination. This property is the key factor that reduces the complexity of the shortest anypath algorithms from exponential to polynomial time. For k neighbors, we do not have to test every one of the $2^k - 1$ possible forwarding candidate sets. Instead, we only need to check at most k forwarding candidate sets.

Lemma 2.7 *Assume for a node n_i, $\Phi_i = \langle n_1, n_2, \ldots, n_r \rangle$ with shortest distances $\delta_1 \leq \delta_2 \leq \ldots \leq \delta_r$. If D_i^j is the distance from n_i via forwarding candidate set $\langle n_1, n_2, \ldots, n_j \rangle$ $(1 \leq j \leq r)$, we always have $D_i^1 \geq D_i^2 \geq \ldots \geq D_i^r = \delta_i$.*

Lemma 2.7 explains another important property necessary for the shortest anypath algorithm to converge. Assuming now that the best forwarding set $\Phi_i = \langle n_1, n_2, \ldots, n_r \rangle$ with distances $\delta_1 \leq \delta_2 \leq \ldots \leq \delta_r$, the distance D_i monotonically decreases as each of the forwarding candidate sets $\langle n_1 \rangle$, $\langle n_1, n_2 \rangle$, $\ldots$, $\langle n_1, n_2, \ldots, n_r \rangle$ is used.

2.3.4   Dijkstra-based algorithm

We now introduce the Dijkstra-based algorithm, the shortest anypath first algorithm, proposed in Laufer and Kleinrock (2008b). Given a graph $G = (V, E)$, where V is the node set and E is the edge set, the algorithm calculates the shortest anypaths from all nodes to a destination node n_d. For every node $n_i \in V$, the algorithm keeps an estimate D_i which is an upper bound of the distance of the shortest anypath from n_i to n_d. In addition, this algorithm also keeps a forwarding candidate set $\mathcal{F}_i$ for every node, which stores the set of nodes used as the next hops to reach n_d. Finally, two data structures, S and Q, are used. The S set stores the set of nodes for which there is already a shortest anypath defined. Each node $n_i \in V - S$ in which a shortest anypath has not found is stored in a priority queue Q keyed by their D_i values.

Algorithm 2.1 consists of $|V|$ rounds, dictated by the number of elements initially in Q. At each round, the EXTRACT-MIN procedure extracts the node with the minimum distance to the destination from Q. Let this node be n_j. At this point, n_j is settled and inserted into S, since the shortest anypath from n_j to the destination is now known. For each incoming edge $n_i \rightarrow n_j \in E$, if the distance D_i is larger than the distance D_j, node n_j is added to the forwarding candidate set F_i and the distance D_i is updated.

Algorithm 2.1 Shortest Anypath First

Input: G, n_d
for each node $n_i \in V$ **do**
    $D_i \leftarrow \infty$
    $\mathcal{F}_i \leftarrow \emptyset$
end for
  $D_d \leftarrow 0$
  $S \leftarrow \emptyset$
  $Q \leftarrow V$
  while $Q \neq \emptyset$ **do**
    $n_j \leftarrow$ EXTRACT-MIN(Q)
    $S \leftarrow S \cup \{n_j\}$
    for each edge $n_i \rightarrow n_j \in E$ **do**
      $J \leftarrow F_i \cup n_j$
      if $D_i > D_j$ **then**
        $D_i \leftarrow d_{iJ} + D_J$
        $\mathcal{F}_i \leftarrow J$
      end if
    end for
  end while

The optimality of Algorithm 2.1 is proved in Laufer and Kleinrock (2008b). The complexity of this algorithm is $O(|V|log|V| + |E|)$ which is the same complexity as Dijkstra's algorithm (Cormen *et al.* 2001).

2.3.5   Bellman – Ford-based algorithm

Algorithm 2.1 solves the shortest anypath problem in polynomial time, but it requires that the node knows the global information of the network (i.e. the topology of the network and link state). Next, we introduce a Bellman–Ford-based algorithm (Laufer *et al.* 2010), which can solve the shortest anypath problem in a distributed way. By applying Lemma 2.6 presented in Section 2.3.3, this algorithm reduces the complexity of the Bellman–Ford based anypath generalization proposed in (Dubois-Ferriere *et al.* in press) from exponential to polynomial time. As in the regular Bellman–Ford algorithm for single-path routing (Cormen *et al.* 2001), this algorithm takes at most $|V| - 1$ rounds. At each round, every node n_i stores its neighbors in a priority queue Q keyed by their cost. This algorithm then checks each neighbor n_j in ascending order of cost D_j, and verifies whether D_i is larger than D_j. If that is the case, node n_i includes n_j in its forwarding candidate set and updates its distance accordingly. Intuitively, this algorithm works in the same expanding-ring fashion as the regular Bellman–Ford algorithm, settling at each round the costs of the nodes one hop further away from the destination. Since an anypath can not be longer than $|V| - 1$ hops, the algorithm converges after at most $|V| - 1$ iterations.

The running time of Algorithm 2.2 depends on how Q is implemented. Assuming a Fibonacci heap, each of the EXTRACT-MIN operations takes at most $log(|V|)$ running time. The for loop in lines 6–14 runs once for each link, for a total aggregated time of $O(|E|log|V|)$. The total complexity of this algorithm is

---

Algorithm 2.2 Bellman–Ford-Based Shortest Anypath Algorithm

Input: G, n_d
for each node $n_i \in V$ **do**
    $D_i \leftarrow \infty$
    $\mathcal{F}_i \leftarrow \emptyset$
end for
$D_d \leftarrow 0$
for $k \leftarrow 1$ to $|V| - 1$ **do**
    for each node $n_i \in V$ **do**
        $J \leftarrow \emptyset$
        $Q \leftarrow$ GET-NEIGHBORS(n_i)
        while $Q \neq \emptyset$ **do** $n_j \leftarrow$ EXTRACT-MIN(Q)
            $J \leftarrow J \cup \{n_j\}$
            if $D_i > D_j$ **then**
                $D_i \leftarrow d_{iJ} + D_J$
                $\mathcal{F}_i \leftarrow J$
            end if
        end while
    end for
end for

---

then $O(|V||E|log|V|)$, which is only a factor of $log|V|$ higher than the regular Bellman–Ford algorithm.

2.4 Conclusions

In this chapter, we generalized the definition of EPA for an arbitrary number of forwarding candidates in GOR. Through theoretical analysis, we first showed that the maximum EPA can only be achieved by following a relay priority rule – giving the forwarding candidates closer to the destination higher relay priorities when a forwarding candidate set is given. We gave the analytical result of the upper bound of the EPA that any GOR can achieve. We found that the node set achieving the maximum EPA with r nodes is contained in at least one node set achieving the maximum EPA with $r + 1$ nodes. We also showed that giving an available next-hop neighbor set with M nodes, the maximum EPA achieved by selecting r nodes is a strictly increasing and concave function of r and we showed how the candidates should be selected to achieve the maximum EPA. We further showed that increasing the maximum EPA is consistent with increasing the one-hop reliability. These unveiled properties of the local behavior of GOR will enable us to design efficient local routing metric and candidate selection and prioritization algorithms to approach the global optimum performance.

We further introduce the least cost opportunistic routing and important properties about it. Two polynomial algorithms that find shortest anypath for LCOR are described. These two algorithms are based on the proved properties.

References

Cormen TH, Stein C, Rivest RL and Leiserson CE 2001 *Introduction to Algorithms* 2nd edn. McGraw-Hill Higher Education, New York, NY, USA.

Couto DD, Aguayo D, Bicket J and Morris R 2003 A high-throughput path metic for multi-hop wireless routing *ACM MobiCom'03*, San Diego, California.

Dubois-Ferriere H, Grossglauser M and Vetterli M 2007 Least-cost opportunistic routing. Technical Report LCAV-REPORT-2007-001, School of Computer and Communication Sciences, EPFL.

Dubois-Ferriere H, Grossglauser M and Vetterli M In press Valuable detours: Least-cost anypath routing. *IEEE/ACM Transactions on Networking*.

Laufer RP, Dubois-Ferriere H and Kleinrock L 2010 Polynomial-time algorithms for multirate anypath routing in wireless multihop networks. Technical Report UCLA-CSD-TR100034, UCLA Computer Science Department.

Laufer R and Kleinrock L 2008a Multirate Anypath Routing in Wireless Mesh Networks. Technical report, UCLA Computer Science Department.

Laufer RP and Kleinrock L 2008b Multirate anypath routing in wireless mesh networks. Technical Report UCLA-CSD-TR080025, UCLA Computer Science Department.

Lee S, Bhattacharjee B and Banerjee S 2005 Efficient geographic routing in multihop wireless networks *MobiHoc*.

Reis C, Mahajan R, Rodrig M, Wetherall D and Zahorjan J 2006 Measurement-based models of delivery and interference in static wireless networks *Proceedings of the 2006 Conference on Applications, Technologies, Architectures, and Protocols for Computer Communications*, pp. 51–62 SIGCOMM '06. ACM, New York, NY, USA.

Seada K, Zuniga M, Helmy A and Krishnamachari B 2004 Energy efficient forwarding strategies for geographic routing in wireless sensor networks *ACM Sensys'04*, Baltimore, MD.

3

Energy efficiency of geographic opportunistic routing

Wireless sensor networks (WSNs) are characterized by multihop lossy wireless links and severely resource-constrained nodes. Among the resource constraints, energy is probably the most crucial one since sensor nodes are typically battery powered and the lifetime of the battery imposes a limitation on the operation hours of the sensor network. Unlike the microprocessor industry or the communication hardware industry, where computation capability or the line rate has been continuously improved (regularly doubled every 18 months), battery technology has been relatively unchanged for many years. Energy efficiency has been a critical concern in wireless sensor network protocol design. Researchers are investigating energy conservation at every layer in the traditional protocol stack, from the physical layer up to the network layer and application layer.

Among the energy consumption factors, communication has been identified as the major source of energy consumption and costs significantly more than computation in WSNs (Pottie and Kaiser 2000). Opportunistic routing has shown its advantage on energy efficiency (Zeng *et al*. 2007; Zorzi and Rao 2003) comparing to traditional routing. However, the existing opportunistic routing schemes like GeRaF (Zorzi and Rao 2003) typically include all the available next-hop neighbors as forwarding candidates, which does not lead to optimal energy efficiency.

In this chapter, we propose an energy-efficient geographic opportunistic routing (EGOR) framework, which is based on opportunistic routing but more judiciously selects a subset of the available next-hop neighbors as the forwarding candidates

Multihop Wireless Networks: Opportunistic Routing, First Edition.
Kai Zeng, Wenjing Lou and Ming Li.
© 2011 John Wiley & Sons, Ltd. Published 2011 by John Wiley & Sons, Ltd.

to strike a good balance between the packet advancement and energy cost. The analysis of how to achieve maximum EPA in Chapter 2 provides useful insights on the selection of the forwarding candidate set. Based on which, we propose two localized candidate selection algorithms with $\mathbf{O}(M^3)$ and $\mathbf{O}(M^2)$ running time in the worst case respectively and $\Omega(M)$ in the best case, where M is the number of available next-hop neighbors of the transmitter. The algorithms efficiently determine the optimal forwarding candidate set with respect to the EPA per unit of energy consumption. The performance of EGOR is justified through extensive simulations and comparisons with those of the existing geographic routing and opportunistic routing schemes. The simulation results show that EGOR strikes a good balance between energy consumption and routing efficiency in terms of EPA, and achieves the best energy efficiency among the three schemes in all the cases. Our simulation results also show that, under a realistic lossy channel model, the best energy efficiency can be achieved with only a very small number of forwarding candidates (around two), even when the energy consumption of reception is negligible to that of transmission.

The rest of the chapter is organized as follows. We formulate the EGOR problem in Section 3.1. Two efficient localized candidate selection algorithms are proposed in Section 3.2. In Section 3.3, we propose and analyze our EGOR scheme. Simulation results are presented in Section 3.4. Conclusions are drawn in Section 3.5.

3.1 EGOR problem formulation

3.1.1 Energy consumption model

Here we do not assume the promiscuous mode in which every node "overhears" the transmission within its range. Instead, being energy efficient, we assume nodes/sensors only listen to the transmissions intended for themselves. To achieve this, a second low-power radio (Vaidya and Miller 2005) can be used to wake up nodes that should participate in the EGOR or to inform the neighbors (including nodes giving negative advancement), which are not selected as forwarding candidates to shut down their data radios. Nodes can also only read the headers of packets for early rejection (Seada *et al*. 2004). For simplicity, we also ignore the energy consumption of the control packets,[1] as control packets are usually much smaller than data packets. We only consider the energy consumption of packet transmission and reception.[2] So the total energy consumption for one opportunistic forwarding attempt is:

$$E_t(r) = E_{tx} + r \cdot E_{rx} \tag{3.1}$$

---

[1] For different MAC protocols, the energy consumption of control packets may be different. However, the energy consumption is likely to be a nondecreasing function of the number of forwarding candidates. So ignoring it will not affect the upper bound analysis of the energy efficiency in this book.

[2] In sensor networks, the energy consumption of reception is comparable to that of transmission see the MICA2 datasheet at https://www.eol.ucar.edu/rtf/facilities/isa/internal/CrossBow/DataSheets/mica2.pdf, so is non-negligible.

where E_{tx} and E_{rx} are the packet transmission and reception energy consumption, respectively. Recall that r is the number of candidates in the forwarding candidate set $\mathcal{F}$.

3.1.2   Tradeoff between EPA and energy consumption

As we proved in Lemma 2.1, the more nodes get involved in GOR, the larger the EPA can be. So the GOR that involves all the nodes in $\mathcal{C}$ will achieve the largest EPA. This fact has been implicitly used in the existing opportunistic routing approaches. However, it is not always the most energy-efficient way to forward packets by involving all the nodes in $\mathcal{C}$. As from Equation (3.1), we know one transmission from the transmitter is accompanied by r receptions of the r forwarding candidates, so involving all the nodes in $\mathcal{C}$ consumes the most energy. On the other hand, conventional geographic routing involving only one forwarding candidate has the least energy cost of one transmission and one reception but it achieves the least EPA per hop. This indicates lower routing efficiency as more hops (transmissions) might be necessary to reach the final destination. Clearly there is a tradeoff between per-hop routing efficiency and overall energy efficiency.

This tradeoff is illustrated in Figure 3.2, which corresponds to the example in Figure 3.1 by assuming $E_{tx} = 1$ unit of energy, $E_{rx} = 0.5$ unit. Note that although $\mathrm{EM}(\mathcal{C}, r)$ (defined in Definition 2.1) and $E_t(r)$ are both strictly increasing function of r, the ratio $\frac{\mathrm{EM}(\mathcal{C},r)}{E_c(r)}$ reaches its maximum at $r = 2$, and the corresponding ordered node set is $\langle i_1, i_4 \rangle$ with node i_1 having higher relay priority than i_4.

Based on the analysis above, we propose a new local metric that aims to strike a good balance between the routing efficiency and energy efficiency. The new metric is denoted as $G(\pi(\mathcal{F}))$ and defined in Equation (3.2) as follows.

$$G(\pi(\mathcal{F})) = \frac{\mathrm{EPA}(\pi(\mathcal{F})) \cdot L_{pkt}}{E_t(|\mathcal{F}|)} \tag{3.2}$$

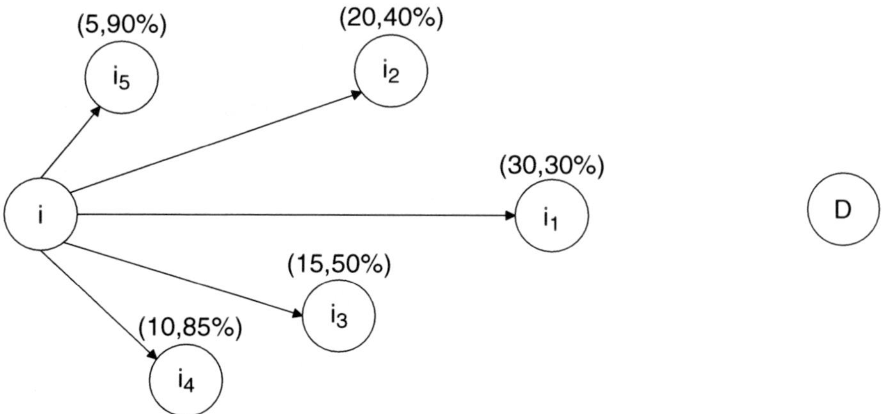

Figure 3.1   Example in which node i is forwarding a packet to a remote destination D. Reproduced by permission of © 2007 IEEE.

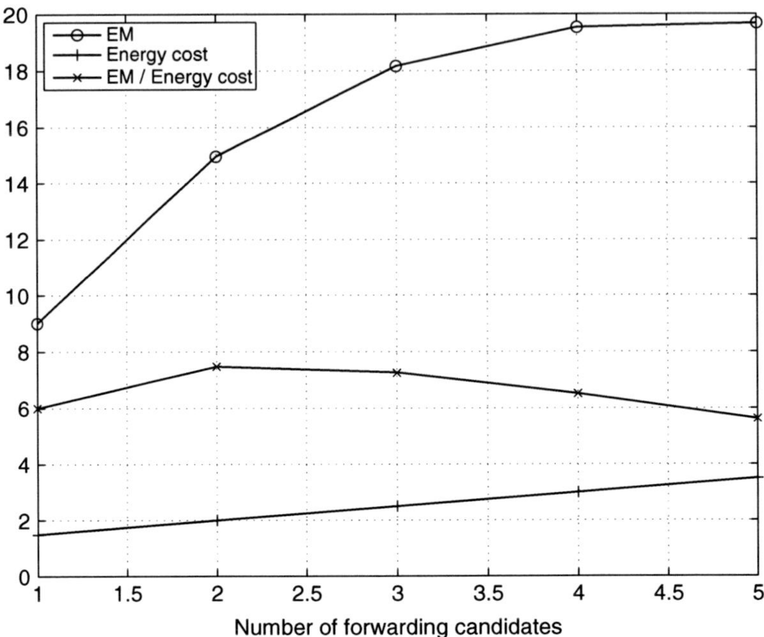

Figure 3.2   EM, energy cost and their ratio as functions of number of forwarding candidates. Reproduced by permission of © 2007 IEEE.

where L_{pkt} is the packet length in bits, and $\text{EPA}(\pi(\mathcal{F}))$ is defined in Equation (2.1). If the unit of $\text{EPA}(\pi(\mathcal{F}))$ is the meter, and $E_t(|\mathcal{F}|)$ is the joule, the unit of $G(\pi(\mathcal{F}))$ is bmpJ. The physical meaning of $G(\pi(\mathcal{F}))$ is the expected bit advancement to the destination by consuming 1 J of energy per packet forwarding attempt.

Now, our goal is to find a way to select a $\mathcal{F}^*$ which maximizes the metric $G(\mathcal{F}^*)$, which can be formulated as the following optimization problem:

$$\mathcal{F}^* = \text{argmax}_{\mathcal{F} \subseteq \bigcup_{S \subseteq 2^\mathcal{C}} Sym(S)} G(\mathcal{F}) \tag{3.3}$$

where $2^\mathcal{C}$ is the powerset of $\mathcal{C}$ and $Sym(S)$ is the set of all the permutations of S. A solution to this optimization problem needs to answer the following two questions: 1. How many and which nodes should be involved in the local forwarding? 2. What priority should they follow to forward a packet?

3.2   Efficient localized node-selection algorithms

3.2.1   Reformulate the node-selection optimization problem

We know that when the number of neighbors involved in GOR is given, the denominator of the function $G(\pi(\mathcal{F}))$ defined in Equation (3.2) is fixed, then maximizing $G(\pi(\mathcal{F}))$ is equivalent to maximize its numerator. So we can find the suboptimal solution for each $r = 1, 2, \cdots, N$, then get a global optimal solution by picking the largest one of the suboptimal solutions. From this analysis, as the packet length L_{pkt} is fixed, combining Equation (3.1), the optimization problem in (3.3) is equivalent to

$$Maximize\ M(r) := \frac{\text{EM}(\mathcal{C},r)}{E_{tx}+r\cdot E_{rx}}\ s.t.\ 1 \leq r \leq |\mathcal{C}| \tag{3.4}$$

We now introduce the following corollary that can help us solve this optimization problem more efficiently.

Corollary 3.1 *(**Local maximum of M(r) is global maximum**) Given the available next-hop node set $\mathcal{C}$ with $|\mathcal{C}| = M(M \geq 1)$, the receiving energy consumption $E_{rx} > 0$ and transmission energy consumption $E_{tx} > 0$, the local maximum of the objective function $M(r)$ defined in (3.4) is the global maximum. That is, if $M(k - 1) < M(k)$ and $M(k) \geq M(k + 1)$ $(1 \leq k \leq M)$, $M(k) \geq M(k + n)$, $\forall$ $1 \leq n \leq M - k$.*

Proof. As

$$M(k) \geq M(k + 1)$$

that is

$$\frac{\text{EM}(\mathcal{C},k)}{E_{tx}+k\cdot E_{rx}} \geq \frac{\text{EM}(\mathcal{C},k+1)}{E_{tx}+(k+1)E_{rx}} \Rightarrow n \times \frac{\text{EM}(\mathcal{C},k+1)-\text{EM}(\mathcal{C},k)}{\text{EM}(\mathcal{C},k)} \leq n \times \frac{E_{rx}}{E_{tx}+k\cdot E_{rx}} \tag{3.5}$$

Since $\text{EM}(\mathcal{C}, r)$ is concave and positive, we have

$$\frac{\text{EM}(\mathcal{C},k+n)-\text{EM}(\mathcal{C},k)}{\text{EM}(\mathcal{C},k)} \leq n \times \frac{\text{EM}(\mathcal{C},k+1)-\text{EM}(\mathcal{C},k)}{\text{EM}(\mathcal{C},k)} \tag{3.6}$$

From inequalities (3.5) and (3.6), we have

$$\frac{\text{EM}(\mathcal{C},k+n)-\text{EM}(\mathcal{C},k)}{\text{EM}(\mathcal{C},k)} \leq n \times \frac{E_{rx}}{E_{tx}+k\cdot E_{rx}} \Rightarrow \frac{\text{EM}(\mathcal{C},k)}{E_{tx}+k\cdot E_{rx}} \geq \frac{\text{EM}(\mathcal{C},k+n)}{E_{tx}+(k+n)E_{rx}}$$

that is $M(k) \geq M(k + n)$, $\forall$ $1 \leq n \leq M - k$.

3.2.2   Efficient node-selection algorithms

3.2.2.1   Algorithm based on Lemma 2.2 and Corollary 3.1

Based on the **containing property** in Lemma 2.2, a straightforward way to find an optimal node set containing r nodes is to add a new node into the optimal node set containing r − 1 nodes. Furthermore, when a local maximum is found, it is the global maximum based on Corollary 3.1. The algorithm GetM-A in Table 3.1 finds an optimal forwarding candidate set $\mathcal{F}^*$ and the corresponding energy-efficiency value M^* of the objective function defined in (3.4). Note that $\mathcal{F}^*$, $\mathcal{F}_c^*$ and $\mathcal{F}$ are all ordered node sets with nodes closer to the destination having higher relay priorities. For feasible sets having the same maximum EPA, we choose the one that achieves higher one-hop reliability (line 6).

Table 3.1   Pseudocode for finding the maximum energy efficiency value M^* and an optimal forwarding candidate set $\mathcal{F}^*$ based on Lemma 2.2

```
GetM-A(C, E_tx, E_rx)
1    M* ← M*_c ← 0; A* ← A*_c ← 0;
2    F* ← F*_c ← ∅; P_F* ← P_F*_c ← 0; B ← C; /* Initialization */
3    while (B ≠ ∅) do /* B is the remained node set */
4      for each node i_j ∈ B
5        F ← F* ∪ {i_j}; P_F ← 1 − (1 − P_F*)(1 − p_j); A ← EPA(F);
6        if A > A*_c || (A = A*_c & P_F > P_F*_c)
7          A*_c ← A; F*_c ← F; P_F*_c ← P_F;
8      end for
9      M*_c ← A*_c / (E_tx + |F*_c| · E_rx);
10     if M*_c ≤ M* /* Local maximum is found */
11       return(M*, F*);
12     else
13       B ← C \ F*_c; A* ← A*_c; M* ← M*_c; F* ← F*_c; P_F* ← P_F*_c;
14   end while
15   return(M*, F*);
```


It is not difficult to find an algorithm to calculate EPA($\mathcal{F}$) (in line 5) in $\mathbf{O}(|\mathcal{F}|)$ running time. Then the algorithm GetM-A costs $\mathbf{O}(M^3)$ running time in the worst case and in the best case it only costs $\Omega(M)$.

Table 3.2 shows the procedure of finding the M^* and an $\mathcal{F}^*$ by applying the algorithm GetM-A on the example in Figure 3.1 with $E_{tx} = 1$ and $E_{rx} = 0.5$. The procedure runs from Round 1 to Round 3 and in each round it runs from the top to the bottom. In the first round, $\langle i_1 \rangle$ is found as the node achieves the maximum EPA by selecting one forwarding candidate; in the second round, $\langle i_1, i_4 \rangle$ is found

Table 3.2   The procedure for finding the maximum energy efficiency value M^* and an optimal forwarding candidate set $\mathcal{F}^*$ by applying the algorithm GetM-A on the example in Figure 3.1 with $E_{tx} = 1$ and $E_{rx} = 0.5$

Round 1		Round 2		Round 3	
$\mathcal{F}$	EPA($\mathcal{F}$)	$\mathcal{F}$	EPA($\mathcal{F}$)	$\mathcal{F}$	EPA($\mathcal{F}$)
$\langle i_1 \rangle$	9	$\langle i_1, i_2 \rangle$	14.6	$\langle i_1, i_2, i_4 \rangle$	18.17
$\langle i_2 \rangle$	8	$\langle i_1, i_3 \rangle$	14.25	$\langle i_1, i_3, i_4 \rangle$	17.225
$\langle i_3 \rangle$	7.5	$\langle i_1, i_4 \rangle^*$	14.95	$\langle i_1, i_4, i_5 \rangle$	15.4225
$\langle i_4 \rangle$	8.5	$\langle i_1, i_5 \rangle$	12.15		
$\langle i_5 \rangle$	4.5				
M(1) = 6		$M^*(2) = 7.475$		M(3) = 7.268	

as the optimal node set by selecting two forwarding candidates; in the third round, $\langle i_1, i_2, i_4 \rangle$ is found as the optimal node set by selecting three forwarding candidates, and $M(3) < M(2)$; so searching is terminated, and $M(2)$ is the maximum energy efficiency value and $\langle i_1, i_4 \rangle$ is an optimal forwarding candidate set.

3.2.2.2 Dynamic programing algorithm

We now propose another efficient dynamic programing algorithm, which is not based on the **containing property**, and only costs $\mathbf{O}(M^2)$ in the worst case and $\Omega(M)$ in the best case.

Recall that nodes i_j ($1 \le j \le M$) in $\mathcal{C}$ are ordered according to the advancements as $d_1 > d_2 > \ldots > d_M$. Denote the set $\langle i_q, i_{q+1}, \ldots, i_M \rangle$ ($1 \le q \le M$) as $\mathcal{C}_q$. Following the denoting, $\mathcal{C}_1 = \mathcal{C}$. According to the **relay priority rule** in Theorem 2.1 and the definition of $\mathrm{EM}(\mathcal{C}, r)$, we then have

$$\mathrm{EM}(\mathcal{C}_q, r) = \begin{cases} 0 & r = 0 \ or \ M - q + 1 < r \\ Max\{d_q p_q + (1 - p_q)\mathrm{EM}(\mathcal{C}_{q+1}, r - 1), \mathrm{EM}(\mathcal{C}_{q+1}, r)\} \end{cases} \quad (3.7)$$

$\mathrm{EM}(\mathcal{C}, r)$ ($1 \le r \le M$) can be efficiently calculated by applying Equation (3.7) recursively using dynamic programing (Cormen *et al.* 2001).

The pseudocode of the dynamic programing algorithm GetM-B is given in Table 3.3, where $|\mathcal{C}| = M$, $F_{(q,r)}$ is the ordered node set corresponding to $\mathrm{EM}(\mathcal{C}_q, r)$, $P_{(q,r)}$ is the corresponding one-hop reliability, and d_is are sorted in descending order ($d_1 > d_2 \ldots > d_M$). We also choose the feasible set that achieves a higher one-hop reliability when two feasible sets have the same EPA (line 9). Based on Corollary 3.1, if a local maximum is found (line 16), the searching is terminated and the optimal solution is returned (line 17). The algorithm GetM-B costs $\mathbf{O}(M^2)$ running time in the worst case and $\Omega(M)$ in the best case.

Table 3.4 shows the procedure for finding the M^* and an $\mathcal{F}^*$ by applying the algorithm GetM-B on the example in Figure 3.1 with $E_{tx} = 1$ and $E_{rx} = 0.5$. The procedure runs from Round 1 to Round 3, and in each round it runs from the bottom to the top. It can be seen that although it finds the same M^* ($M(2)$) and $\mathcal{F}^*$ ($\langle i_1, i_4 \rangle$) as in Table 3.2, most of the tested node sets are different from the ones in Table 3.2.

3.3 Energy-efficient geographic opportunistic routing

The EGOR that applies the local forwarding candidates selection algorithms GetM-A or GetM-B to get $\mathcal{F}^*$ is described in Table 3.5, where node i (i $\ne$ destination D) is routing a packet and the forwarding candidates' i_js are trying to relay the packet collaboratively. Here we do not consider any mechanism to route around voids (when $\mathcal{C} = \emptyset$). If the packet becomes stuck due to no node being available for forwarding, it is dropped (line 14 in Procedure A). Mechanisms such as FACE routing (Bose *et al.* 1999) or perimeter forwarding in GPSR (Karp and Kung 2000) can be applied here to deal with the communication void problem but

Table 3.3 Pseudocode of dynamic programming algorithm finding the maximum energy efficiency value M^* and an optimal forwarding candidate set $\mathcal{F}^*$

```
GetM-B(C, E_tx, E_rx)
1  for i ← 1 to N
2    EM(C_{i+1}, 0) ← 0; F_{(i+1,0)} ← ∅; P_{(i+1,0)} ← 0;
3  end for
4  M^* ← 0; F^* ← ∅; EM(C_{N+1}, 1) ← 0; P_{(N+1,1)} ← 0;
5  for r ← 1 to N
6  for q ← N-r+1 down to 1
7    A ← d_q p_q + (1 − p_q)EM(C_{q+1}, r − 1);
8    P ← 1 − (1 − P_{(q+1,r−1)})(1 − p_q);
9    if A > EM(C_{q+1}, r)||(A = EM(C_{q+1}, r) & P > P_{(q+1,r)})
10     EM(C_q, r) ← A; F_{(q,r)} ← F_{(q+1,r−1)} ∪{i_q}; P_{(q,r)} ← P;
11   else
12     EM(C_q, r) ← EM(C_{q+1}, r); F_{(q,r)} ← F_{(q+1,r)};
13     P_{(q,r)} ← P_{(q+1,r)};
14   end for
15   M(r) ← EM(C_1, r)/(E_tx + r · E_rx);
16   if M(r) ≤ M^*  /* Local maximum is found */
17   return(M^*, F^*)
18   else
19     M^* ← M(r); F^* ← F_{(1,r)};
20   end for
21   return(M^*, F^*);
```

Table 3.4 The procedure for finding the maximum energy efficiency value M^* and an optimal forwarding candidate set $\mathcal{F}^*$ by applying the algorithm GetM-B on the example in Figure 3.1 with $E_{tx} = 1$ and $E_{rx} = 0.5$

	Round 1		Round 2		Round 3	
q	$F_{(q,1)}$	$EM(C_q, 1)$	$F_{(q,2)}$	$EM(C_q, 2)$	$F_{(q,3)}$	$EM(C_q, 3)$
1	$\langle i_1 \rangle$	9	$\langle i_1, i_4 \rangle^*$	14.95	$\langle i_1, i_2, i_4 \rangle$	18.17
2	$\langle i_4 \rangle$	8.5	$\langle i_2, i_4 \rangle$	13.1	$\langle i_2, i_3, i_4 \rangle$	15.05
3	$\langle i_4 \rangle$	8.5	$\langle i_3, i_4 \rangle$	11.75	$\langle i_3, i_4, i_5 \rangle$	12.0875
4	$\langle i_4 \rangle$	8.5	$\langle i_4, i_5 \rangle$	9.175		
5	$\langle i_5 \rangle$	4.5				
	M(1) = 6		$M^*(2) = 7.475$		M(3) = 7.268	

Table 3.5   The procedure of EGOR when node i is forwarding the packet

---

Procedure A
run by transmitter i:
1   $RN \leftarrow 0$
2   **while** $(RN \leq RL)$ **do**
3   **if** $C \neq \emptyset$
4   **if** $D \notin C$
5   Get $\mathcal{F}^*$ from C, broadcast the packet to the nodes in $\mathcal{F}^*$.
6   **else**
7   Get $\mathcal{F}^*$ from $C \setminus \{D\}$,
8   broadcast the packet to the nodes in $\mathcal{F}^* \bigcup D$.
9   **if** None of candidates received packet correctly
10      $RN \leftarrow RN + 1$
11   **if** $RN > RL$
12   Drop the packet
13   **else break**
14   **else**
15   Drop the packet

Procedure B
run by forwarding candidate i_j receiving the packet correctly:
1 **if** $(i_j \neq D)$
2   **if** No candidates having higher priorities received packet correctly
3   i_j becomes the actual forwarder and runs Procedure A
4   **else**
5   i_j drops the packet
6 **else** The packet is arriving at D and routing is terminated.

---

it is beyond the scope of this chapter. Retransmission limitation is applied. If the retransmission number (RN) reaches the limitation (RL) (line 11 in Procedure A), the packet will also be dropped (line 12 in Procedure A). It is worth mentioning that there is a last-hop behavior (line 7 and 8 in Procedure A) in EGOR. When the sink D is in the available next-hop node set C, we calculate the forwarding set by eliminating D from C. Because sink D is not energy constrained, its receiving energy cost should not be counted when maximizing the energy efficiency. After calculating the $\mathcal{F}^*$, D should be added into the forwarding candidate set, since the packet always has a chance to reach D whatever the link quality from i to D is when D is the neighbor of i.

3.4   Performance evaluation

We evaluate the performance of EGOR through extensive Monte-Carlo simulations. We compare EGOR with geographic routing, which only has one forwarding candidate that achieves the maximum EPA, and opportunistic routing, which involves all the available next-hop nodes as forwarding candidates. When the packet becomes stuck, all the three protocols just drop the packet. Various situations are simulated by varying node densities, transmission-to-reception energy ratios and retransmission limits.

3.4.1   Simulation setup

Channel Model: To simulate a realistic channel model for lossy WSNs, we use the log-normal shadowing path loss model derived in (Zuniga and Krishnamachari 2004):

$$PRR(L_f, d) = \left(1 - \frac{1}{2}exp^{-\frac{\gamma(d)}{2 \times 0.64}}\right)^{8\rho L_f} \qquad (3.8)$$

where d is the transmitter–receiver distance, $\gamma(d)$ is the signal-to-noise ratio (SNR), ρ is the encoding ratio and L_f is the frame length in bytes. This model considers several environmental and radio parameters,[3] such as the path-loss exponent (α) and log-normal shadowing variance (σ) of the environment, and the modulation and encoding schemes of the radio. This particular equation resembles a MICA2 mote see the MICA2 datasheet at www.eol.ucar.edu/rtf/facilities/isa/internal/CrossBow/DataSheets/mica2.pdf, which has data rate of 19.2 kbps, and the noise bandwidth 30 kHz. Non-coherent FSK and Manchester are used as the modulation and encoding schemes $(\rho = 2)$, respectively. The environmental parameters are set to $\alpha = 3.5$ and $\sigma = 4$.

Energy Model: The energy consumption is obtained by multiplying the power consumption and the packet transmission time. The transmission power consumption P_{tx} is the summation of the power of power amplifier (P_{PA}) and electronic (P_{elec}), and P_{rx} is equal to P_{elec}. We assume that P_{PA} is proportional to the transmit power P_{trans}. Then P_{tx} is

$$P_{tx} = P_{PA} + P_{elec} = \frac{P_{trans}}{\eta} + P_{elec} \qquad (3.9)$$

where η is the PA power efficiency, which is set to be 0.3 in our simulation.

Evaluation Metrics: We define the following metrics to evaluate the performance of the three protocols.

- *Packet Delivery Ratio (PDR)*: percentage of packets sent by the source that actually reach the sink. This is a measure for reliability.

- *Energy efficiency $\eta(S, D)$*: this metric is measured in bit-meters per joule (bmpJ). It is calculated as in Equation (3.10),

$$\eta(S, D) = \frac{L_{pkt} \cdot N_r \cdot Dist(S, D)}{E_{total}} \qquad (3.10)$$

where N_r denotes the number of packets received at the destination, L_{pkt} is the packet length in bits, and E_{total} is the (transmission and reception) energy

---

[3] Please refer to Zuniga and Krishnamachari (2004) for a complete description of the model.

consumed by all the nodes involved in the routing procedure excluding the sink. We account for the distance factor, because the energy efficiency is indeed relevant to the distance between the communication pair due to the lossy property of multihop wireless links in WSNs.

- *Hop count*: it is measured as the number of hops a successfully delivered packet travels from source to destination.

The simulated sensor network has stationary nodes uniformly distributed in a $60 \times 60\,m^2$ square region, with nodes having identical fixed transmission power of 0 dbm. The frame length is fixed on 50 bytes with preamble of 20 bytes. The source and the sink node are fixed at two corners across the diagonal of the square area. All simulations are run for 5000 iterations. For each iteration, node locations are randomly reassigned and PRRs between nodes are recalculated.

3.4.2 Simulation results and analysis

3.4.2.1 Impact of node density

We use different node numbers (100, 144, 196, 225) to achieve various node densities corresponding to the average number of neighbors per node of 9.5, 14, 20, 23 respectively. The reception power consumption is fixed on 2 MW, so the reception to transmission energy ratio is $\frac{3}{8}$. No retransmission is allowed.

Figure 3.3 shows that EGOR achieves better energy efficiency than the other two routing protocols. This result can be explained as follows: for every forwarding decision, EGOR chooses the forwarding set that maximizes the EPA per unit energy consumption. This local optimal behavior can achieve a good global performance under a uniformly randomly distributed node deployment, where any intermediate-hop forwarding can be viewed as a similar new first-hop packet forwarding. Statistically, every hop may make similar progress in such a homogeneous environment. The overall energy cost for successfully delivering one packet from source S to destination D can be approximated by $\frac{Dist(S,D)}{EPA/Local\ energy\ cost}$, so when we maximize the numerator, the total energy consumption is minimized. Opportunistic routing involving all the available next-hop nodes in the routing has the worst performance, since it has the lowest EPA/*Local energy cost* ratio.

Another observation from Figure 3.3 is that the energy efficiency of EGOR and geographic routing is increased as the network becomes denser, while opportunistic routing shows the opposite trend. This result is related to the PDR performance shown in Figure 3.4 and the hop count performance shown in Figure 3.5. As we can see, the PDR of the opportunistic routing remains as high as nearly 1 under all the different node densities. That is to say, higher node density does not bring much gain to opportunistic routing on successfully delivering packets. Although the hop count of opportunistic routing decreases when node density increases, the energy consumption due to unnecessarily involving more nodes in forwarding overwhelms the benefit of hop count decreasing. For geographic routing, the PDR is increased and hop count is also decreased when the network is denser, so the energy efficiency

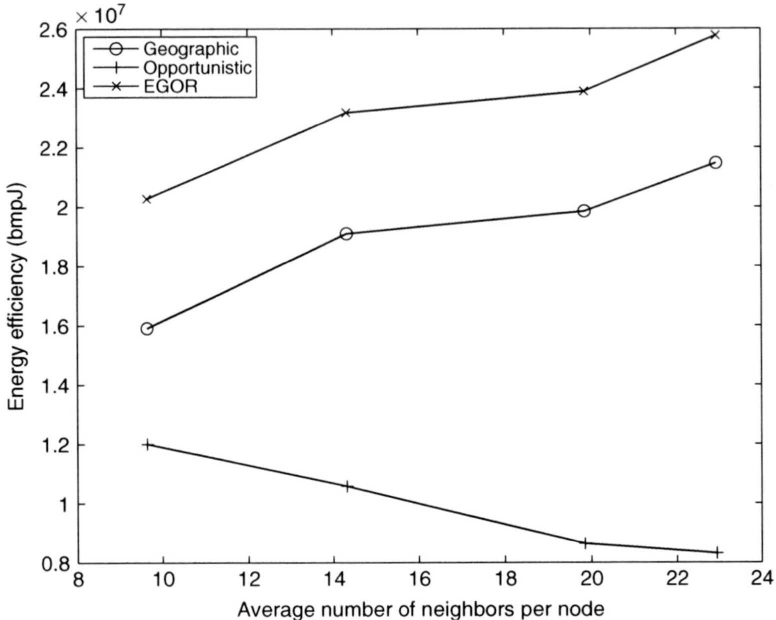

Figure 3.3   *Energy efficiency versus network density. Reproduced by permission of © 2007 IEEE.*

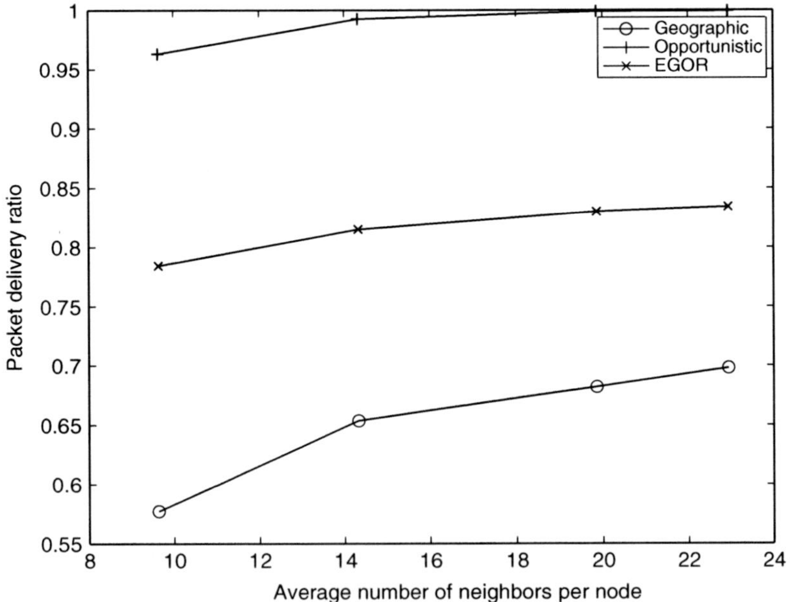

Figure 3.4   *Packet delivery ratio versus network density. Reproduced by permission of © 2007 IEEE.*

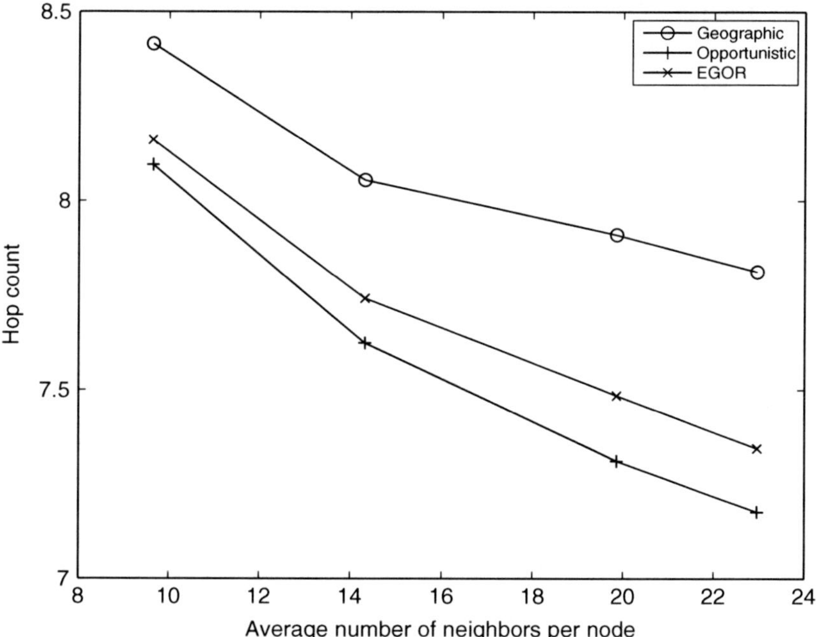

Figure 3.5 Hop count versus network density. Reproduced by permission of © 2007 IEEE.

is increased. For EGOR, PDR is not increased much but hop count is decreased when node density increases, so the energy efficiency of EGOR is also increased.

Figure 3.5 also shows that the hop counts decrease when network is denser for all the three protocols, because involving more available next-hop nodes brings more chance for packets to make larger advancement when nodes are uniformly distributed. Opportunistic routing has the smallest hop count under all the node densities, while geographic routing has the largest hop count. Hop count of EGOR is between two of them. From Lemma 2.1 in Section 2.2.1, we know that the maximum EPA, by choosing r nodes from a given set, is strictly increasing with r. So opportunistic routing archives the largest EPA by selecting all the available next-hop nodes as forwarding candidates; geographic routing gets the smallest EPA by choosing only one forwarding candidate, and EGOR archives larger EPA than geographic routing and smaller EPA than opportunistic routing by selecting some (not all) available next-hop nodes as forwarding candidates. Actually, in this simulation, EGOR selects 1.2 nodes on average as the candidate forwarders under each node density. This observation suggests that under such settings, only a few nodes are necessary in order to take advantage of opportunistic routing efficiently. Adding one more node in forwarding can get much better energy efficiency and reliability than geographic routing. Involving all the available next-hop nodes in opportunistic routing is an energy wasteful method.

3.4.2.2 Impact of reception to transmission energy ratio (RTER)

We study the performance of the three protocols under different RTERs in this section. In the simulations, no retransmission is allowed and the available next-hop node set size is 6.7 on average. The reception power consumption is varied from 10^{-3} to 10 MW, so the corresponding RTER is in the range $[3 \times 10^{-4}, 0.75]$.

Figure 3.6 shows that the energy efficiency of EGOR is always the best of the three protocols and the RTER is a crucial parameter affecting the energy efficiency of the opportunistic routing. There is a watershed on RTER, smaller than which the energy efficiency of the opportunistic routing is better than the geographic routing, while greater than which the geographic routing surpasses the opportunistic routing. The reason is that when the energy consumption of reception is negligible compared to that of the transmission, the opportunistic routing achieves a larger EPA than the geographic routing while consuming nearly the same energy as geographic routing. So opportunistic routing is more energy efficient. However, when the energy consumption of reception is comparable to transmission, involving all the available next-hop nodes in the opportunistic routing consumes much more energy than the geographic routing and the cost of the increased energy consumption overwhelms the benefit of the increased EPA. Thus, the energy efficiency of the opportunistic routing is less than the geographic routing when RTER is greater than the watershed. For these two protocols, RTER does not affect the forwarding candidate(s) selecting criteria, so it does not affect the PDR.

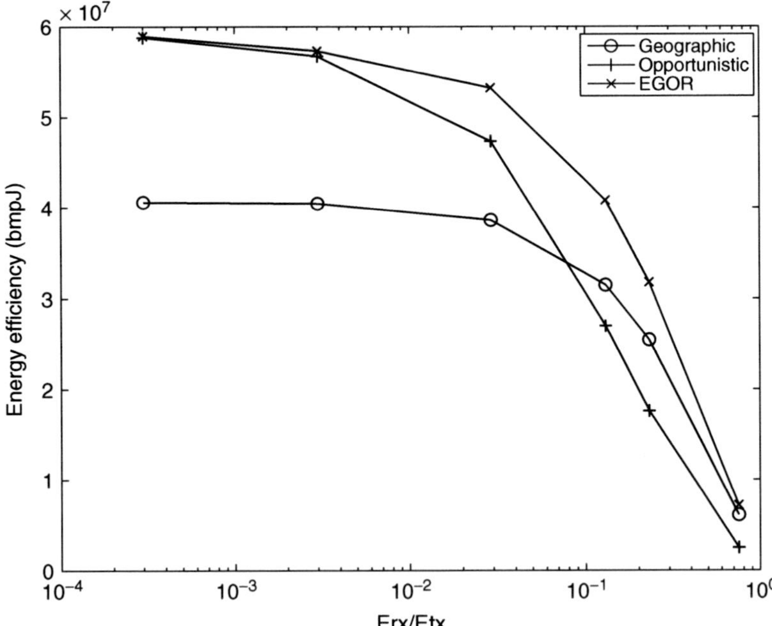

Figure 3.6   Energy efficiency versus reception to transmission power ratio.

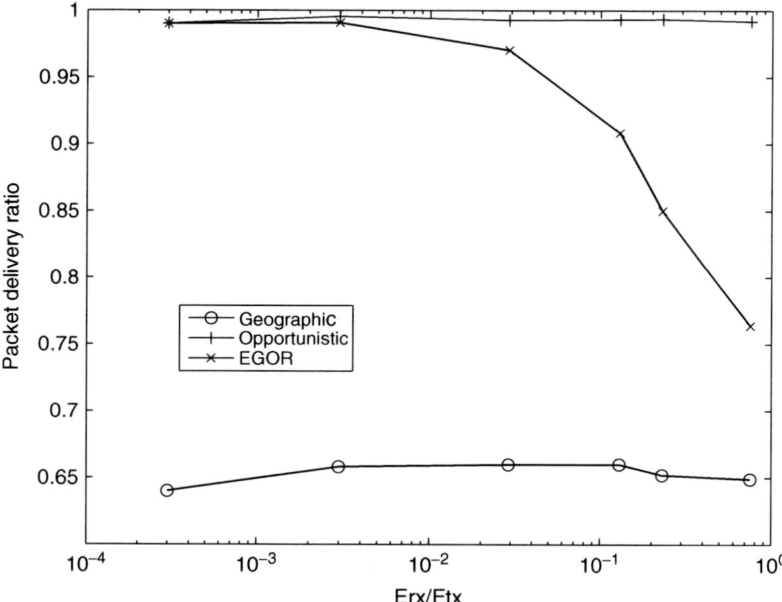

Figure 3.7   Packet delivery ratio versus reception to transmission power ratio.

Figure 3.7 shows that the PDR of the geographic routing and opportunistic routing does not change according to the RTER. For EGOR, the PDR decreases when RTER increases because EGOR takes the energy consumption into account. When the reception energy cost increases, fewer nodes are selected as forwarding candidates, then the packet is more likely to be lost without retransmission. An interesting observation here is that, even when RTER is very small, EGOR only selects a very small number of nodes as the forwarding candidate, but achieves nearly the same energy efficiency as opportunistic routing. For example, when RTER = 0.03%, EGOR only selects 2.2 forwarding candidates on average, while it has the same energy efficiency and PDR as the opportunistic routing which selects 6.7 candidates on average. This result again suggests that only a small number of nodes need to be involved in opportunistic routing to achieve a good balance between energy efficiency and routing efficiency.

The hop-count performance shown in Figure 3.8 indicates that the RTER does not affect the hop count of the opportunistic routing and geographic routing, the reason is as the same as the PDR performance of these two protocols. For EGOR, the hop count increases after the RTER is larger than 10% because fewer forwarding candidates are selected and the EPA is decreased.

3.4.2.3   Impact of retransmission limit

In this section we study how the retransmission limit affects the performance of the three protocols. The reception power consumption is fixed at 2 MW and the available next-hop node set size is also 6.7 on average.

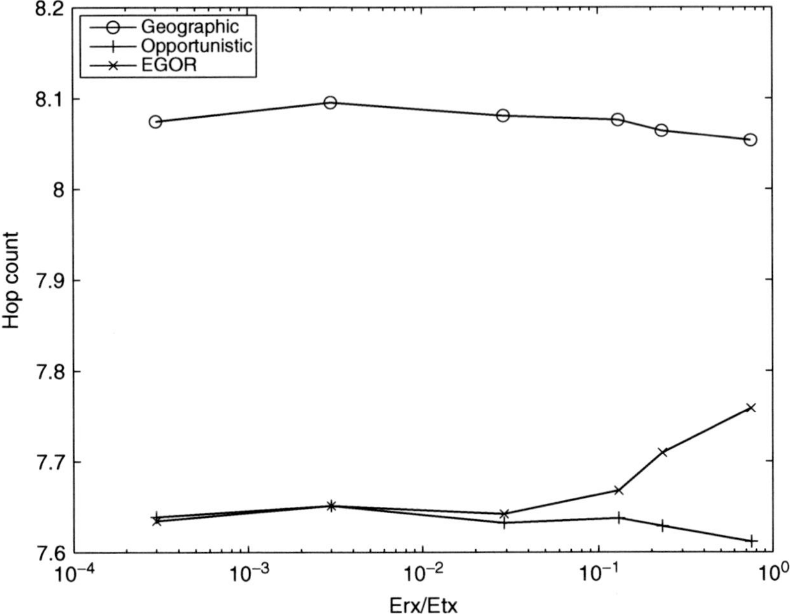

Figure 3.8   Hop count versus reception-to-transmission power ratio.

Intuitively, increasing retransmission limit will increase the reliability, say PDR. Figure 3.9 shows this trend for all the three protocols. It is worth mentioning that the benefit of increasing the retransmission limit (which can be seen as the slopes of the curves) for the opportunistic routing is trivial but for the geographic routing is obvious (especially when retransmission limit is less than 4). The reason is that the opportunistic routing has already achieved a high PDR (nearly 1) by involving all the available next-hop nodes in forwarding even when there is no retransmissions allowed. For the geographic routing, however, there is only one next-hop node involved in the forwarding, then the packet is more likely to be lost in one transmission than in the opportunistic routing. For EGOR, the PDR increasing rate is less than that of the geographic routing because EGOR already achieves higher PDR than the geographic routing when no retransmissions are allowed. When the retransmission limit is larger than 1 the PDR gains become less and less for both EGOR and the geographic routing, and when the limit is larger than 3, the PDRs of both are approaching to 1.

Figure 3.10 shows that, for opportunistic routing, the energy efficiency has not changed much with the change of retransmission limit. The reason is that the retransmission does not play a role for the PDR in opportunistic routing and almost the same packets can be delivered to the destination whether retransmission is allowed or not. For geographic routing and EGOR, the retransmission does play a role for the energy efficiency when retransmission limit is less than 3. As

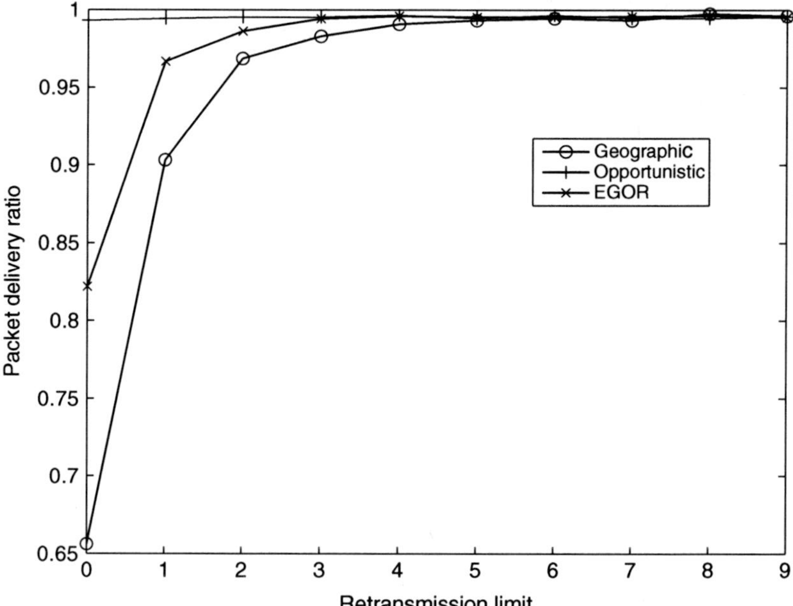

Figure 3.9   *Packet delivery ratio versus retransmission limit.*

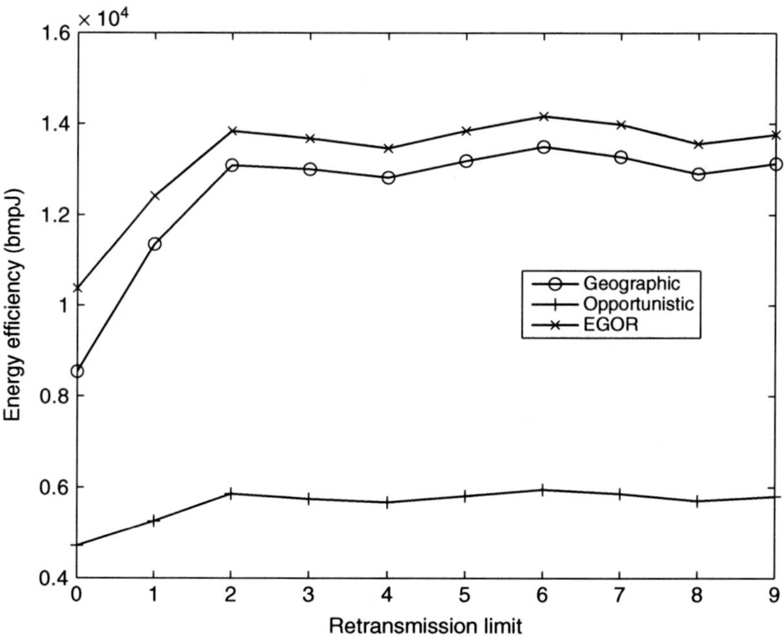

Figure 3.10   *Energy efficiency versus retransmission limit.*

we have analyzed, retransmission affects the PDR, especially from allowing no retransmission to one and from one to two. When retransmission limit is larger than 3, the energy efficiency of these two protocols do not change much as the PDRs are already approaching 1.

As retransmission does not affect EPA much, Figure 3.11 shows that the hop count remains almost the same when the retransmission limit varies for each of the three protocols.

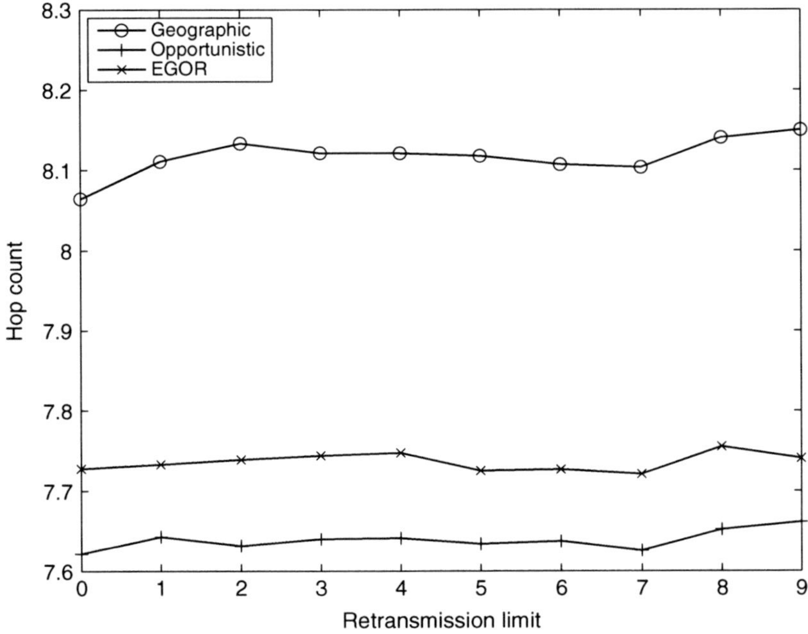

Figure 3.11   Hop count versus retransmission limit.

3.4.2.4   Number of local forwarding candidates involved

We study the number of local forwarding candidates involved in opportunistic routing and EGOR under various node densities and RTERs. In the simulations, we use different node numbers (100, 144, 196, 225) to achieve various node densities corresponding to the available next-hop candidate set sizes of 4.6, 6.7, 9.2, 10.5 on average respectively. The RTER is in the range $[3 \times 10^{-4}, 0.75]$.

Figure 3.12 shows the simulation result. Opportunistic routing uses all the available next-hop nodes as the forwarding candidates, whereas EGOR only uses a very small number of forwarding candidates (around two or fewer). For example, even when the RTER is as small as 0.03%, EGOR only chooses 2.2 forwarding candidates on average under various node densities. This means even when the energy consumption of reception is far less than that of transmission, in order to achieve maximum energy efficiency we still only need to involve two forwarding

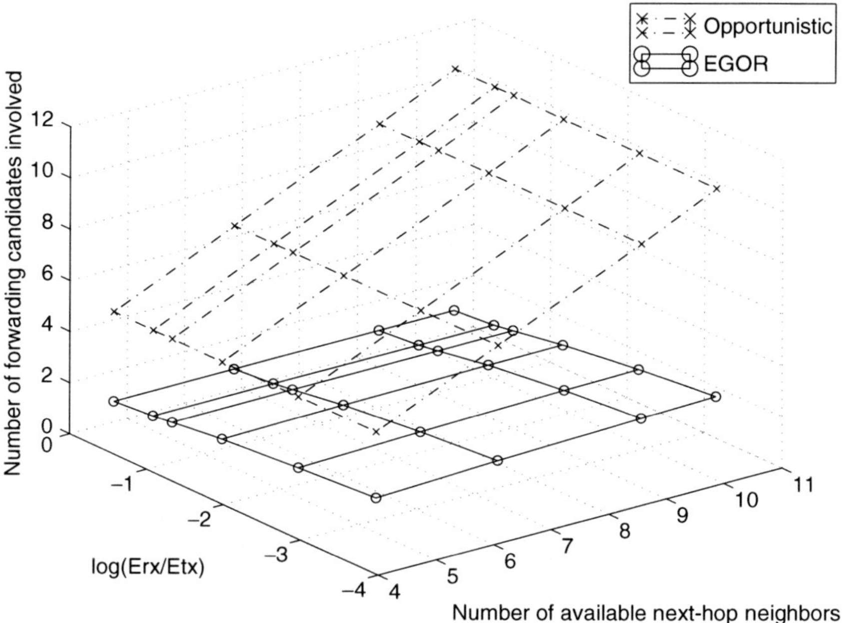

Figure 3.12   Number of forwarding candidates involved under different node densities and RTERs.

candidates. Figure 3.12 also shows that the number of forwarding candidates is affected mainly by the RTER but not by the node density under a uniform node distribution. For example, when RTER = 0.75, the forwarding candidates in EGOR are nearly unchanged as 1.1 under different node densities. This is an important result as it indicates that involving more forwarding candidates will not bring much more expected packet advancement. A small number of forwarding candidates is sufficient to strike a good balance between EPA and energy consumption. This is a very desirable result because the cost incurred due to assuring one final forwarder from multiple forwarding candidates at the MAC layer is expected to grow when involving more forwarding candidates (Shah *et al.* 2005). Involving fewer candidates introduces less rendezvous and contention cost.

3.4.2.5   Concavity of maximum EPA and its slope

In this section, we study the concavity of the maximum EPA function and its slope in one hop under various node densities. The nodes are uniformly distributed and the next-hop available node number varies from 6 to 12. From Figure 3.13 we can see that the maximum EPA increases when the number of the forwarding candidates increases, and when nodes are denser, the EPA is larger. A very interesting result is that under different node densities, the slopes of each curve in Figure 3.13 are nearly the same, which is shown in Figure 3.14. Notice that, when the forwarding candidate number is 3, the slope is already decreased to below 0.01.

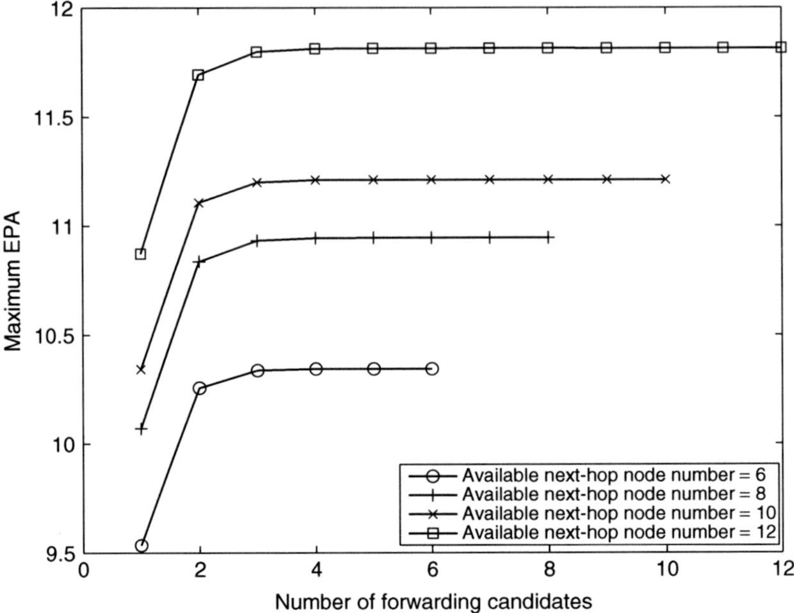

Figure 3.13   *Maximum EPA versus forwarding candidate number under different node densities.*

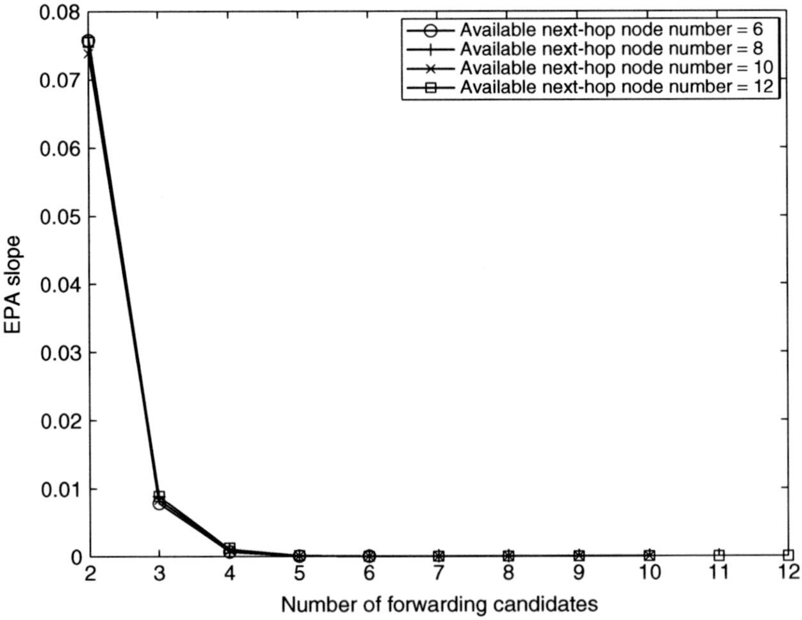

Figure 3.14   *EPA slope under different node densities.*

When the number of forwarding candidates is larger than 4, the slope is near to zero. Figures 3.13 and 3.14 manifest that no matter what the node density is, the EPA gain of involving more forwarding candidates becomes very small when the number of forwarding candidates is larger than 3. These results are consistent with Figure 3.12 where the optimal energy efficiency is achieved when the number of forwarding candidates is around 2.2.

3.5 Conclusion

In this chapter, we studied the geographic opportunistic routing strategy with both routing and energy efficiencies as the major concerns. We proposed a new routing metric that evaluates EPA per unit of energy consumption so that the energy efficiency can be taken into consideration in routing. By leveraging the proved findings in Chapter 2, we proposed two localized candidate selection algorithms with $O(M^3)$ and $O(M^2)$ running time in the worst case, respectively, and $\Omega(M)$ in the best case, where M is the number of available next-hop neighbors of the transmitter. The algorithms efficiently determine the forwarding candidate set that maximizes the proposed new metric for energy efficiency, namely the EPA per unit of energy consumption. We further propose an EGOR framework applying the node selection algorithms to achieve the energy efficiency. The performance of EGOR is studied through extensive simulations and compared with those of the existing geographic routing and opportunistic routing protocols. The results show that EGOR achieves the best energy efficiency among the three protocols in all the cases while maintaining very good routing performance. Our simulation results also show that the number of forwarding candidates necessary to achieve the maximum energy efficiency is mainly affected by the reception to transmission energy ratio but not by the node density under a uniform node distribution. Although the EPA can be maximized by involving the most number of nodes in GOR, in terms of energy efficiency, only a very small number of forwarding candidates (around 2) is needed on average. This is true even when the energy consumption of reception is far less than that of transmission.

References

Bose P, Morin P, Stojmenovic I and Urrutia J 1999 Routing with guaranteed delivery in ad hoc wireless networks *3rd International Workshop on Discrete Algorithms and Methods for Mobile Computing and Communications*, Seattle, WA, New York, NY.

Cormen TH, Leiserson CE, Rivest RL and Stein C 2001 *Introduction to Algorithms, Second Edition*. MIT Press and McGraw-Hill.

Karp B and Kung H 2000 Gpsr: Greedy perimeter stateless routing for wireless networks *ACM MOBICOM*, Boston.

Pottie GJ and Kaiser WJ 2000 Wireless integrated network sensors. *Communications of the ACM* **43**(5), 51–58.

Seada K, Zuniga M, Helmy A and Krishnamachari B 2004 Energy efficient forwarding strategies for geographic routing in wireless sensor networks *ACM Sensys'04*, Baltimore, MD.

Shah RC, Wietholter S, Wolisz A and Rabaey JM 2005 When does opportunistic routing make sense? *IEEE PerSens*.

Vaidya NH and Miller MJ 2005 A mac protocol to reduce sensor network energy consumption using a wakeup radio. *IEEE Transactions on Mobile Computing* **4**(3), 228–242.

Zeng K, Lou W, Yang J and Brown DR 2007 On geographic collaborative forwarding in wireless ad hoc and sensor networks *WASA'07*, Chicago, IL.

Zorzi M and Rao RR 2003 Geographic random forwarding (geraf) for ad hoc and sensor networks: energy and latency performance. *IEEE Transactions on Mobile Computing* **2**(4), 349–365.

Zuniga M and Krishnamachari B 2004 Analyzing the transitional region in low power wireless links *IEEE Secon'04*.

4

Capacity of multirate opportunistic routing

The existing works on OR mainly focused on a single-rate system. Researchers have proposed several candidate selection and prioritization schemes to improve throughput or energy efficiency. However, there is a lack of theoretical analysis on the performance limit or the throughput bounds achievable by OR. In addition, one of the current trends in wireless communication is to enable devices to operate using multiple transmission rates. For example, many existing wireless networking standards such as IEEE 802.11a/b/g include this multirate capability. The inherent rate–distance tradeoff of multirate transmissions has shown its impact on the throughput performance of traditional routing (Awerbuch *et al*. 2006; Zhai and Fang 2006a,b). Generally, low-rate communication covers a long transmission range, while high-rate communication must occur at short range. It is intuitive to expect that this rate–distance tradeoff will also affect the throughput of OR because different transmission ranges also imply different neighboring node sets, which results in different spacial diversity opportunities. These rate–distance–diversity tradeoffs will no doubt affect the throughput of OR, which deserves careful study. To the best of our knowledge, there is no existing work addressing the throughput problem of OR in a multirate network.

In this chapter, we bridge these two gaps by studying the throughput bound of OR and the performance of OR in a multirate scenario. First, for OR, we propose the concept of concurrent transmission sets, which captures the transmission conflict constraints of OR. Then, for a given network with a given opportunistic routing strategy (i.e., forwarder selection and prioritization), we formulate the maximum end-to-end throughput problem as a maximum-flow linear programming problem subject to the constraints of transmitter conflict. The solution of the optimization

Multihop Wireless Networks: Opportunistic Routing, First Edition.
Kai Zeng, Wenjing Lou and Ming Li.
© 2011 John Wiley & Sons, Ltd. Published 2011 by John Wiley & Sons, Ltd.

problem provides the performance bound of OR. The proposed method establishes a theoretical foundation for the evaluation of the performance of different variants of OR with various forwarding candidate selection, prioritization policies, and transmission rates. We also propose two OR metrics: *expected medium time* (EMT) and *expected advancement rate* (EAR), and the corresponding distributed and local rate and candidate set selection schemes, one of which is Least Medium Time OR (LMTOR) and the other is Multirate Geographic OR (MGOR). Simulation results show that, for OR, by incorporating our proposed multirate OR schemes, systems operating at multirates achieve higher throughput than systems operating at any single rate. Several observations are made about OR: 1. the end-to-end capacity gained decreases when the number of forwarding candidates is increased; 2. there exists a node-density threshold, higher than which 24 mbps GOR performs better than 12 mbps GOR, and lower than which the opposite is the case.

The rest of this chapter is organized as follows. We propose the framework of computing the throughput bounds of OR in Section 4.1. Section 4.2 studies the impact of multirate capability and forwarding strategy on the throughput bound of OR. We then propose the OR metrics, and rate and candidate selection schemes for multirate systems in Section 4.3. Simulation results are presented and analyzed in Section 4.4. Conclusions are drawn in Section 4.5.

4.1   Computing throughput bound of OR

The first fundamental issue to address is the maximum end-to-end throughput when OR is used. Any traffic load higher than the throughput capacity is not supported and even causes performance to deteriorate as a result of excessive medium contention. The knowledge of throughput capacity can be used to reject any excessive traffic in the admission control for real-time services. It can also be used to evaluate the performance of different OR variants. Furthermore, the derivation of throughput of OR may suggest novel and efficient candidate selection and prioritization schemes.

In this section we present our methodology to compute the throughput bound between two end nodes in a given network with a given OR strategy (i.e., given each node's forwarding candidate set, node relay priority, and transmission/broadcast rate at each node). We first introduce two concepts, transmitter based conflict graph and concurrent transmission set, which are used to represent the constraints imposed by the interference among wireless transmissions in a multihop wireless network. We then present methods for computing bounds on the optimal throughput that a network can support when OR is used.

4.1.1   Transmission interference and conflict

Wireless interference is a key issue affecting throughput. Existing wireless interference models generally fall into two categories: *protocol model* and *physical model* (Gupta and Kumar 2000). Under the protocol model, a transmission is considered successful when both of the following conditions hold: 1. the receiver is in the effective transmission range of the transmitter; and 2. no other node that is in the carrier sensing range of the receiver is transmitting. This kind of protocol model requires only the receiver to be free of interference. To model a 802.11 like bidirectional communications we can extend the protocol model by adding the requirement of interference free also at the transmitter side. Under the physical model, for a successful transmission, the aggregate power at the receiver from all other ongoing transmissions plus the noise power must be less than a certain threshold so that the SNR requirement at the ongoing receiver is satisfied. In this book, we use the term "**usable**" to describe a link when it is able to make a successful transmission based on either the protocol model or the physical model. When two (or more) links are not able to be usable at the same time, they are having a "**conflict**".

Link conflict graphs have been used to model such interference (Jain *et al.* 2003; Zhai and Fang 2006a). As shown in Figure 4.2(b), in a link conflict graph, each vertex corresponds to a link in the original connectivity graph. There is an edge between two vertices if the corresponding two links may not be active simultaneously due to interference (e.g., having a "conflict"). However, this link-based conflict graph cannot be directly applied to study capacity problem of OR networks because by the nature of opportunistic routing, for one transmission, throughput may take place on any one of the links from the transmitter to its forwarding candidates. The throughput dependency among multiple outgoing links from the same transmitter makes the subsequent maximum-flow optimization problem very difficult (if it is still possible). Therefore, in this paper, we propose a new construction of conflict graph to facilitate the computation of throughput bounds of OR. Instead of creating link conflict graph, we study the conflict relationship by transmitters (or nodes) associated with their forwarding candidates. As shown in Figure 4.2(c) in the node conflict graph, each vertex corresponds to a node in the original connectivity graph. Each vertex is associated with a set of links, e.g., the links to its selected forwarding candidates. There is an edge (conflict) between two vertices if the two nodes cannot be transmitting simultaneously due to a conflict caused by one or more unusable links, which we will define in Section 4.1.2.

4.1.2   Concurrent transmission sets

We define the concepts of **concurrent transmission sets** (CTSs) for OR as follows. These concepts capture the impact of interference of wireless transmissions and OR's opportunistic nature. They are the foundation of our method of computing the end-to-end throughput.

1. **Conservative CTS:** According to a specific OR policy, when one node is transmitting, the packet is broadcast to all the nodes in its forwarding candidate set. The links from a transmitter to all its forwarding candidates are defined as links associated with the transmitter. We define a conservative CTS (CCTS) as a set of transmitters, when all of them are transmitting simultaneously, all links associated with them are still usable. If adding any one more node into a CCTS will result in a non-CCTS, the CCTS is called a maximum CCTS.

The conservative CTS actually requires all the opportunistic receivers to be interference free for one transmission. This is probably true for certain protocols (Fussler *et al*. 2003) where RTF (Request To Forward) and CTF (Clear To Forward) control packets are used to clear certain ranges within transmitter and forwarding candidates or confirm a successful reception. But this is a stricter requirement than necessary and will only give us a lower bound of end-to-end capacity. We define the following greedy CTS to compute the maximum end-to-end throughput.

2. **Greedy CTS:** In order to maximize the throughput, we permit two or more transmitters to transmit at the same time even when some links associated with them become unusable. The idea is to allow a transmitter to transmit as long as it can deliver some throughput to one of the next-hop forwarder(s). Therefore, we define a greedy CTS as a set of transmitters, when all of them are transmitting simultaneously, at least one link associated with each transmitter is usable. If adding any one more node into a GCTS will result in changes in the usability status of any link associated with nodes in that set, the GCTS is called a maximum GCTS.

4.1.3   Effective forwarding rate

After we find a CTS, we need to identify the capacity on every link associated with a node in the CTS. We introduce the concept of **effective forwarding rate** on each link associated with a transmitter according to a specified OR strategy. Assume node n_i's forwarding candidate set $\mathcal{F}_i = \langle n_{i_1}, n_{i_2} \ldots n_{i_r} \rangle$, with relay priorities $n_{i_1} > n_{i_2} > \ldots > n_{i_r}$. Let ψ_q denote the indicator function on link l_{ii_q} when it is in a particular CTS: $\psi_q = 1$ indicating link l_{ii_q} is usable, and $\psi_q = 0$ indicating that link l_{ii_q} is not usable. Then the effective forwarding rate of link l_{ii_q} in that particular CTS is defined in Equation (4.1):

$$\widetilde{R}_{ii_q} = R_i \cdot \psi_q \cdot p_{ii_q} \prod_{k=0}^{q-1}(1 - \psi_k \cdot p_{ii_k}) \tag{4.1}$$

where R_i is the broadcast rate of transmitter i, and $p_{ii_0} := 0$. $p_{ii_q} \prod_{k=0}^{q-1}(1 - \psi_k \cdot p_{ii_k})$ is the probability of candidate n_{i_q} receiving the packet correctly but all the higher-priority candidates not. Note that the candidate (with $\psi_q = 0$), which is interfered by other transmissions, is not involved in the opportunistic forwarding, and has no effect on the effective forwarding rate from the transmitter to lower-priority candidates, as $(1 - \psi_k \cdot p_{ii_k}) = 1$.

In a conservative CTS, all the receptions are interference free. Therefore, in each CCTS, every link associated with a transmitter is usable, i.e. $\psi = 1$, and the effective forwarding rate on each link is non zero. And the effective forwarding rate for a particular link remains same when the link is in a different CCTS. The effective forwarding rate indicates that according to the relay priority, only when a usable higher forwarding candidate did not receive the packet correctly, a usable lower priority candidate may have a chance to relay the packet if it received the packet correctly. Note that this definition generalizes the effective rate for unicast in traditional routing, that is, when there is only one forwarding candidate, the effective forwarding rate reduces to the unicast effective data rate.

For the greedy mode, some link(s) associated with one transmitter may become unusable, thus having zero effective forwarding rate. Furthermore, the effective forwarding rate of the links may be different when they are in different GCTSs. To indicate this possible difference, we use $\widetilde{R}^\alpha_{ii_w}$ to denote the effective forwarding rate of link l_{ii_w} when it is in the αth GCTS.

4.1.4 Lower bound of end-to-end throughput of OR

Assume we have found all the maximum CCTSs $\{T_1, T_2 \ldots T_M\}$ in the network. At any time, one CTS at most can be scheduled to transmit. When one CTS is scheduled to transmit, all the nodes in that set can transmit simultaneously. Let λ_α denote the time fraction scheduled to CCTS T_α $(1 \le \alpha \le M)$. Then the maximum throughput problem can be converted to an optimal scheduling problem that schedules the transmission of the maximum CTSs to maximize the end-to-end throughput. Therefore, considering communication between a single source, n_s, and a single destination, n_d, with opportunistic routing, we formulate the maximum achievable throughput problem between the source and the destination as a linear programming problem corresponding to a maximum-flow problem under additional constraints in Figure 4.1.

$$Max \sum_{l_{si} \in \mathbf{E}} f_{si}$$

$$s.t.$$

$$\sum_{l_{ij} \in \mathbf{E}} f_{ij} = \sum_{l_{ji} \in \mathbf{E}} f_{ji} \ \forall \ n_i \in \mathbf{V} - \{n_s, n_d\} \tag{4.2}$$

$$\sum_{l_{is} \in \mathbf{E}} f_{is} = 0 \tag{4.3}$$

$$\sum_{l_{di} \in \mathbf{E}} f_{di} = 0 \tag{4.4}$$

$$f_{ij} \ge 0 \ \forall \ l_{ij} \in \mathbf{E} \tag{4.5}$$

$$f_{ij} = 0 \ \forall \ l_{ij} \in \mathbf{E}, n_j \notin \mathcal{F}_i \tag{4.6}$$

$$\sum_{\alpha=1}^{M} \lambda_\alpha \le 1 \tag{4.7}$$

$$\lambda_\alpha \ge 0, 1 \le \alpha \le M \tag{4.8}$$

$$f_{ij} \le \sum_{n_i \in T_\alpha, n_j \in \mathcal{F}_i, 1 \le \alpha \le M} \lambda_\alpha \widetilde{R}^\alpha_{ij} \ \forall \ l_{ij} \in \mathbf{E} \tag{4.9}$$

Figure 4.1 LP formulations to optimize the end-to-end throughput of OR. Reproduced by permission of © 2008 IEEE.

In Figure 4.1, f_{ij} denotes the amount of flow on link l_{ij}, $\mathbf{E}$ is a set of all links in the connected graph G, and $\mathbf{V}$ is the set of all nodes. The maximization states that we wish to maximize the sum of flow out of the source. Constraint (4.2) represents flow conservation, i.e., at each node, except the source and the destination, the amount of incoming flow is equal to the amount of outgoing flow. Constraint (4.3) states that the incoming flow to the source node is 0. Constraint (4.4) indicates that the outgoing flow from the destination node is 0. Constraint (4.5) restricts the amount of flow on each link to be non-negative. Constraint (4.6) says there is no flow from the node to the neighboring nodes that are not selected as the forwarding candidates of it. Constraint (4.7) represents, at any time, at most one CTS will be scheduled to transmit. Constraint (4.8) indicates the scheduled time fraction should be non-negative. The constraint (4.9) states the actual flow delivered on each link is constrained by the total amount of flow that can be delivered in all activity periods of the OR modules which contain this link.

The key difference between our maximum flow formulations and the formulations for traditional routing in Jain *et al.* (2003); and Fang (2006a) lies in the methodology we use to schedule concurrent transmissions. With the construction of concurrent transmission sets, we are able to schedule the transmissions based on node set (with each node associated with a set of forwarding candidates) rather than link set in traditional routing. When we schedule a transmitter, we effectively schedule the links from the transmitter to its forwarding candidates at the same time according to OR strategy. For traditional routing, any two links sharing the same sender cannot be scheduled simultaneously. When a packet is not correctly received by the intended receiver but is opportunistically received by neighboring nodes of the sender, traditional routing will retransmit that packet instead of making use of the correct receptions on the other links. OR takes advantage of the correct receptions. That is why OR achieves higher throughput than traditional routing. Our proposed model accurately captures OR's capability of delivering throughput opportunistically.

A Simple Example: Next, we give an example to show how our formulation helps us to find the end-to-end throughput bound of OR, and we compare this result with the maximum throughput derived from multipath traditional routing based on results in Jain *et al.* (2003).

For simplicity, in the four-node network shown in Figure 4.2(a), we assume each node transmits at the same rate R, and each link is associated with a *PRR* indicated in the pair on each link. Assume every node is in the carrier sensing range of any other nodes. We are going to find the maximum end-to-end throughput from node a to d for traditional routing and OR.

For traditional routing, we first construct the link-conflict graph as shown in Figure 4.2(b). In the conflict graph, each vertex corresponds to each link in the original connectivity graph. There is an edge between two vertices when these two links conflict with each other. According to the protocol model, no two links in Figure 4.2(a) can be scheduled simultaneously. So the link-conflict graph for traditional routing is a complete graph (clique). There are four independent sets, each containing one node in the conflict graph. Each independent set corresponds to one concurrent schedulable link set. By running the linear programming formulated in Jain *et al.* (2003) we can find an optimal schedule on links to maximize the throughput. Assuming the whole communication period is τ, one feasible solution is to assign $\frac{3}{10}\tau$, $\frac{3}{10}\tau$, $\frac{2}{10}\tau$, $\frac{2}{10}\tau$ to l_{ab}, l_{ac}, l_{bd}, l_{cd}, respectively. So the maximum end-to-end throughput between a and d is $\frac{2(\frac{3}{10}R\cdot 0.5\tau)}{\tau} = \frac{3}{10}R$ for the traditional routing.

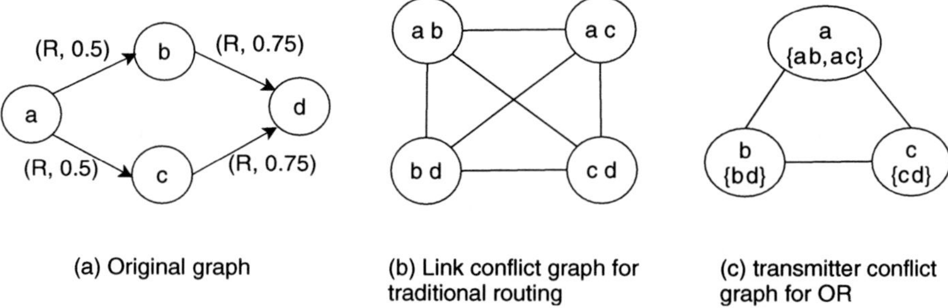

(a) Original graph

(b) Link conflict graph for traditional routing

(c) transmitter conflict graph for OR

Figure 4.2   Conflict graph. Reproduced by permission of © 2008 IEEE.

For OR, we construct the node conflict graph. Assume a chooses nodes b and c as its forwarding candidates, and b and c's forwarding candidate is just the destination d. According to the protocol model, the node conflict graph is constructed in Figure 4.2(c), which only contains three vertices and is also a clique. So the three conservative transmission sets are $T_1 = \{a\}$, $T_2 = \{b\}$, and $T_3 = \{c\}$. Assume node b has higher relay priority than node c, then we have $\widetilde{R}_{ab}^1 = 0.5R$, $\widetilde{R}_{ac}^1 = 0.25R$, and $\widetilde{R}_{bd}^2 = \widetilde{R}_{cd}^3 = 0.75R$. By running the linear programming formulated in Figure 4.1, we get an optimal schedule that assigns $\frac{1}{2}\tau$, $\frac{1}{3}\tau$ and $\frac{1}{6}\tau$, to nodes a, b and c respectively. So the throughput of OR under this optimal schedule is $\frac{0.25R\tau+0.125R\tau}{\tau} = \frac{3}{8}R$, which is 25% higher than that of the traditional routing.

4.1.5   Maximum end-to-end throughput of OR

The throughput bound we find based on the maximum conservative CTSs in Section 5.3.4 is a lower bound of maximum end-to-end throughput. The CCTSs can be constructed based on either the protocol model or the physical model. However, the interference freedom at every intended receiver is a stricter requirement than necessary. It may be applicable under some protocol scenarios but it fails to take full advantage of opportunistic nature of OR because it excludes the situations where concurrent transmission is able to deliver throughput on some of the links even though some other links are having conflicts. In order to compute the exact capacity, we apply the same optimization technique to the greedy CTSs. Since greedy CTSs include all the possible concurrent transmission scenarios that generate non-zero throughput, the bound found by the optimization technique based on all greedy CTSs will be the maximum end-to-end throughput of OR.

Like the construction of CCTSs, GCTSs can be constructed based on either the protocol model or the physical model. Under the protocol model, the conflict between two links is binary, either conflict or no conflict. It is not difficult to construct the GCTSs under the protocol model with the proposed node conflict graph. On the other hand, it is well known that the physical model captures the interference property more accurately. However, it is more complicated to represent the interference when multiple transmitters are active at the same time. In this section, we discuss the construction of GCTSs based on the physical interference model.

Under the physical interference model, a link l_{ij}, from node n_i to n_j, is usable if and only if the signal to noise ratio at receiver n_j is no less than a certain threshold, e.g., $\frac{Pr_{ij}}{P_N} \geq SNR_{th}$, where Pr_{ij} is the average signal power received at n_j from n_i's transmission, P_N is the interference + noise power, and SNR_{th} is the SNR threshold, under which the packet cannot be correctly received and above which the packet can be received at least with probability p_{td}. Note that, SNR_{th} is different for different data rates.

Under the physical model, the interference gradually increases as the number of concurrent transmitters increases and becomes intolerable when the interference+noise level reaches a threshold. We define a weight function w_{ij_q}, to capture the impact of a transmitter n_i's transmission on a link l_{jj_q}'s reception. Link l_{jj_q} represents the data forwarding from node n_j its forwarding candidate n_{j_q}.

$$w_{ij_q} = \frac{Pr_{ij_q}}{\frac{Pr_{jj_q}}{SNR_{th}} - P_{noise}} \tag{4.10}$$

where Pr_{ij_q} and Pr_{jj_q} are the received power at node n_{j_q} from the transmissions of nodes n_i and n_j, respectively, P_{noise} is the ambient noise power, and $\frac{Pr_{jj_q}}{SNR_{th}} - P_{noise}$ is the maximum allowable interference at node n_{j_q} for keeping link l_{jj_q} usable.

Then given a transmission set S and $n_j \in S$, a link l_{jj_q} is usable if and only if $\sum_{n_i \in S, i \neq j} w_{ij_q} < 1$. It means that link l_{jj_q} is usable even when all the transmitters in set S are simultaneously transmitting. For conservative mode, if this condition is true for every link associated with each transmitter in S, this set S is a CCTS. For greedy mode, if this condition is true for at least one link associated with each transmitter in S, the set S is a GCTS.

After finding all the GCTSs, we can apply the same optimization technique to the maximum-flow problem based on all the GCTSs. The result is the exact bound of maximum end-to-end throughput.

When each node has only one forwarding candidate, OR degenerates to the traditional routing. Therefore, finding all the concurrent transmission sets is at least as hard as the NP-hard problem of finding the independent sets in Jain *et al.* (2003) and Zhai and Fang (2006a) for traditional routing. However, it may not be necessary to find all of them to maximize an end-to-end throughput. A heuristic algorithm similar to that in Tang *et al.* (2007), or column generation technique (Zhang *et al*. 2005) can be applied to find a good subset of all the CTSs. In addition, complexity can be further reduced by taking into consideration the fact that interferences/conflicts always happen for nodes within a certain range. How to efficiently find all the CTSs is beyond the scope of this book. We simply apply a greedy algorithm to find all the CTSs, say each time we add new transmitters into the existing CTSs to create new CTSs, until no any additional transmitter can be added into any of the existing CTSs.

4.1.6   Multi-flow generalization

Our formulations in Figure 4.1 can be extended from a single source–destination pair to multiple source–destination pairs using a multi-commodity flow formulation (Chavtal 1983) augmented with OR transmission constraints. By assigning a unique connection identifier to each source–destination pair, we introduce the variable f_{ij}^k to denote the amount of flow for connection k on link l_{ij}. For each flow k, according to some OR routing strategy, the corresponding transmitters and their forwarding candidates can be decided. Then the CCTS or GCTS can be constructed over the union of all the OR modules. Referring to Figure 4.1, the objective is now to maximize the summation of all the flows out of all the sources; the flow conservation constraints at each node apply on a per-connection basis (constraint (4.2)); the total incoming flow into a source node is zero only for the connection(s) originating at that node (constraint (4.3)); similarly, the total outgoing flow from a destination node is zero only for the connection(s) terminating at that node (constraint (4.4)); f_{ij}^k is non-negative (constraint (4.5)); f_{ij}^k is equal to zero if the flow k is not routed by any link (constraint (4.6)); and the sum of all the flows traversing on a link is constrained by the total amount of flow that can be delivered in all activity periods of the OR modules that contain this link (constraint (4.9)).

4.2   Impact of transmission rate and forwarding strategy on throughput

The impact of the transmission rate on the throughput of OR is twofold. On the one hand, different rates have different transmission ranges, which lead to different neighborhood diversity. A high rate usually has short transmission range. In one hop, there are few neighbors around the transmitter, which presents low neighborhood diversity. A low rate is likely to have long transmission range and therefore achieves high neighborhood diversity. From the diversity point of view, a low rate may be better. On the other hand, although low rate brings the benefit of larger one-hop distance, which results in higher neighborhood diversity and fewer hop counts to reach the destination, it may still end up with a low effective end-to-end throughput because the low rate disadvantage may overwhelm all other benefits. It is nontrivial to decide which rate is indeed better.

We now use a simple example in Figure 4.3 to illustrate that transmitting at lower rate may achieve higher throughput than transmitting at higher rate for OR. In this example, we assume all the nodes operate on a common channel, but each node

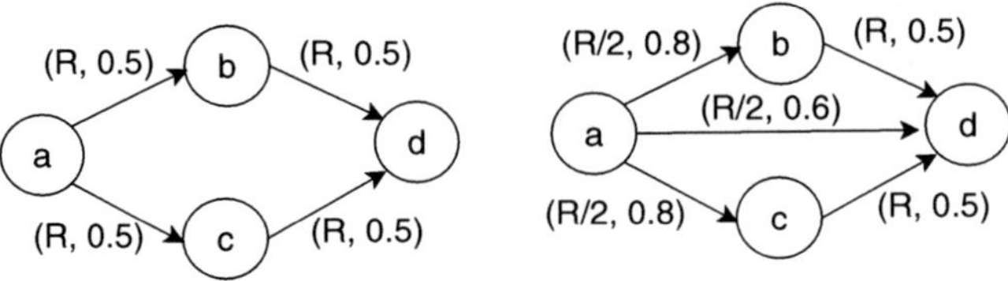

Figure 4.3   End-to-end throughput comparison at different transmission rates. Reproduced by permission of © 2008 IEEE.

can transmit at two different rates R and $R/2$. We compare the throughput from source a to destination d when the source transmits the packets at the two different rates. Figure 4.3(a) shows the case when all the nodes transmit at rate R, and the packet delivery ratio on each link is 0.5. So the effective data rate on each link is $0.5R$. There is no link from a to d because d is out of a's effective transmission range when a operates on rate R. Assume the four nodes are in the carrier sensing range of each other, so they cannot transmit at the same time. Assuming b and c are the forwarding candidates of a, and b has higher relay priority than c. Then link l_{ac} has effective forwarding rate of $0.25R$. By using the formulations in Figure 4.1, we obtain an optimal transmitter schedule such that a, b and c are scheduled to transmit for a fraction of time 0.4, 0.4 and 0.2, respectively. So the maximum end-to-end throughput from a to d is $0.3R$. While in Figure 4.3(b), when a is transmitting at a lower rate $R/2$, it can reach d directly with packet delivery ratio of 0.6. Additionally, we get higher packet deliver ratio from a to b and c as 0.8. In this case, the lower rate achieves longer effective transmission range and brings more spacial diversity chances. Assume d, b, and c are forwarding candidates of a, and with priority $d > b > c$. Similarly, we calculate the maximum throughput from a to d as 0.36R, which is 20% higher than the scenario in Figure 4.3(a) where the system operates on a single rate.

Besides the inherent rate–distance, rate–diversity and rate–hop tradeoffs, which affect the throughput of OR, the forwarding strategy will also have an impact on the throughput. For example, different forwarding candidates may achieve very different throughput, and even for the same forwarding candidate set, different forwarding priority will also result in different throughput, etc. We refer readers to Zeng *et al.* (2007c) for detail analysis of the impact of forwarding strategy on the OR throughput.

4.3   Rate and candidate selection schemes

How to select the transmission rates and forwarding strategy for each node efficiently so that the network capacity can be globally optimized is still an open research issue. We have shown the example in Figure 4.3 that nodes transmitting at a lower rate may lead to a higher end-to-end throughput than when nodes are transmitting at a higher rate. Then, what criteria should node a follow to select transmission rate, forwarding candidates and candidate priority to approach the capacity? It is nontrivial to answer this question. Towards the development of distributed and localized OR protocol that maximize the capacity, in this section, we propose two rate and candidate selection schemes, one is enlightened by least-cost opportunistic routing (LCOR) as proposed in Dubois-Ferriere *et al.* (2007), and the other is inspired by geographic opportunistic routing (GOR) (Fussler *et al.* 2003; Zeng *et al.* 2007a,b,c; Zorzi and Rao 2003).

4.3.1   Least medium time opportunistic routing

In traditional routing, the medium time metric (MTM) (Awerbuch *et al.* 2006) and expected transmission time (ETT) (Draves *et al.* 2004) have shown to be good metrics to achieve high throughput. For OR, we define the opportunistic ETT (OETT) as the expected transmission time to send a packet from n_i to any node in its forwarding candidate set $\mathcal{F}_i$.

$$OETT_{n_i}^{\mathcal{F}_i} = \frac{L_{pkt}}{R_i P_{\mathcal{F}_i}} \tag{4.11}$$

where L_{pkt} is the packet length, R_i is the data transmission rate at node n_i, and $P_{\mathcal{F}_i}$ is the probability of at least one candidate in $\mathcal{F}_i$ correctly receiving the packet sent by n_i:

$$P_{\mathcal{F}_i} = 1 - \prod_{q=1}^{r}(1 - p_{ii_q}) \tag{4.12}$$

Note that this metric actually generalizes the unicast ETT, that is, for $|\mathcal{F}_i| = 1$, the OETT reduces to the unicast ETT.

Denote by D_i the expected medium time (EMT) to reach the destination n_d from a node n_i. Assume that n_i's forwarding candidates are prioritized according to their expected medium time D_{i_q}, such that $D_{i_1} < D_{i_2} \ldots < D_{i_r}$. Then we define the remaining EMT to the destination n_d when node n_i chooses the forwarding candidate set $\mathcal{F}_i$ as follows:

$$EMT_{\mathcal{F}_i}^{n_d} = \frac{1}{P_{\mathcal{F}_i}} \sum_{q=1}^{r} D_{i_q} p_{ii_q} \prod_{k=0}^{q-1}(1 - p_{ii_k}) \tag{4.13}$$

where $p_{ii_0} := 0$.

$p_{ii_q} \prod_{k=0}^{q-1}(1 - p_{ii_k})$ is the probability of candidate n_{i_q} receiving the packet correctly but all the higher priority candidates do not. That is, it is the probability of n_{i_q} becoming the actual forwarder. So the summation $\sum_{q=1}^{r} D_{i_q} p_{ii_q} \prod_{k=0}^{q-1}(1 - p_{ii_k})$ is the expected remaining medium time needed for a packet to travel to the destination for one transmission from node n_i. $\frac{1}{P_{\mathcal{F}_i}}$ is the expected transmission count n_i needs to make in order to deliver the packet to one of its forwarding candidates.

Note that like the OETT, the EMT generalizes the single-path case: when $|\mathcal{F}_i| = 1$, it simply becomes the delay from the next hop to the destination. We should also notice that any two different transmitters, n_i and n_j, even if $\mathcal{F}_i = \mathcal{F}_j$, may have different EMTs, since this EMT is affected by the delivery probabilities from the transmitter to its each forwarding candidate. In other words, the remaining EMT from a forwarding candidate set to the destination depends not only on the candidate set itself, but also on the predecessor node of this set.

We now define the least EMT of node n_i to the destination n_d in a multirate scenario:

$$D_i = \min_{\mathcal{F}_i^j \in 2^{C_i^j}, 1 \le j \le J} (OETT_{n_i}^{\mathcal{F}_i^j} + EMT_{\mathcal{F}_i^j}^{n_d}) \tag{4.14}$$

where C_i^j is the neighboring node set of node n_i when n_i transmits at rate R^j, $\mathcal{F}_i^j$ is the corresponding forwarding candidate set.

Equation (4.14) represents the steady state of the least medium time OR (LMTOR), that selects the forwarding candidates and transmission rate for each node to achieve the minimum end-to-end EMT.

Multirate Bellman–Ford based algorithm   A distributed algorithm running like Bellman–Ford can solve the LMTOR problem. That is, in one iteration, each node n_i updates its value D_i^k, where k is the iteration index. This D_i^k is the estimated EMT from n_i to the destination at the kth iteration; it converges toward D_i. $D_d^k = 0$, $\forall\ k$. One iteration step consists of updating the estimated EMT to the destination from each node:

$$D_i^{k+1} = \min_{\mathcal{F}_i^j \in 2^{C_i^j}, 1 \le j \le J} (OETT_{n_i}^{\mathcal{F}_i^j} + EMT_{\mathcal{F}_i^j}^{n_d}(k))\ \forall\ n_i \ne n_d \tag{4.15}$$

where $EMT_{\mathcal{F}_i^j}^{n_d}(k)$ is the remaining EMT computed using the costs $D_{i_q}^k$ $(n_{i_q} \in \mathcal{F}_i^j)$ from the previous iteration.

At each step, according to Lemma 2.6 described in Section 2.3.3, we do not need to enumerate all the possible $\mathcal{F}_i^j$ to find the optimal forwarding candidate and transmission rate that achieve D_i^{k+1}. For each node n_i, at rate R^j, suppose there are r neighbors denoted as $1, 2, \ldots, r$ and sorted as $D_1^k \le D_2^k \le \ldots \le D_r^k$, we only need to test the candidate sets in the form of $\langle 1 \rangle$, $\langle 1, 2 \rangle$, $\ldots$, $\langle 1, 2, \ldots, r \rangle$. For each rate, we select an optimal candidate set that maximizes the D_i^{k+1}, then the final optimal set should be the one in the optimal sets that achieves the maximum of these maximums. The algorithm terminates when: $D_i^{k+1} = D_i^k\ \forall\ n_i \ne n_d$. Like the proof in Dubois-Ferriere *et al.* (2007), this algorithm converges after at most N iterations, where $N = |V|$ is the number of nodes in the network. The complexity of this algorithm is $O(|V||E|log|V| + |V||E||J|)$.

The pseudo-code of this algorithm (Laufer *et al.* 2010) is presented in Algorithm 4.1, which is the extension of the single rate Bellman–Ford-based algorithm described in Section 2.3.5. The $d_{i\mathcal{S}}$ in the algorithm is the OETT defined in Equation (4.11). The $D_{\mathcal{S}}$ is the remaining EMT defined in Equation (4.13).

Algorithm 4.1 Multirate Bellman–Ford-based Shortest Anypath Algorithm

---

Input: G, n_d
for each node $n_i \in V$ **do**
  $D_i \leftarrow \infty$
  $\mathcal{F}_i \leftarrow \emptyset$
  $T_i \leftarrow$ NIL
  for j $\leftarrow 1$ to J **do**
    $D_{i_j}^{j} \leftarrow \infty$
    $\mathcal{F}_i^{j} \leftarrow \emptyset$
  end for
end for
$D_d \leftarrow 0$
for $t \leftarrow 1$ to $|V| - 1$ **do**
  for each node $n_i \in V$ **do**
    $Q \leftarrow$ GET-NEIGHBORS(n_i)
    while $Q \neq \emptyset$ **do**
      $n_k \leftarrow$ EXTRACT-MIN(Q)
      for $j \leftarrow 1$ to J **do**
        $\mathcal{S} \leftarrow \mathcal{F}_i^{j} \cup \{n_k\}$
        if $D_i^{j} > D_k$ **then**
          $D_{i_j}^{j} \leftarrow d_{i\mathcal{S}} + D_{\mathcal{S}}$
          $\mathcal{F}_i^{j} \leftarrow \mathcal{S}$
          if $D_i > D_i^{j}$ **then**
            $D_i \leftarrow D_{i_j}^{j}$
            $\mathcal{F}_i \leftarrow \mathcal{F}_i^{j}$
            $T_i \leftarrow j$
          end if
        end if
      end for
    end while
  end for
end for

Multirate Dijkstra-based algorithm   The Dijkstra-based algorithm proposed in Section 2.3.4 can also be extended to the multirate case. Algorithm 4.2 presents this algorithm that is proposed in (Laufer *et al.* 2010).

The idea of Algorithm 4.2 is that each node $n_i \in V$ has an independent cost estimate D_i^{j} when the node is transmitting at each rate R^j ($1 \leq j \leq J$). The minimum of these estimates is kept as D_i, the expected medium time (cost) from this node to the destination. At each round of the **while** loop, the node with the minimum cost from Q is settled. At the first round, the destination node will be extracted. In the subsequent each round, suppose n_k is extracted. For each ingress edge $n_i \to n_k$ in the edge set, for every rate R^j, if the cost D_i^j is larger than the cost (D_k) from the settled node n_k to the destination, node n_k is added to the forwarding candidate set $\mathcal{F}_i^j$. The cost D_i^j is then updated accordingly. If the new cost D_i^j is lower than the node cost D_i, D_i is updated as well as the forwarding candidate set $\mathcal{F}_i$ and transmission rate T_i to reflect the new minimum.

Algorithm 4.2 Multirate Bellman-Ford-Based Shortest Anypath Algorithm

---

Input: G, n_d
for each node $n_i \in V$ **do**
    $D_i \leftarrow \infty$
    $\mathcal{F}_i \leftarrow \emptyset$
    $T_i \leftarrow \text{NIL}$
    for $j \leftarrow 1$ to J **do**
        $D_i^j \leftarrow \infty$
        $\mathcal{F}_i^j \leftarrow \emptyset$
    end for
end for
$D_d \leftarrow 0$
$S \leftarrow \emptyset$
$Q \leftarrow V$
while $Q \neq \emptyset$ **do**
    $n_k \leftarrow \text{EXTRACT-MIN}(Q)$
    $S \cup n_k$
    for each ingress edge $n_i \to n_k$ **do**
        for $j \leftarrow 1$ to J **do**
            $S \leftarrow \mathcal{F}_i^j \cup \{n_k\}$
            if $D_i^j > D_k$ **then**
                $D_i^j \leftarrow d_{iS} + D_S$
                $\mathcal{F}_i^j \leftarrow S$
                if $D_i > D_i^j$ **then**
                    $D_i \leftarrow D_i^j$
                    $\mathcal{F}_i \leftarrow \mathcal{F}_i^j$
                    $T_i \leftarrow j$
                end if
            end if
        end for
    end for
end while

Although these two algorithms can find the optimal candidate set and transmission rate to minimize the expected medium time (delay) of multirate opportunistic routing, they either need to know the global network information (Dijkstra-based algorithm), or exchange cost information among node iteratively (Bellman–Ford-based algorithm). In a very large-scale network, these algorithms can still introduce a lot of overheads. We propose another local rate and candidate selection scheme by leveraging on the node's location information as in GOR. Each node only needs to know its neighborhood information and does not need to propagate any cost information. The complexity of the candidate selection procedure can be reduced to linear.

4.3.2 Per-hop greedy: most advancement per unit time

A local metric: Expected Advancement Rate. The location information is available to the nodes in many applications of multihop wireless networks, such as sensor networks for monitoring and tracking purposes (Zorzi and Rao 2003) and vehicular networks (Fussler *et al.* 2003). Geographic opportunistic routing has been proposed as an efficient routing scheme in such networks. In GOR, nodes are aware of the location of itself, its one-hop neighbors, and the destination. A packet is forwarded to neighbor nodes that are geographically closer to the destination. In Chapter 2, we proposed a local metric, *expected packet advancement (EPA)* for GOR to achieve efficient packet forwarding. EPA for GOR is a generalization of EPA for traditional geographic routing (Lee *et al.* 2005; Seada *et al.* 2004). It represents the expected packet advancement achieved by opportunistic routing in one transmission without considering the transmission rate. In this chapter, we extend it into a bandwidth adjusted metric, *expected advancement rate (EAR)*, by taking into consideration various transmission rates.

We define the EAR as follows:

$$EAR_{n_i}^{\mathcal{F}_i} = R_i \sum_{q=1}^{r} a_{ii_q} p_{ii_q} \prod_{k=0}^{q-1}(1 - p_{ii_k}) \tag{4.16}$$

The physical meaning of EAR is the *expected bit advancement per second* towards the destination when the packet is forwarded according to the opportunistic routing procedure introduced in Section 8.1.

The definition of EAR is the rate R_i multiplying the EPA proposed in (Zeng *et al.* 2007b). According to the proved **relay priority rule** for EPA in Section 2.2.2, we have the following theorem for EAR:

Theorem 4.1 (Relay priority rule) *For a given transmission rate at n_i and $\mathcal{F}_i$, the maximum EAR can only be achieved by giving the candidates closer to the destination higher relay priorities.*

This theorem indicates how to prioritize the forwarding candidates when a transmission rate and the forwarding candidate set are given. From the definition of EAR, it is also not difficult to find that adding more neighboring nodes with positive advancement into the existing forwarding candidate set will lead to a larger EAR.

Therefore, we conclude that *an OR strategy that includes all the neighboring nodes with positive advancement into the forwarding candidate set and gives candidates with larger advancement higher relay priorities will lead to the maximum EAR* for a given rate.

Then a straightforward way to find the best rate is: for node n_i, at each transmission rate R^m ($1 \leq m \leq J$), we calculate the largest EAR according to the above conclusion, then we pick the rate that yields the maximum EAR. This would be the local optimal transmission rate and the corresponding forwarding candidate set. Note that for a node n_i, it is possible that no neighboring nodes are closer to the destination than itself. In this case we need some mechanism like face routing (Karp and Kung 2000) to contour the packet around the void. However, solving the communication voids problem is beyond the scope of this chapter.

Note that the above discussion does not take protocol overheads into consideration. As we have shown in (Zeng *et al*. 2007a, b, c), including as many as possible nodes might not be the optimal strategy when overheads, such as the time used to coordinate the relay contention at MAC layer, are taken into consideration. To consider the protocol overhead, the EAR can be extended to the metric EOT (expected one-hop throughput), which we will study in Chapter 8. However, in this chapter, as our goal is on studying the end-to-end throughput bound of OR, we apply EAR as the local metric, which is the upper bound of the packet advancement rate that can be made by any GOR.

4.4   Performance evaluation

In this section, we use Matlab to investigate the impact of different factors on the end-to-end throughput bound of opportunistic routing, such as source–destination distances, node densities, and number of forwarding candidates. Both line and square topologies are studied for each factor. We also compare the performance of single rate opportunistic routing and multirate ones, and the performance of OR with traditional routing (TR). We call a routing scheme "traditional" when there is only one forwarding candidate selected for each packet relay at each hop.

The OR schemes we investigate include single-rate ExOR (Biswas and Morris 2005), single/multirate GOR and single/multirate LMTOR introduced in Section 4.3.1. For ExOR (Biswas and Morris 2005), each transmitter selects the neighbors with lower ETF (Expected number of Transmissions over Forward links) to the destination than itself as the forwarding candidates, and neighbors with lower ETF have higher relay priorities. For GOR, the forwarding candidates of a transmitter are those neighbors that are closer to the destination, and candidates with larger advancement to the destination have higher relay priorities. The EAR metric proposed in Section 4.2 is used to select the transmission rate for each node in the multirate scenario. For multirate LMTOR, the algorithm and metric proposed in Section 4.3.1 is used to choose transmission rate and forwarding candidates at each node. All the evaluations are under protocol model.

4.4.1 Simulation setup

The simulated network has 20 stationary nodes randomly uniformly distributed on a line with length L or in a $W \times W m^2$ square region. The data rates 24, 12, and 6 mbps (chosen from 802.11a) are studied. We use one of the most common models–log-normal shadowing fading model (Rappaport 1996) to characterize the signal propagation. The received signal power is:

$$P_r(d)_{dB} = P_r(d_0)_{dB} - 10\beta log\left(\frac{d}{d_0}\right) + X_{dB} \tag{4.17}$$

where $P_r(d)_{dB}$ is the received signal power at distance d from the transmitter, β is the path loss exponent, and X_{dB} is a Gaussian random variable with zero mean and standard deviation σ_{dB}. $P_r(d_0)_{dB}$ is the receiving signal power at the reference distance d_0, which is calculated by Equation (4.18):

$$P_r(d_0)_{dB} = 10log\left(\frac{P_t G_t G_r c^2}{(4\pi)^2 d_0^2 f^2 L}\right) \tag{4.18}$$

where P_t is the transmitted signal power, G_t and G_r are the antenna gains of the transmitter and the receiver respectively, c is velocity of light, f is the carrier frequency, and L is the system loss.

In our simulation, $d_0 = 1\,m$, $\beta = 3$, $\sigma_{dB} = 6$, G_t, G_r and L are all set to 1, $P_t = 15\,dbm$, $c = 3 \times 10^8\,m/s$, and $f = 5\,GHz$.

We assume a packet is received successfully if the received signal power is greater than the receiving power threshold (P_{Th}). According to Yee and Pezeshki-Esfahani (2002), for 802.11a, the P_{Th} for 24, 12, and 6 mbps is -74, -79, and -82 dbm, respectively. Then according to Equations (4.17) and (4.18), the packet reception ratio for each rate at a certain distance d can be derived. The PRR versus distance for each data rate is shown in Figure 4.4. We set the PRR threshold p_{td} as 0.1, so the effective transmission radius for each rate (24, 12, and 6 mbps) is 47, 70 and 88 m, respectively. As discussed in Zhai and Fang (2006b), 802.11 systems have very close interference ranges for different channel rates, so we use a single interference range 120 m for all channel rates for simplicity.

4.4.2 Impact of source–destination distances

In this subsection, we evaluate the impact of the source–destination distance on the end-to-end throughput bound of OR and TR in line and square topologies. For line topology, the length L is set as 400 m. We fix the left-end node as the destination, and calculate the throughput bounds from all other nodes to it under different OR and TR variants. For square topology, the side length is set as 150 m. We fix the node nearest to the lower left corner as the destination, and calculate the throughput bounds from all other nodes to it. Therefore, there are 19 different source–destination pairs considered in the evaluation for each topology. We evaluate the performance under both single-rate and multirate scenarios. The average numbers of neighbors per node (indicated as ρ) under different topologies and data rates are summarized in Tables 4.1.

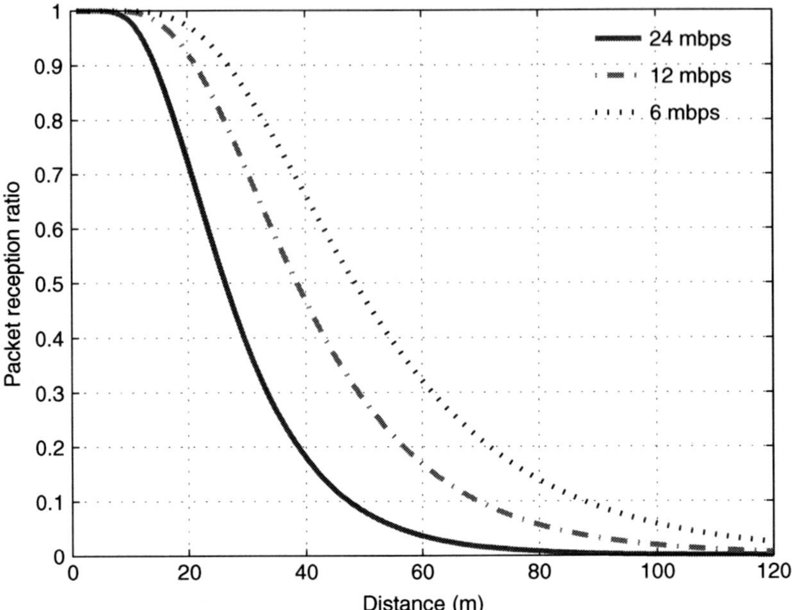

Figure 4.4   Packet reception ratio versus distance at different data rates.

Table 4.1   Average number of neighbors per node under different topologies and data rates.

Rate (mbps)	ρ	
	Line	Square
24	3.5	3.5
12	5.5	7.0
6	6.8	10.0

In the single-rate scenario, for TR, we compute the exact end-to-end throughput bound between the source–destination pairs according to the LP formulations in (Jain *et al.* 2003), which normally result in multiple paths from the source to the destination. So we call it "Multipath TR". We also compute the end-to-end throughput of a single path that is found by minimizing the medium time (delay), and we call it "Single-path TR". The bound of single-path TR is calculated according to the formulations in (Zhai and Fang 2006a). For the three OR variants, we compute the throughput bounds under both conservative (indicated as "c") and greedy (indicated as "g") modes as we discussed in Section 4.1.2.

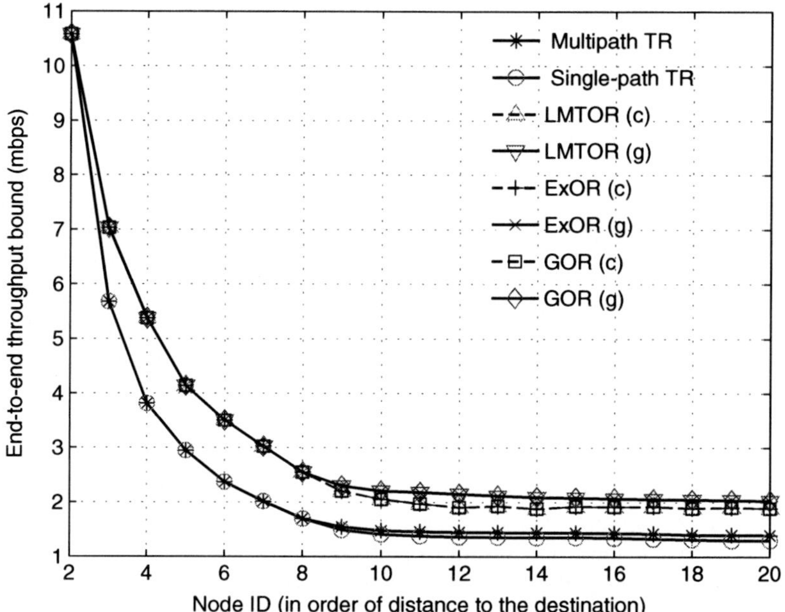

Figure 4.5   End-to-end throughput bound of OR and TR in a single rate (12 mbps) network under line topology. Reproduced by permission of © 2008 IEEE.

Figure 4.5 shows the simulation results of LMTOR, ExOR, GOR and TR in a single rate (12 mbps) system under line topology. We have the following observations: 1. when the distance between the source and destination increases, the end-to-end throughput bound of each routing scheme decreases. 2. the OR achieves higher throughput bound than TR under different source–destination distances. 3. all the OR variants achieve the same performance under the same mode. 4. when source–destination distance is larger than two hops, OR in greedy mode results in higher end-to-end throughput than that in conservative mode, whereas when the source–destination distance is smaller than two hops they represent the same performance. 5. the multipath TR achieves almost the same throughput bound as single-path TR.

In the line topology, the throughput gain of OR over TR mainly comes from the opportunistic property. That is, for each packet transmission, multiple forwarding candidates help on forwarding the packet. The reliability of at least one forwarding candidate correctly receiving the packet is increased comparing to TR. The increased reliability reduces the retransmission overhead and saves the medium time for each packet forwarding, thus improving the throughput.

By tracing into the simulation, we find that the three OR variants result in the same forwarding candidate selection and prioritization at each forwarding node, although they follow different criteria to select the candidates and prioritize them.

That's why we have the observation 3, which indicates that in the line topology the per-hop greedy behavior in GOR can approach the same end-to-end performance as that obtained by a distributed scheme like LMTOR.

For the observation 4, when the source is near to the destination, all the nodes along the paths are in the interference range of each other, thus there is no concurrent transmission allowed in either greedy or conservative mode. Therefore, OR in both modes achieves the same performance when the source–destination distance is smaller than two hops. When the source–destination distance is lager than two hops, concurrent transmission in the network becomes possible. Since the conservative mode requires interference-free communication at all the forwarding candidates, for each transmission, it consumes more space than the greedy mode, which only needs interference-free communication at least at the forwarding candidate. That is, the greedy mode achieves a higher spacial reuse ratio than conservative mode and allows more concurrent transmissions in the network, thus resulting in higher throughput.

The observation 5 indicates that multipath TR does not really improve the wireless network throughput over the single-path TR in the line topology. The reason is that even when there are multiple paths between the source and destination, the links on different paths cannot be scheduled at the same time due to interference. Opportunistic routing does make real use of multiple paths in the sense that throughput can take place on any one of the outgoing links from the sender to its forwarding candidates.

Figure 4.6 shows the simulation results of LMTOR, ExOR, GOR and TR in a single rate (12 mbps) system under square topology. One interesting observation is that the multipath TR achieves (up to 60%) higher throughput bound than single-path TR, and it can achieve comparable or even higher throughput than OR in conservative mode when the source–destination distance is larger than two hops. In the square topology, when the source and destination are far apart, real multipath

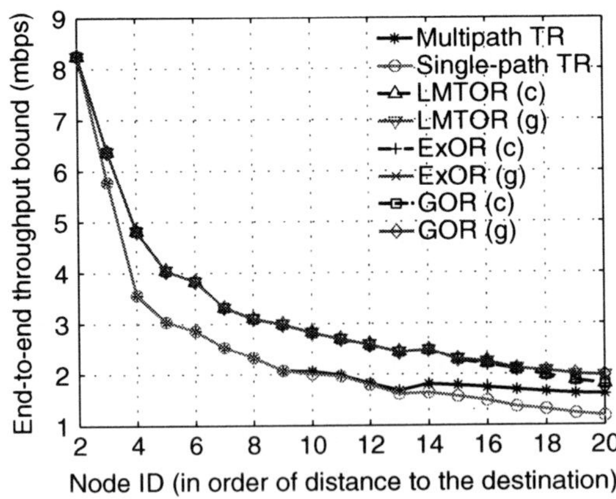

Figure 4.6   End-to-end throughput bound of OR and TR in a single rate (12 mbps) network under square topology. Reproduced by permission of © 2008 IEEE.

routing becomes feasible. That is, different links on different paths can be activated at the same time and this improves the throughput. This observation also indicates that it is not a good idea to include as many forwarding candidates as possible into opportunistic routing when a protocol requires freedom from interference at all the forwarding candidates. As we can see in Figure 4.6 OR in greedy mode still achieves higher throughput than OR in conservative mode and multipath TR. So the advantage of OR over TR is still validated.

Opportunistic routing in greedy mode always achieves higher throughput bound than that in conservative mode, so in the following evaluation, the throughput bound of OR is only calculated under greedy mode. As the performance of ExOR is nearly the same as that of GOR, we will not show the simulation result of ExOR in the following figures. Now, we compare the throughput bounds of OR in multirate and single-rate systems.

Figure 4.7 shows the simulation results of multirate LMTOR, multirate GOR, and single-rate GOR under line topology. We can see that generally multirate OR achieves better performance than any single-rate OR. When the distance between the source and destination is shorter than the interference range (corresponding to node ID 7), the system operating on 24 mbps achieves better performance than that on 12 mbps. However, the difference becomes smaller and smaller when the source–destination distance becomes larger, since more forwarding candidates are involved for 12 mbps and the spacial diversity is increased. When the

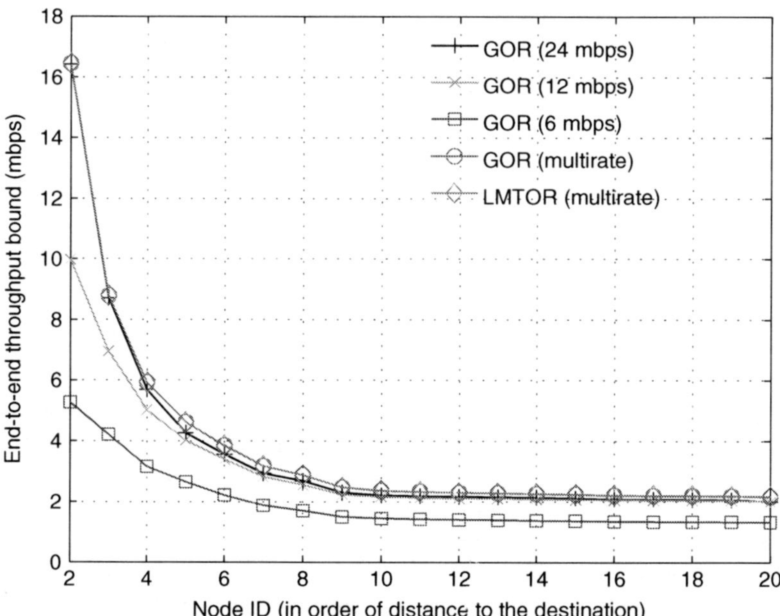

Figure 4.7 End-to-end throughput bound of OR in single-rate and multirate networks under line topology. Reproduced by permission of © 2008 IEEE.

source–destination distance is larger than the interference range, the performance of 24 mbps is the same as that of 12 mbps. Figure 4.8 shows the simulation results under square topology. An interesting difference from line topology is that the system operating at 24 mbps shows lower throughput bound than those operating at 12 mbps and 6 mbps for most of the source–destination pairs. The disadvantage of short transmission range and lower spacial diversity of 24 mbps overwhelms its higher data-rate advantage in the square topology.

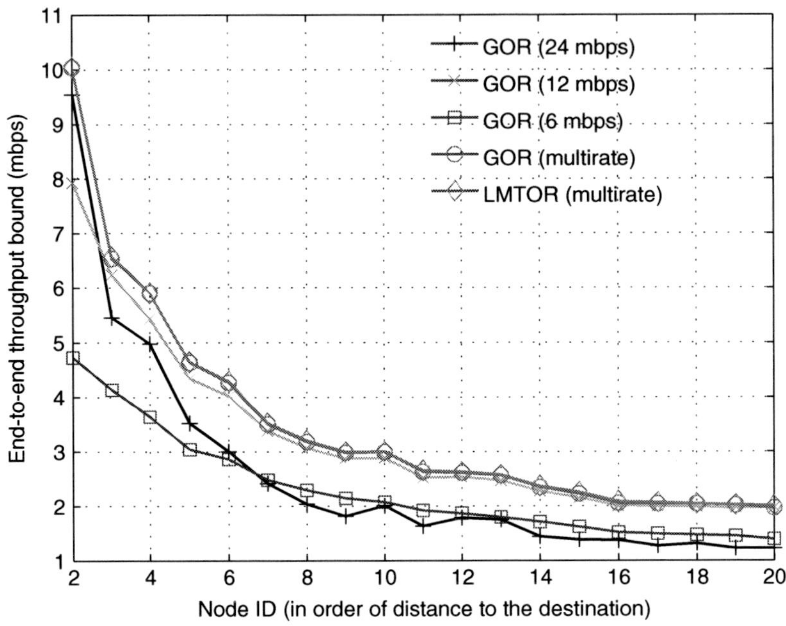

Figure 4.8   End-to-end throughput bound of OR in single-rate and multirate networks under square topology. Reproduced by permission of © 2008 IEEE.

4.4.3   Impact of forwarding candidate number

In this subsection, we study the impact of the number of forwarding candidates on the performance of OR. For line topology, we examine the bound between the two end nodes on the line. For square topology, we examine the throughput bound between the two end nodes on the diagonal. The topology sizes are set as the same as those in the previous simulation. For a transmitter, given a maximum number of forwarding candidates, the single-rate GOR selects the forwarding candidates as follows: first, it finds all the neighbors that are closer to the destination than the transmitter; second, if the number of neighbors found is less than or equal to the maximum number of forwarding candidates, GOR just involves all the found neighbors and gives the neighbors closer to the destination higher relay priorities. If the number of the found neighbors is greater than the maximum number of forwarding candidates, we apply the algorithm proposed in (Zeng *et al.* 2007b) to select the forwarding candidates which maximizes the EPA. For multirate GOR, we select the forwarding candidates for each single-rate GOR and calculate its corresponding EAR, then select the data rate with the highest EAR. For LMTOR, we apply the distributed algorithm proposed in Section 4.3.1. For the local search in Equations (4.14) and (4.15), we only test a subset of all the neighbors with cardinality no larger than the maximum number of forwarding candidates.

Figures 4.9 and 4.10 show the simulation results under line and square topologies, respectively. Generally, multirate OR achieves better performance than any single-rate OR, and multirate LMTOR achieves better performance than multirate GOR. In the square topology (Figure 4.10), GOR on 12 mbps is always the best among all the single-rate GOR for all the different candidate sizes. The 24 mbps GOR performs even worse than the 6 mbps GOR in square topology when the maximum forwarding candidate number is larger than 3. Since 24 mbps has the shortest transmission range, which results in the lowest node density as shown in Table 4.1, GOR on 24 mbps actually does not have three or more forwarding candidates to choose. Note that, the maximum number of forwarding candidates being equal to 1 corresponds to the TR. Although 6 mbps geographic TR (GTR) achieves lower throughput bound than 24 mbps GTR, it is not necessarily the truth for GOR. Lower data rates have longer transmission ranges, so this yields higher neighborhood diversities, which can help to increase the effective forwarding rate for each transmission

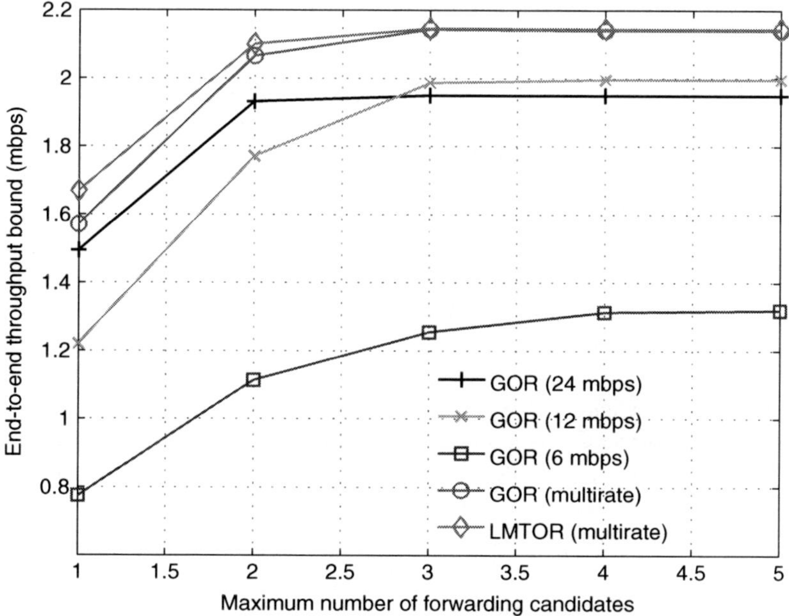

Figure 4.9   End-to-end throughput bound of OR with different number of forwarding candidates under line topology. Reproduced by permission of © 2008 IEEE.

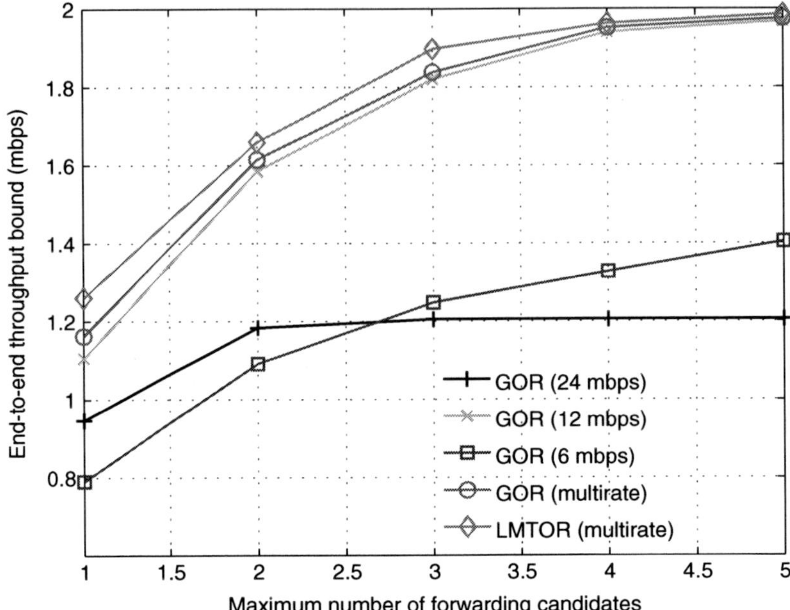

Figure 4.10   End-to-end throughput bound of OR with different number of for-warding candidates under square topology. Reproduced by permission of © 2008 IEEE.

when OR is used. In the line topology (Figure 4.9), when the forwarding candidate number is greater than 3, GOR on 12 mbps achieves better performance than that on 24 mbps which can be explained by the same reason. However, in the line topology, the disadvantage of low data rate of 6 mbps overwhelms its advantage on higher spacial diversity, GOR on 6 mbps shows the worst performance.

An interesting observation in both Figure 4.9 and Figure 4.10 is the concavity of each curve, which indicates that although involving more forwarding candidates improves the end-to-end throughput bound of OR, the capacity gained becomes marginal when we keep doing so. We can see that when the number of forwarding candidates is larger than 3, the end-to-end throughput bound remains almost unchanged. This end-to-end throughput observation is consistent with the local behavior found in Chapter 2. For a realistic MAC for OR, the coordination overhead is likely to increase when more forwarding candidates are involved. Since the throughput gain decreases when the number of forwarding candidates is increased, considering the MAC overhead, it may not be wise or necessary to involve as many as forwarding candidates in OR.

4.4.4   Impact of node density

The impact of the node density on the performance of OR is investigated in this subsection. Instead of single flow, we investigate the multiflow case by randomly

Table 4.2   Average number of neighbors per node at each rate under square topology with different side lengths.

Data rate (mbps)	Line length			
	300	400	500	600
24	4.7	3.4	2.6	2.2
12	7.1	5.5	4.3	3.5
6	9.0	6.8	5.4	4.4

Table 4.3   Average number of neighbors per node at each rate under square topology with different side lengths.

Data rate (mbps)	Square side length			
	100	120	140	180
24	7.7	5.5	4.1	2.8
12	13.8	10.9	8.7	5.8
6	17	14.5	11.9	8.6

Reproduced by permission of © 2008 IEEE.

selecting four source–destination pairs in the network. The settings of the network terrain size and the corresponding number of neighbors per node under different data rates are summarized in Table 4.2 and 4.3.

Figures 4.11 and 4.12 show the simulation results under line and square topologies, respectively. They show the same trend. There exists a threshold on the node density, higher than which the GOR on 24 mbps performs better than that on 12 mbps, and lower than which the opposite is the case. The threshold is about 5.5 and 10.9 neighbors per node on 12 mbps for line and square topologies, respectively. Our proposed multirate GOR and LMTOR can adapt to the different node densities, and choose the proper transmission rate and forwarding candidate set to achieve a better performance than any single-rate GOR.

4.5   Conclusion

In this chapter, we studied the impact of multiple rates, interference, candidate selection and prioritization on the maximum end-to-end throughput of OR. Taking into consideration wireless interference, we proposed a new method of constructing transmission conflict graphs, and present a methodology for computing the end-to-end throughput bounds (capacity) of OR. We formulated the maximum end-to-end throughput problem of OR as a maximum-flow linear programming problem subject to the transmission conflict constraints and effective forwarding rate on each link.

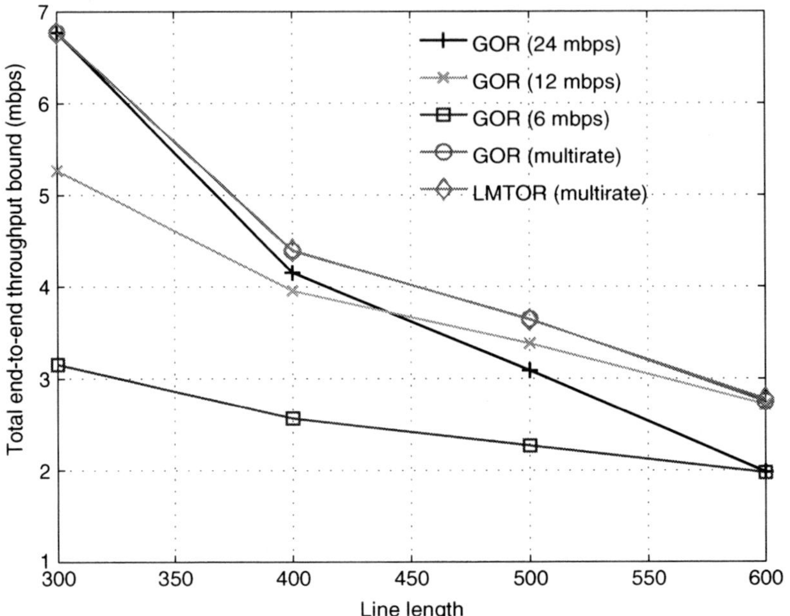

Figure 4.11 *Total end-to-end throughput bound of OR under line topology with different lengths in multiflow case.*

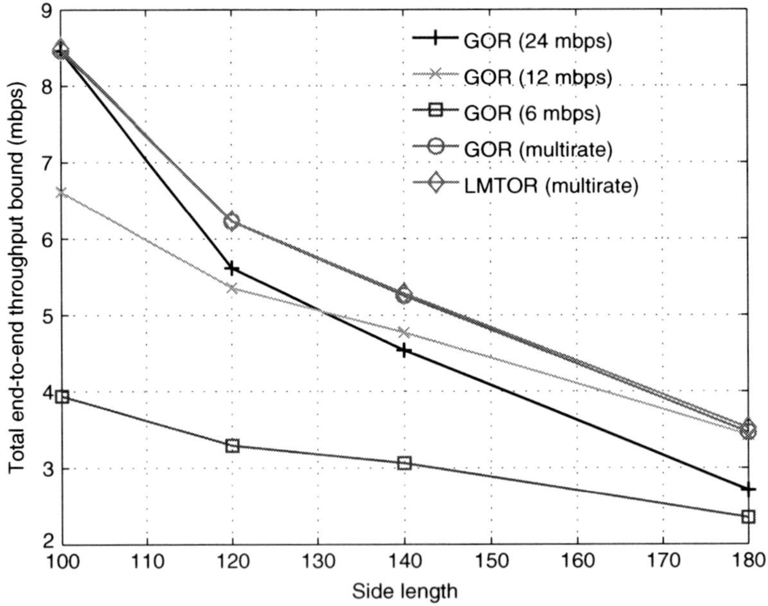

Figure 4.12 *Total end-to-end throughput bound of OR under square topology with different side lengths in multiflow case. Reproduced by permission of © 2008 IEEE.*

To the best of our knowledge, this is the first theoretical work on capacity problem of OR for multihop and multirate wireless networks.

We also proposed two metrics for OR under the multirate scenario. One is *expected medium time* (EMT) and the other is *expected advancement rate* (EAR). Based on these metrics, we proposed the distributed and local rate and candidate selection schemes: LMTOR and MGOR, respectively. We validated the analysis results by simulation, and compared the throughput capacity of multirate OR with single-rate ones under different settings, such as different topologies, source–destination distances, number of forwarding candidates and node densities. We showed that OR has great potential to improve the end-to-end throughput under different settings, and our proposed multirate OR schemes achieve higher throughput bound than any single-rate GOR. We observed some insights of OR: 1. The end-to-end capacity gained decreases when the number of forwarding candidates is increased. When the number of forwarding candidates is larger than 3, the end-to-end throughput bound remains almost unchanged. 2. There exists a node density threshold, higher than which 24 mbps GOR performs better than 12 mbps GOR, and lower than which the opposite is the case. The threshold is about 5.5 and 10.9 neighbors per node on 12 mbps for line and square topologies, respectively.

References

Awerbuch B, Holmer D and Rubens H 2006 The medium time metric: High throughput route selection in multi-rate ad hoc wireless networks. *MONET* 11(2), 253–266.

Biswas S and Morris R 2005 Exor: Opportunistic multi-hop routing for wireless networks *SIGCOMM'05*, Philadelphia, Pennsylvania.

Chavtal V 1983 *Linear Programming*. W.H. Freeman, New York, NY, USA.

Draves R, Padhye J and Zill B 2004 Routing in multi-radio, multi-hop wireless mesh networks *MobiCom '04*.

Dubois-Ferriere H, Grossglauser M and Vetterli M 2007 Least-cost opportunistic routing. Technical Report LCAV-REPORT-2007-001, School of Computer and Communication Sciences, EPFL.

Fussler H, Widmer J, Kasemann M, Mauve M and Hartenstein H 2003 Contention-based forwarding for mobile ad-hoc networks. *Elsevier's Ad Hoc Networks* 1(4), 351–369.

Gupta P and Kumar PR 2000 The capacity of wireless networks. *Trans. Inform. Theory* 46(2), 388–404.

Jain K, Padhye J, Padmanabhan VN and Qiu L 2003 Impact of interference on multi-hop wireless network performance *MobiCom '03: Proceedings of the 9th Annual International Conference on Mobile Computing and Networking*, pp. 66–80. ACM Press, New York, NY, USA.

Karp B and Kung H 2000 Gpsr: Greedy perimeter stateless routing for wireless networks *ACM MOBICOM*, Boston.

Laufer RP, Dubois-Ferriere H and Kleinrock L 2010 Polynomial-time algorithms for multirate anypath routing in wireless multihop networks. Technical Report UCLA-CSD-TR100034, UCLA Computer Science Department.

Lee S, Bhattacharjee B and Banerjee S 2005 Efficient geographic routing in multihop wireless networks *MobiHoc*.

Rappaport TS 1996 *Wireless Communications: Principles and Practice*. Prentice Hall, New Jersey.

Seada K, Zuniga M, Helmy A and Krishnamachari B 2004 Energy efficient forwarding strategies for geographic routing in wireless sensor networks *ACM Sensys'04*, Baltimore, MD.

Tang J, Xue G and Zhang W 2007 Cross-layer design for end-to-end throughput and fairness enhancement in multi-channel wireless mesh networks. *IEEE Transactions on Wireless Communications* 6(10), 3482–3486.

Yee J and Pezeshki-Esfahani H 2002 Understanding wireless lan performance trade-offs. *CommsDesign.com*, http://i.cmpnet.com/commsdesign/csd/2002/nov02/feat3-nov02.pdf.

Zeng K, Lou W and Zhang Y 2007a Multi-rate geographic opportunistic routing in wireless ad hoc networks *IEEE Milcom*, Orlando, FL.

Zeng K, Lou W, Yang J and Brown DR 2007b On geographic collaborative forwarding in wireless ad hoc and sensor networks *WASA'07*, Chicago, IL.

Zeng K, Lou W, Yang J and Brown DR 2007c On throughput efficiency of geographic opportunistic routing in multihop wireless networks *QShine'07*, Vancouver, British Columbia, Canada.

Zhai H and Fang Y 2006a Impact of routing metrics on path capacity in multirate and multihop wireless ad hoc networks *IEEE ICNP*.

Zhai H and Fang Y 2006b Physical carrier sensing and spatial reuse in multirate and multihop wireless ad hoc networks *IEEE Infocom*.

Zhang J, Wu H, Zhang Q and Li B 2005 Joint routing and scheduling in multi-radio multi-channel multi-hop wireless networks *IEEE Broadnets*.

Zorzi M and Rao RR 2003 Geographic random forwarding (geraf) for ad hoc and sensor networks: energy and latency performance. *IEEE Transactions on Mobile Computing* **2**(4), 349–365.

5

Multiradio multichannel opportunistic routing

Two major factors that limit the throughput in multihop wireless networks are the co-channel interference and unreliability of wireless transmissions. Multiradio multichannel technology and opportunistic routing (OR) have shown their promise to significantly improve the network capacity by combating these two limits. It raises an interesting problem concerning the tradeoff between multiplexing and spacial diversity when integrating these two techniques for throughput optimization. It is not known what the capacity of the network could be when nodes have multiple radios and OR capability. In this chapter, we present our study on optimizing an end-to-end throughput of the multiradio multichannel network when OR is available. First, we formulate the end-to-end throughput bound as a linear programming (LP) problem, which solves the radio-channel assignment, transmission scheduling, and forwarding candidate selection problems together. Second, we propose an LP approach and a heuristic algorithm to find a feasible scheduling of opportunistic forwarding priorities to achieve the capacity. Simulations show that the heuristic algorithm achieves desirable performance under various number of forwarding candidates. Leveraging our analytical model, we find that 1. OR can achieve better performance than traditional routing (TR) under different radio/channel configurations, however, in particular scenario (e.g. bottleneck links exist between the sender and relays), TR is preferable; 2. OR can achieve comparable or better performance than TR by using less radio resource.

Multihop Wireless Networks: Opportunistic Routing, First Edition.
Kai Zeng, Wenjing Lou and Ming Li.
© 2011 John Wiley & Sons, Ltd. Published 2011 by John Wiley & Sons, Ltd.

5.1 Introduction

With the spur of modern wireless technologies, a promising way to improve the system throughput is to allow more concurrent transmissions by installing multiple radio interfaces on one node with each radio tuned to a different orthogonal channel (Alicherry *et al*. 2005; Kodialam and Nandagopal 2005; Zhang *et al*. 2005). Other than the multiradio multichannel technology, opportunistic routing (OR) also shows its potential for significantly improving network throughput (Biswas and Morris 2005; Chachulski *et al*. 2007; Dubois-Ferriere *et al*. 2007; Fussler *et al*. 2003; Shah *et al*. 2004; Zeng *et al*. 2008; 2007a,b,c; Zhong *et al*. 2006). Opportunistic routing is a network-MAC cross-layer design, which involves multiple forwarding candidates at each hop, and the actual forwarder is selected *after* packet transmission according to the instant link reachability and availability. It is quite different from the traditional routing (TR) in that only one *pre-selected* next-hop node is involved to forward packets at each hop.

When integrating these two techniques, an interesting question arises: "what is the end-to-end throughput bound of the multi-radio multi-channel network when OR is available?" In this chapter, we will propose a methodology to answer this question. However, it is a nontrivial task.

First, unlike TR, OR has a unique quality in that for each packet transmission, any one of the forwarding candidates of the transmitter can become the actual forwarder. Thus, effective throughput can take place from a transmitter to any one of its forwarding candidates at any instant. However, for TR, throughput can only happen from a transmitter to a predefined next-hop node even if other neighboring nodes overhear the transmission. Therefore, previous work (Alicherry *et al*. 2005; Kodialam and Nandagopal 2005; Zhang *et al*. 2005) on the throughput optimization in multiradio multichannel systems based on traditional routing (TR) cannot be directly applied to OR.

Second, multiradio multichannel capability raises challenging issues on radio-channel assignment for OR. In a single-radio single-channel system, OR naturally takes advantage of the redundant receptions on multiple neighboring nodes without consuming or sacrificing any extra channel resources. When a node is sending packets, all of its one-hop neighbors usually cannot send or receive other packets at the same time due to co-channel interference. That is, these one-hop neighbors have no other choices but listen to the transmission. However, in multiradio/channel systems, the one-hop neighbors have two choices: 1. they can operate on the same channel as the transmitter to improve the diversity gain on the receiver side, then more effective traffic can flow out of the transmitter and the system throughput can be increased; or 2. they can operate on other orthogonal channels. Thus they have chances to transmit/receive packets to/from other nodes, which may result in more concurrent effective traffic flowing in the network and can also increase the system throughput. This can be considered as a tradeoff between multiplexing and spacial diversity. Which choice the neighboring nodes should make is nontrivial. The radio-channel assignment for optimizing the end-to-end throughput in multiradio, multichannel systems when OR is available deserves careful study.

Third, due to the broadcast nature of the wireless medium, a transmission may interfere with neighboring links operating on the same channel. Therefore, a node's transmission should be optimally scheduled in order to maximize the throughput. Finally, even when the radio-channel assignment and transmission scheduling are given, we still need to optimally (often dynamically) select forwarding candidates and assign relay priorities among them in order to maximize the end-to-end throughput. How to dynamically assign and schedule the forwarding priority among forwarding candidates has not been well studied in the existing literature.

In summary, in order to maximize the end-to-end throughput of the multiradio, multichannel network when OR is available, we should jointly address multiple issues: radio-channel assignment, transmission scheduling, and forwarding candidate selection and forwarding priority scheduling. In this chapter, we carry out a comprehensive study on these issues. First, we formulate the end-to-end throughput bound between a source–destination pair in multiradio, multichannel, multihop wireless networks with OR capability as a linear programming (LP) problem which jointly solves the radio-channel assignment, transmission scheduling, and forwarding candidate selection. Second, we propose an LP approach and a heuristic algorithm to find a feasible scheduling of opportunistic forwarding priority to achieve the throughput bound. The proposed heuristic algorithm achieves desirable performance under different number of forwarding candidates. Leveraging our analytical model, we gain the following two insights: 1. OR can achieve better performance than TR under different radio/channel configurations, however, in some scenarios (e.g. when bottleneck links exist between the sender and relays), TR is preferable; 2. OR can achieve comparable or even better performance than TR by using fewer radio resources.

The rest of this chapter is organized as follows. We introduce the system model and opportunistic routing in Section 5.2. We propose the framework of computing the throughput bound between a source–destination pair in multiradio, multichannel, multihop wireless networks with OR capability in Section 5.3. The scheduling of opportunistic forwarding priorities is studied in Section 5.4. Examples and simulation results are presented and analyzed in Section 5.5. Conclusions are drawn in Section 5.6.

5.2 System model and opportunistic routing primer

We consider a multihop wireless network with N nodes. Each node n_i ($1 \leq i \leq N$) is equipped with one or more wireless interface cards, referred to as radios in this work. Denote the number of radios in each node n_i as t_i ($i = 1 \ldots N$). Assume K orthogonal channels are available in the network without any interchannel interference. We consider the system with channel-switching capability, such that a radio can dynamically switch across different channels. We assume there is no performance gain to assigning the same channel to the different radios on the same node (i.e. we do not consider MIMO). For simplicity, we assume each node n_i transmits at the same data rate R_i among all its radios and channels. However, our model can be easily extended to the multirate case. We also assume half-duplex on each radio-that is, a radio cannot transmit and receive packets at the same time. This is usually true in practice. There is a unified transmission range and interference range for the whole network. The transmission range and interference range are largely dependent on the transmission power, which is fixed in our model. Typically, the interference range is larger than the transmission range. Two nodes, n_i and n_j, can communicate with each other if the Euclidean distance between them is less than the transmission range and they are operated on the same channel. Due to the unreliability of the wireless links, a packet reception ratio (PRR) is associated with each transmission link. In this chapter, we assume that the link quality on each channel is independent and can be obtained by the existing measurement schemes (Aguayo *et al.* 2004; Kim and Shin 2006). In order to analyze the throughput bound, we assume that packet transmission/forwarding at an individual node and radio/channel allocation can be perfectly scheduled by an omniscient and omnipotent central entity. Thus, we do not concern ourselves with issues such as MAC contention or coordination overheads that may be unavoidable in a distributed network. This is a very commonly used assumption for theoretical studies (Jain *et al.* 2003; Zeng *et al.* 2008; Zhai and Fang 2006).

5.2.1   Opportunistic routing primer

Different from TR, OR basically runs in such a way that, for each local packet forwarding, a set of next-hop forwarding candidates are selected at the network layer and one of them is chosen as the actual relay at the MAC layer according to their instantaneous availability and reachability at the time of transmission. Using Figure 5.1 as an example, the **one-hop neighbor set** of a transmitter n_i is $C_i = \{n_{i_1}, \ldots, n_{i_5}\}$, which consists of nodes that operate on the same channel as node n_i and are in its transmission range. A subset $\mathcal{F}_i = \{n_{i_1}, \ldots, n_{i_3}\}$ of C_i is selected as the **forwarding candidate set** of n_i. We name $(n_i, \mathcal{F}_i)$ as an **opportunistic module**.

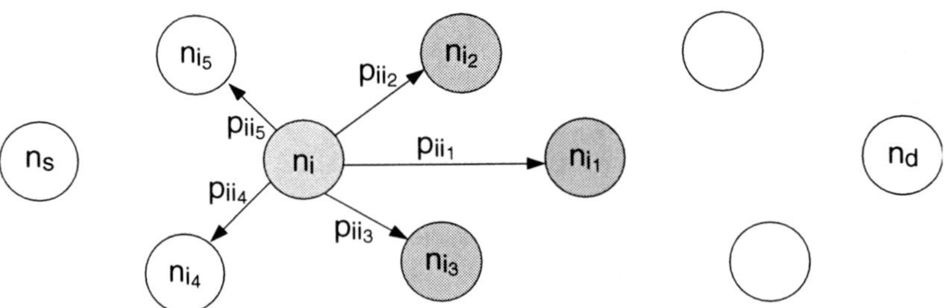

Figure 5.1   An example of opportunistic module $(n_i, \mathcal{F}_i)$, where $\mathcal{F}_i = \{n_{i_1}, n_{i_2}, n_{i_3}\}$. Reproduced by permission of © 2010 IEEE.

To avoid packet duplication, only one of the forwarding candidates becomes the actual forwarder of each packet. There is a forwarding priority among these forwarding candidates to decide which should forward the packet if multiple forwarding candidates correctly receive the same packet. We use $\mathcal{P}$ to represent the forwarding priority among the forwarding candidate, such that $\mathcal{P}(i_j) > \mathcal{P}(i_k)$ indicates n_{i_j} has higher forwarding priority than n_{i_k}.

To send a packet from the source n_s to a destination n_d, OR works by the source n_s sending the packet to the receivers in its forwarding candidate set $\mathcal{F}_s$. One of the candidate nodes continues the forwarding based on their relay priorities. If the first-priority node in the set has received the packet successfully, it forwards the packet towards the destination while all other nodes suppress themselves from duplicate forwarding. Otherwise, the second-priority node in the set is arranged to forward the packet if it has received the packet correctly. Otherwise the third-priority node, the fourth-priority node, etc. A forwarding candidate will forward the packet only when all the other candidates with higher priorities failed to do so.

Several MAC protocols (Biswas and Morris 2005; Fussler *et al.* 2003; Zorzi and Rao 2003) have been proposed to ensure the relay priority among the candidates. For example, in Biswas and Morris (2005) a batch map is used to indicate the packets known to have been received by higher priority candidates, thus prohibiting the lower priority candidates from relaying duplicate copies of the packets. Only when none of the forwarding candidates has successfully received the packet will the sender retransmit the packet if retransmission is enabled. The forwarding reiterates until the packet is delivered to the destination n_d.

5.3 Problem formulation

In this section we present our methodology to compute the throughput bound between two end nodes in a multiradio, multichannel, multihop wireless network when OR is available. We first study which opportunistic modules can coexist at the same time under the constraints of wireless interference and radio interface limits. We then formulate the end-to-end throughput bound as an LP problem, which solves the radio-channel assignment, transmission scheduling and forwarding candidate selection problems together. We further propose an LP approach and a heuristic algorithm to find a feasible scheduling of opportunistic forwarding priorities to achieve the capacity.

5.3.1 Concurrent transmission sets

In this subsection, we will discuss which opportunistic modules in the network can be activated at the same time. The set of opportunistic modules that can be activated at the same time is called the **concurrent transmission set** (CTS). The motivation for building a concurrent transmission set is similar that for building an independent set in Jain *et al.* (2003) and concurrent transmission patterns in Zhang *et al.* (2005) – that is, taking the benefit of time-sharing scheduling of different concurrent transmission sets, we could achieve a collection of capacity graphs, associated with capacity constraint on each link. Opportunistic routing can be performed on the underlying capacity graph to achieve the maximum throughput. However, the methodology of constructing CTS for OR is quite different from those in Jain *et al.* (2003) and Zhang *et al.* (2005) for TR. Because for OR, any of the forwarding candidates can become the actual forwarder for each transmission, and the instantaneous throughput can take place on any link from the transmitter to any forwarding candidate. So the CTS is constructed based on opportunistic modules (involving multiple links sharing the same transmitter) instead of individual links. Furthermore, besides the co-channel interference, radio interface limits in the multiradio system also impose constraints on concurrent transmissions in the network.

We introduce the concept of **transceiver configuration**, v_i^k, which indicates node n_i operating on channel k ($1 \leq k \leq K$). Each transceiver configuration can be in either the transmission or the reception state and we call it transmitter or receiver, respectively. We say there is a wireless link l_{ij}^k ($i \neq j$) when v_i^k is a transmitter and v_j^k is a receiver and v_j^k is in the transmission range of v_i^k. Link l_{ij}^k is **usable** when v_j^k is not in the interference range of any other transmitters; otherwise, it is **unusable**. When a link is usable, its transmitter and receiver are also usable. Let $V = \{v_i^k | i = 1 \ldots N, k = 1 \ldots K\}$, and $E = \{l_{ij}^k | i, j = 1 \ldots N, i \neq j, k = 1 \ldots K\}$.

A CTS T_α can be represented by an indicator vector on all wireless links, written as $T_\alpha = \{\psi_{ij}^{k\alpha} | l_{ij}^k \in E\}$.

$$\psi_{ij}^{k\alpha} = \begin{cases} 1, & l_{ij}^k \text{ is usable in CTS } T_\alpha \\ 0, & \text{otherwise} \end{cases} \tag{5.1}$$

Denote the following indicator variable to represent the transceiver configuration status in CTS T_α:

$$\eta_i^{k\alpha} = \begin{cases} 1, & v_i^k \text{ is usable in CTS } T_\alpha \\ 0, & \text{otherwise} \end{cases} \tag{5.2}$$

where v_i^k can be a transmitter or receiver.

An opportunistic module in a CTS T_α can be represented as $(v_i^k, \{v_j^k | l_{ij}^k \in E, \psi_{ij}^{k\alpha} == 1\})$. Note that according to the unique property of OR, when a transmitter v_i^k is usable, its multiple receivers can be usable at the same time, while a usable receiver can only correspond to one transmitter. This can be represented formally by:

$$\eta_i^{k\alpha} = \min\left(1, \sum_{l_{ij}^k \in E} \psi_{ij}^{k\alpha}\right) + \sum_{l_{ji}^k \in E} \psi_{ji}^{k\alpha}, \forall\, i = 1 \ldots N, k = 1 \ldots K \tag{5.3}$$

Although any two active links operating on different channels do not interfere with each other, due to radio interface constraints, the number of channels being used on one node cannot exceed the number of radios installed on this node. To satisfy this constraint, we have

$$\sum_{k=1}^{K} \eta_i^{k\alpha} \leq t_i, \forall i = 1 \ldots N \tag{5.4}$$

If two wireless links are concurrently usable on the same channel, they should either share the same transmitter or not interfere with each other. This can be represented by

$$\psi_{ij}^{k\alpha} + \psi_{pq}^{k\alpha} \leq 1 + I(l_{ij}^k, l_{pq}^k), \forall k = 1 \ldots K \tag{5.5}$$

where

$$I(l_{ij}^k, l_{pq}^k) = \begin{cases} 1, & i == p, \ or \ l_{ij}^k \ and \ l_{pq}^k \ do \ not \ interfere \\ 0, & otherwise \end{cases} \tag{5.6}$$

According to Equations (5.3), (5.4), and (5.5), we can enumerate all the CTSs. One CTS represents one radio-channel assignment. Note that the number of all the CTS's is exponential in the number of nodes, radios and channels. However, it may not be necessary to find all of them to maximize an end-to-end throughput. A heuristic algorithm similar to that in Tang *et al.* (2007), or a column-generation technique (Zhang *et al.* 2005) can be applied to find a subset of all the CTSs to approach the throughput bound. Applying these technologies to find CTSs is beyond the scope of this chapter. In this chapter, we simply enumerate all the CTSs. Next, we discuss which link rate (or rate vector) is supportable by OR from a transmitter to its forwarding candidates.

5.3.2   Effective forwarding rate

A fundamental difference of OR from TR is that effective throughput can take place from a transmitter to any of its forwarding candidates at any instant. To capture the unique property of OR we apply the definition of **effective forwarding rate** in Zeng *et al.* (2008) to represent the throughput on each link from a transmitter to each of its forwarding candidate according to a forwarding strategy. For a given transmitter n_i and its forwarding candidate set $\mathcal{F}_i$, under a forwarding priority $\mathcal{P}$, the effective forwarding rate on link l_{ii_q} is defined in Equation (5.7):

$$\widetilde{R}_{ii_q} = R_i \cdot p_{ii_q} \prod_{\mathcal{P}(i_k) > \mathcal{P}(i_q), n_{i_k} \in \mathcal{F}_i} (1 - p_{ii_k}) \tag{5.7}$$

where R_i is the data transmission rate at transmitter n_i.

The effective forwarding rate indicates that according to the relay priority, only when higher-priority forwarding candidates do not receive the packet correctly, a lower-priority candidate may have a chance to relay the packet if it does. A similar methodology is used to define the remaining path cost for a forwarding candidate set in (Dubois-Ferriere *et al.* 2007) and compute the expected number of packets a transmitter must forward in (Chachulski *et al.* 2007).

Then the effective forwarding rate from a transmitter n_i to its forwarding candidate set $\mathcal{F}_i$ is the summation of the effective forwarding rate to each forwarding candidate in the sequence:

$$\widetilde{R}_{i\mathcal{F}_i} = \sum_{n_{i_q} \in \mathcal{F}_i} \widetilde{R}_{ii_q} = R_i \cdot \left(1 - \prod_{n_{i_q} \in \mathcal{F}_i} (1 - p_{ii_q}) \right) \tag{5.8}$$

Note that, the effective forwarding rate from a transmitter to a set of its forwarding candidates only depends on the transmission rate and the PRRs on the corresponding links but does not depend on priority among the forwarding candidates. In Section 5.3.4 we will show that this property eases the LP formulation by avoiding enumerating all the possible prioritizations among the forwarding candidates. It will also be used to design a heuristic scheduling of opportunistic forwarding priorities to satisfy a rate vector in Section 5.4.2.

5.3.3 Capacity region of an opportunistic module

In this subsection, we study the capacity region of an opportunistic module $(n_i, \mathcal{F}_i)$. This capacity region will serve as a bound of a rate vector corresponding to the links in the opportunistic module.

By applying the result proved in Tassiulas and Ephremides (1993), we have the capacity region of $(n_i, \mathcal{F}_i)$ indicated in the following inequality (5.9).

$$\sum_{q=1}^{r} \mu_{ii_q} \cdot \phi_{i_q} \leq R_i \left(1 - \prod_{q=1}^{r} (1 - p_{ii_q} \cdot \phi_{i_q}) \right), \forall [\phi_{i_1}, \ldots, \phi_{i_r}] \in \{0, 1\}^r \qquad (5.9)$$

where μ_{ii_q} $(1 \leq q \leq r)$ is the rate from n_i to n_{i_q} in $\mathcal{F}_i$, and $r = |\mathcal{F}_i|$.

The physical meaning of inequality (5.9) is that any subset summation of the rates on the outgoing links from a transmitter to its forwarding candidates must be bounded by the effective forwarding rate from the transmitter to the corresponding forwarding candidate subset. Now we are ready to formulate the end-to-end throughput bound of OR in multiradio, multichannel systems by making use of the CTS and the capacity region of the opportunistic module.

5.3.4 Maximum end-to-end throughput in multiradio, multichannel, multihop networks with OR capability

Assume we have found all the CTSs $\{T_1, T_2 \ldots T_M\}$ in the network. At any time, we activate all the transmitters in one CTS. Let λ_α denote the time fraction scheduled for CTS T_α $(\alpha = 1 \ldots M)$. The maximum throughput problem can then be converted to an optimal scheduling problem, which schedules the activation of the CTSs to maximize the end-to-end throughout. Therefore, considering communication between a single source, n_s, and a single destination, n_d, with opportunistic routing, we formulate the throughput capacity problem between the source and the destination as a linear programming problem corresponding to a maximum-flow problem under additional constraints in Figure 5.2.

In Figure 5.2, $\mu_{ij}^{k\alpha}$ and $\mu_{ii_q}^{k\alpha}$ denote the rate on link l_{ij}^k and $l_{ii_q}^k$ in the CTS T_α, respectively. Recall that $\mathbf{E}$ is the set of all the wireless links, and $\mathbf{V}$ is the set of all the transceiver configurations. The maximization states that we wish to maximize the sum of the flow rates out of the source, which is the accumulated flow rates on all outgoing links and all channels from the source in all CTSs. The constraint (5.11) represents flow-conservation, i.e., at each node, except the source and the destination, the accumulated incoming flow rate is equal to the accumulated

$$Max \sum_{k=1}^{K} \sum_{l_{si}^{k} \in \mathbf{E}} \sum_{\alpha=1}^{M} \mu_{si}^{k\alpha} \tag{5.10}$$

$$s.t.$$

$$\sum_{k=1}^{K} \sum_{l_{ij}^{k} \in \mathbf{E}} \sum_{\alpha=1}^{M} \mu_{ij}^{k\alpha} = \sum_{k=1}^{K} \sum_{l_{ji}^{k} \in \mathbf{E}} \sum_{\alpha=1}^{M} \mu_{ji}^{k\alpha}$$

$$\forall \, i = 1 \ldots N, i \neq s, i \neq d \tag{5.11}$$

$$\sum_{k=1}^{K} \sum_{l_{is}^{k} \in \mathbf{E}} \sum_{\alpha=1}^{M} \mu_{is}^{k\alpha} = 0 \tag{5.12}$$

$$\sum_{k=1}^{K} \sum_{l_{di}^{k} \in \mathbf{E}} \sum_{\alpha=1}^{M} \mu_{di}^{k\alpha} = 0 \tag{5.13}$$

$$\mu_{ij}^{k\alpha} \geq 0, \quad \forall \, k = 1 \ldots K, l_{ij}^{k} \in \mathbf{E}, \alpha = 1 \ldots M \tag{5.14}$$

$$\sum_{\alpha=1}^{M} \lambda_{\alpha} \leq 1 \tag{5.15}$$

$$\lambda_{\alpha} \geq 0, \quad \forall \, \alpha = 1 \ldots M \tag{5.16}$$

$$\sum_{\mathcal{C}} \mu_{ii_q}^{k\alpha} \cdot \phi_{i_q} \leq \lambda_{\alpha} R_i \left(1 - \prod_{\mathcal{C}} (1 - p_{ii_q}^{k} \cdot \phi_{i_q}) \right)$$

$$\mathcal{C} = \{ n_{i_q} | l_{ii_q}^{k} \in \mathbf{E}, \psi_{ii_q}^{k\alpha} == 1 \}$$

$$\forall v_i^{k} \in \mathbf{V}, \alpha = 1 \ldots M, \forall \, \Phi(\mathcal{C}) \in \{0, 1\}^{|\mathcal{C}|} \tag{5.17}$$

Figure 5.2   LP formulations to optimize the end-to-end throughput between two end nodes in multiradio, multichannel, multihop wireless networks with OR capability. Reproduced by permission of © 2010 IEEE.

outgoing flow rate. The constraint (5.12) states that the incoming accumulated flow rate to the source node is 0. The constraint (5.13) indicates that the outgoing accumulated flow rate from the destination node is 0. The constraint (5.14) restricts the amount of flow rate on each link to be non-negative. The constraint (5.15) shows that at any time, at most one CTS will be scheduled to be active. The constraint (5.16) indicates that the scheduled time fraction should be non-negative.

In the constraint (5.17), $\Phi(\mathcal{C})$ is an indicator vector of ϕ_j with length $|\mathcal{C}|$. The constraint (5.17) states that no matter which forwarding candidates are selected, the flow rates from a transmitter v_i^k to its usable one-hop neighbors in $\mathcal{C}$ should be in the capacity region of the opportunistic module $(v_i^k, \mathcal{C})$. That is, in any CTS T_α, any subset-summation of the flow rates from a transmitter to its usable one-hop neighbors is bounded by the effective forwarding rate from the transmitter to the corresponding neighbor set. So constraint (5.17) actually contains $2^{|C|}$ inequalities.

The solution of the objective function (5.10) is the upper bound of the throughput between two nodes in multiradio, multichannel, multihop wireless networks when OR is available. The byproduct of the LP in Figure 5.2 is the radio-channel assignment (CTS's $\{T_\alpha | \alpha = 1 \ldots M\}$) and transmission scheduling ($\{\lambda_\alpha | \alpha = 1 \ldots M\}$). We also get the rate $\mu_{ij}^{k\alpha}$ on each link l_{ij}^{k} in each CTS T_α. When $\mu_{ij}^{k\alpha} \neq 0$, node j operating on channel k is selected as a forwarding candidate of v_i^k in the CTS T_α; otherwise, it is not. Candidate selection is therefore also solved by the LP. Note that only one forwarding candidate being selected indicates the usage of TR. So our model is general for OR and TR cases. However, the LP does not tell us how to achieve the link rate $\mu_{ij}^{k\alpha}$ in the opportunistic module, which is a forwarding priority scheduling problem.

In the following section we propose an LP approach and a heuristic algorithm to satisfy the flow rate $\frac{\mu_{ij}^{k\alpha}}{\lambda_\alpha}$ on each link l_{ij}^{k} by scheduling the forwarding priorities among the forwarding candidates in an opportunistic module in a CTS T_α.

5.4 Forwarding priority scheduling

In this section, we will answer the question "in the time fraction λ_α assigned to T_α, how can we schedule the forwarding priorities among the forwarding candidates $\mathcal{F}_{v_i^k} = \{v_j^k | \mu_{ij}^{k\alpha} \neq 0\}$ of the transmitter v_i^k to satisfy $\mu_{ij}^{k\alpha}$?" Note that $\mu_{ij}^{k\alpha}$ is the normalized link rate over the whole scheduling period, thus, during the time fraction λ_α, the link rate on l_{ij}^{k} is $\frac{\mu_{ij}^{k\alpha}}{\lambda_\alpha}$. For simplicity, we denote v_i^k as n_i, and v_j^k as n_{i_q} in the following discussion. Furthermore, we denote the rate vector $[\frac{\mu_{ij}^{k\alpha}}{\lambda_\alpha} | v_j^k \in \mathcal{F}_{v_i^k} \text{ in } T_\alpha]$ as $\vec{\mu} = [\mu_1, \ldots, \mu_r]$, where $r = |\mathcal{F}_{v_i^k}|$. The **forwarding priority scheduling problem (FPSP)** can therefore be formally defined as follows:

Definition 5.1 *FPSP: Given $\vec{\mu}$, find a forwarding priority scheduling $[(\mathcal{P}_m, \beta_m) | m = 1 \ldots L]$, such that on link l_{ii_q}, the accumulated effective rate $R_{ii_q} \geq \mu_q \; \forall 1 \leq q \leq r$.*

In the definition, $\mathcal{P}_m$ and β_m are the mth forwarding priority $\mathcal{P}_m$ and its time fraction, respectively. So $\beta_m \geq 0 \; \forall \; 1 \leq m \leq L$, and $\sum_{m=1}^{L} \beta_m \leq 1$. L is the total number of different priority assignment. Under the scheduling, R_{ii_q} can be computed as follows:

$$R_{ii_q} = \sum_{m=1}^{L} \beta_m \widetilde{R}_{ii_q}^{m} \tag{5.18}$$

where $\widetilde{R}_{ii_q}^{m}$ is the effective forwarding rate on link l_{ii_q} defined in Equation (5.7) under the forwarding priority $\mathcal{P}_m$.

5.4.1   A scheduling based on LP

One way to get a scheduling of opportunistic forwarding priorities for a rate vector $\overrightarrow{\mu}$ is by solving a linear programming problem in Figure 5.3. The basic idea of this linear programming is to enumerate all possible $r!$ opportunistic forwarding priorities to see if we can find a feasible solution. If the solution of the objective function (5.19) is no greater than 1, then the flow vector is schedulable, and $[(\mathcal{P}_m, \beta_m)|m = 1 \ldots r!]$ is a feasible scheduling; otherwise, the flow vector is not schedulable.

$$Min \sum_{k=1}^{r!} \beta_k \qquad (5.19)$$

$$s.t.$$

$$\mu_q \le \sum_{k=1}^{r!} \beta_k \widetilde{R}_{ii_q}^k, \ \forall \ q = 1 \ldots r \qquad (5.20)$$

$$0 \le \beta_k \le 1, \ \forall \ k = 1 \ldots r! \qquad (5.21)$$

Figure 5.3   LP formulations for finding a forwarding priority scheduling to satisfy a rate vector $[\mu_1, \ldots, \mu_r]$. *Reproduced by permission of © 2010 IEEE.*

The linear programming in Figure 5.3 provides a way to judge the schedulability of a rate vector corresponding to an opportunistic module and find a schedule of forwarding priorities if the rate vector is schedulable. At most, r is the number of all the one-hop neighbors of a transmitter, so it tends to be a relatively small number. However, it may not be necessary to enumerate all the possible forwarding priorities to find a feasible scheduling. In the following subsection, we propose a heuristic algorithm to solve the FPSP in a more efficient way.

5.4.2   A heuristic scheduling

Table 5.1 describes the heuristic recursive algorithm that finds a schedule of opportunistic forwarding priorities satisfying the rate vector $\overrightarrow{\mu}$. The basic idea of this algorithm is to satisfy each rate one-by-one by using two priority settings: assigning the corresponding candidate the highest and lowest priority in the existing subset of the candidates. In the algorithm, we take advantage of the property of OR that the effective forwarding rate of a lower-priority candidate is not affected by the priority relationships among the higher priority candidates. Then we can consider a group of forwarding candidates $\mathcal{F}$ as a virtual candidate, whose PRR is the probability of at least one forwarding candidate receiving the packet correctly, and the rate to this virtual candidate can be computed using Equation (5.8).

In Table 5.1, the input of the prioritizing and scheduling algorithm PS includes: $\overrightarrow{\mu}$, the rate vector; $\overrightarrow{p}$, the corresponding PRR vector; $\overrightarrow{l}$, the corresponding

Table 5.1 Pseudocode of a heuristic recursive algorithm for finding a scheduling of opportunistic forwarding strategies

---

$(S,\Gamma) = \text{PS}(\overrightarrow{\mu}, \overrightarrow{p}, \overrightarrow{I}, r, \beta, \omega)$

1 **if** $r == 1$

2   **return**$(\langle I_1 \rangle, \beta)$

3 **else**

4 **if** $\exists\, \mu_q == \omega R_i p_q \,||\, \mu_q \leq \omega R_i p_q \prod_{j \neq q}(1 - p_j)$

5   swap(μ_1, μ_q); swap(p_1, p_q); swap(I_1, I_q);

6 $P_2 = 1 - \prod_{q=2}^{r}(1 - p_q)$;

7 $\beta_2 = \min(\frac{R_i p_1 \cdot \omega - \mu_1}{R_i P_2 p_1 \cdot \omega}, 1)$; $\beta_1 = 1 - \beta_2$; $\omega' = \omega(1 - p_1\beta_1)$;

8 $(S_{11}, \Gamma_{11}) = \text{PS}(\mu_1, p_1, I_1, 1, \beta\beta_1, \omega)$;

9 $(S_{12}, \Gamma_{12}) = \text{PS}(\mu_1, p_1, I_1, 1, \beta\beta_2, \omega)$;

10 $(S_{21}, \Gamma_{21}) = \text{PS}([\mu_2 \ldots \mu_r], [p_2 \ldots p_r], [I_2 \ldots I_r], r - 1, \beta\beta_2, \omega')$;

11 $(S_{22}, \Gamma_{22}) = \text{PS}([\mu_2 \ldots \mu_r], [p_2 \ldots p_r], [I_2 \ldots I_r], r - 1, \beta\beta_1, \omega')$;

12 $(S_1, \Gamma_1) = \text{Merge}(S_{11}, S_{22}, \Gamma_{11}, \Gamma_{22})$;

13 $(S_2, \Gamma_2) = \text{Merge}(S_{21}, S_{12}, \Gamma_{21}, \Gamma_{12})$;

14 **return**$(S_1 \bigcup S_2, \Gamma_1 \bigcup \Gamma_2)$;

---

forwarding candidate index vector; r, the number of candidates; β, the active time fraction of the links corresponding to candidates in $\overrightarrow{I}$; ω, a scalar on the PRR, which is used to calculate time fraction β_1 and β_2 in line 7. Initially, $\beta = \omega = 1$. The output of this algorithm is a set of opportunistic forwarding priorities, S, and the corresponding time fraction vector, Γ.

Lines 1 and 2 indicate the basic case where there is only one candidate; then the candidate index and the corresponding time fraction β are returned. When the number of candidates is larger than 1, we first pre process the rate vector (in lines 4 and 5) such that if there is a rate equal to its corresponding scaled PRR timing the transmission rate or no greater than the scaled effective forwarding rate when the corresponding candidate is assigned the lowest priority, we put this candidate at the first place of the candidate vector. We then split the candidates into two parts, part 1: I_1 and part 2: $[I_2 \ldots I_r]$. Next, we calculate the accumulated PRR P_2 of candidates $[I_2 \ldots I_r]$. In line 7, we calculate the time fractions β_1 and β_2 corresponding to prioritization $\langle I_1, [I_2 \ldots I_r] \rangle$ and $\langle [I_2 \ldots I_r], I_1 \rangle$, respectively. Note that $\langle I_1, [I_2 \ldots I_r] \rangle$ indicates the candidate I_1 has higher relay priority than the group of candidates $[I_2 \ldots I_r]$, and vice versa. Then we recursively call the function PS on I_1 and $[I_2 \ldots I_r]$ (in lines 8 to 11). The returned S_{ij} is the set of forwarding strategies when part i is in the jth place ($j = 1, 2$ indicates higher and lower priority, respectively). Then we combine the sequences in S_{11} and S_{22} to get S_1 which are sequences of candidates with I_1 having higher priority than $[I_2 \ldots I_r]$ (in line 12). Similarly, we combine S_{21} and S_{12} with group of candidates $[I_2 \ldots I_r]$ having higher priority than I_1 (in line 13). Finally, we return the whole series of prioritizations by taking the union of S_1 and S_2.

Table 5.2   Pseudocode of merging two prioritized sub-sets of candidates

---

(S, Γ)=Merge($S_1,S_2,\Gamma_1,\Gamma_2$);
1   $S = \emptyset$; $\Gamma = \emptyset$;
2   **while** ($S_1 \neq \emptyset \,\|\, S_2 \neq \emptyset$)
3   push(S,top(S_1)|top(S_2));
4   **if** (top(Γ_1)>top(Γ_2))
5   push(Γ,top(Γ_2)); pop(Γ_2); pop(S_2); top(Γ_1)=top(Γ_1)−top(Γ_2);
6   **else if** (top(Γ_2)>top(Γ_1))
7   push(Γ,top(Γ_1)); pop(Γ_1); pop(S_1); top(Γ_2)=top(Γ_2)−top(Γ_1);
8   **else**
9   push(Γ,top(Γ_1)); pop(Γ_1); pop(S_1); pop(Γ_2); pop(S_2);
10 **end while**
11 **return**(S, Γ);

Next, we explain the Merge algorithm in Table 5.2. We assume both input (S_1, S_2, Γ_1 and Γ_2) and output (S and Γ) are stored in stacks. The basic idea of this Merge algorithm is to concatenate the sequence (corresponding to a prioritization) in the top of S_1 with that in the top of S_2 (in line 3) to create a new sequence (prioritization). The time fraction of this new sequence is the minimum of the time fractions of these two subsequences. After creating a new sequence, we pop the sequence with smaller time fraction, and update the time fraction of the other sequence by subtracting the used time fraction (in lines 5, 7, and 9). When all the sequences in S_1 and S_2 are popped out, a series of new sequences S and the corresponding time fraction vector Γ are returned (in line 11).

The computation complexity of Merge algorithm is $\Theta(|S_1| + |S_2|)$, where $|S_i|$ ($i = 1, 2$) is the number of sequences in S_i. For S_i with x elements, we have at most $O(2^x)$ and at least $\Omega(1)$ sequences in it. So the complexity of the algorithm PS is $O(2^{r-1})$ in the worst case and $\Omega(r)$ in the best case, where r is the number of forwarding candidates.

5.4.2.1   Correctness of the heuristic algorithm

This heuristic algorithm does not guarantee to return a feasible schedule of opportunistic forwarding priorities even when the rate vector is schedulable. When this happens, we need to run the LP in Figure 5.3 to get a feasible scheduling. However, we will prove that this heuristic algorithm does return a feasible scheduling for a schedulable rate vector $\overrightarrow{\mu}$ when $r \leq 2$. We will also show numerical results that this heuristic algorithm works well for larger number of forwarding candidates in terms of achieving low unsatisfied rate ratio.

Preposition 5.1 *When* $r = |\overrightarrow{\mu}|$ *is no greater than 2, any rate vector* $\overrightarrow{\mu} = [\mu_1, \ldots, \mu_r]$ *in the capacity region defined in Inequality (5.9) can be satisfied by the scheduling obtained by the heuristic algorithm PS in Table 5.1.*

Proof. First, when $r = 1$, it is obvious that any μ_1, s.t. $\mu_1 \leq R_i p_{ii_1}$, is schedulable. Lines 1 and 2 in Table 5.1 deal with this case.

Second, when $r = 2$, there are two forwarding priorities: $\mathcal{P}_1 : \mathcal{P}_1(i_1) > \mathcal{P}_1(i_2)$ and $\mathcal{P}_2 : \mathcal{P}_2(i_2) > \mathcal{P}_2(i_1)$. Assuming the whole transmission period is unit 1, we allocate β_1 and β_2 time fraction for $\mathcal{P}_1$ and $\mathcal{P}_2$, respectively. Then according to Equation (5.18), we have

$$R_{ii_1} = R_i(\beta_1 \cdot p_{ii_1} + \beta_2 \cdot p_{ii_1}(1 - p_{ii_2})) \tag{5.22}$$

$$R_{ii_2} = R_i(\beta_1 \cdot p_{ii_2}(1 - p_{ii_1}) + \beta_2 \cdot p_{ii_2}) \tag{5.23}$$

Then we only need to prove, for any μ_1 and μ_2, s.t. $0 \leq \mu_1 \leq R_i p_{ii_1}$, $0 \leq \mu_2 \leq R_i p_{ii_2}$, and $\mu_1 + \mu_2 \leq R_i(1 - (1 - p_{ii_1})(1 - p_{ii_2}))$, $\exists \beta_1$ and β_2, s.t. $0 \leq \beta_1 \leq 1$, $0 \leq \beta_2 \leq 1$, and $\beta_1 + \beta_2 = 1$, to make $\mu_1 \leq R_{ii_1}$ and $\mu_2 \leq R_{ii_2}$.

With $\mu_2 \leq R_{ii_2}$, $\mu_2 \leq R_i p_{ii_2}$, Equation (5.23) and $\beta_1 = 1 - \beta_2$, we have

$$0 \leq \beta_1 \leq \frac{R_i p_{ii_2} - \mu_2}{R_i p_{ii_1} p_{ii_2}} \tag{5.24}$$

With $\mu_1 \leq R_{ii_1}$, $\mu_1 \leq R_i p_{ii_1}$, Equation (5.22) and $\beta_2 = 1 - \beta_1$, we have

$$0 \leq \beta_2 \leq \frac{R_i p_{ii_1} - \mu_1}{R_i p_{ii_1} p_{ii_2}} \tag{5.25}$$

By satisfying μ_1, we set

$$\beta_2 = \min\left(\frac{R_i p_{ii_1} - \mu_1}{R_i p_{ii_1} p_{ii_2}}, 1\right), \beta_1 = 1 - \beta_2 \tag{5.26}$$

By substituting Equation (5.26) into Equations (5.22) and (5.23), we can verify that $\mu_1 \leq R_{ii_1}$ and $\mu_2 \leq R_{ii_2}$. Note that, the setting of β_1 and β_2 makes inequalities (5.24) and (5.25) hold. Equation (5.26) exactly corresponds to the first two equations in line 7 in Table 5.1. So we proved the correctness of the heuristic algorithm PS for $r = 2$.

The proof of the correctness of the heuristic algorithm also indicates that any rate vector in the capacity region shown in Figure 5.4 is schedulable.

5.4.2.2 An example

We use an example to illustrate how the PS algorithm works. Assume n_i has three forwarding candidates $\{n_{i_1}, n_{i_2}, n_{i_3}\}$, the corresponding rate on each link l_{ii_q} ($q = 1, 2, 3$) is $0.2R_i$, $0.3R_i$, and $0.46R_i$, and the corresponding PRR on these links are 0.5, 0.6 and 0.8, respectively. Figure 5.5 shows the running result of algorithm PS. In the first stage, μ_1 is satisfied, and in the second stage μ_2 is satisfied, then μ_3. The time fraction β of each forwarding priority is listed at the right of the priority.

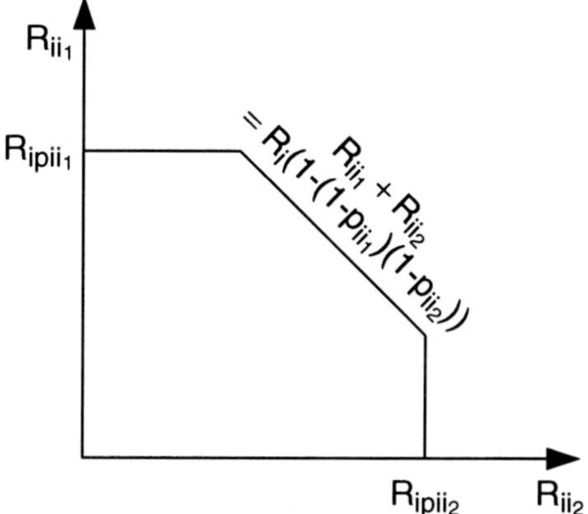

Figure 5.4 Capacity region for two forwarding candidates. Reproduced by permission of © 2010 IEEE.

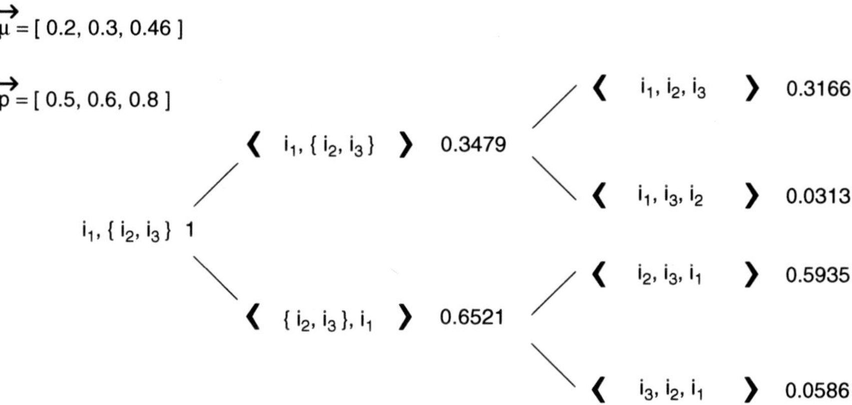

Figure 5.5 An example of opportunistic forwarding strategy scheduling for three forwarding candidates. Reproduced by permission of © 2010 IEEE.

5.4.2.3 Performance of the heuristic algorithm

We conducted numerical simulations to evaluate the performance of the PS algorithm. We propose the metric of **unsatisfied rate ratio** γ to indicate how well the heuristic algorithm can satisfy the rate vector $\vec{\mu}$.

$$\gamma = \frac{\sum_{q=1}^{r}(\mu_q - R_{ii_q})I(R_{ii_q} < \mu_q)}{\sum_{q=1}^{r}\mu_q} \tag{5.27}$$

where R_{ii_q} is the accumulated effective rate on link l_{ii_q} defined in Equation (5.18) under a scheduling of forwarding priorities, and $I()$ is an indicator function. When the input expression of $I()$ is true, $I() = 1$; otherwise $I() = 0$. According to Equation (5.27), $0 \leq \gamma \leq 1$. A smaller γ indicates better performance.

In the simulation, we vary the number of forwarding candidates from 1 to 10. For each number of forwarding candidates, we conducted 10^4 runs. In each run, we randomly assign the PRR on the links in the opportunistic module, and generates a rate vector that reaches the capacity of that opportunistic module. Then we run PS algorithm on the rate vector and opportunistic module, and compute the corresponding γ. Figure 5.6 shows the mean of γ with 95% confidence interval under different number of forwarding candidates. We can see that the PS algorithm works well in terms of having low γ. It satisfies the rate vector almost all the time with the unsatisfied ratio as low as 0.7% when there are no more than five forwarding candidates. When the number of forwarding candidates increases, γ is increased. However, even when there are ten forwarding candidates, γ is below 10%.

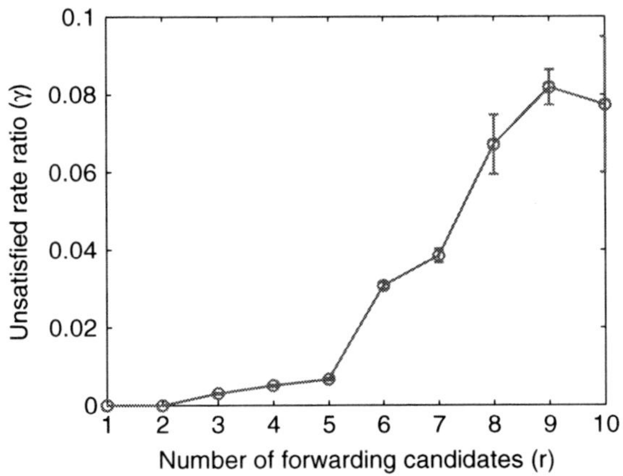

Figure 5.6   Unsatisfied rate ratio versus number of forwarding candidates using PS algorithm for forwarding priority scheduling. Reproduced by permission of © 2010 IEEE.

5.5   Performance evaluation

In this section, we show the results of joint radio-channel assignment, routing, and scheduling for optimizing an end-to-end throughput solved by our methodology for two simple scenarios, and simulation results for more general networks. All the simulations are implemented in Matlab.

5.5.1   Two scenarios with different link qualities

We consider two four-node network scenarios in Figure 5.7 with different link qualities. Suppose each node has one radio that can be operated on two orthogonal channels. The PRR is indicated on each link. For simplicity, we assume the PRR

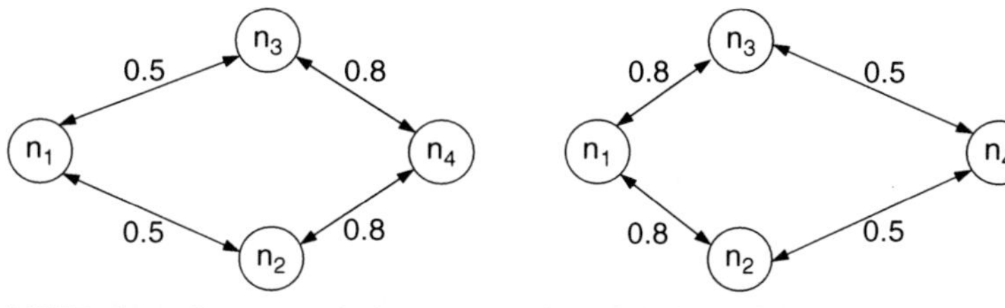

Figure 5.7   Four-node networks under different channel conditions (link PRRs). Reproduced by permission of © 2010 IEEE.

is identical under different channels in each network. We assume each node is in the interference range of each other. So there is only one transmitter can be active on the same channel at any instant in the network. By applying the methodology in Sections 5.3 and 5.4, we solve the joint radio-channel assignment, routing, scheduling problem for maximizing the throughput from n_1 to n_4. We summarize the results for Figure 5.7(a) and Figure 5.7(b) in Table 5.3. The optimal throughputs from n_1 to n_4 for these two scenarios are 0.58 and 0.5, respectively. An interesting observation from Table 5.3 is that the opportunistic routing is not used when n_1 is transmitting packets. Since in Figure 5.7(b), the channel conditions from the source to the relays are better than that from the relays to the destination, the maximum throughput is constrained by the bottleneck links from the relays to the destination. So we should allow more concurrent transmissions to saturate the bottleneck links instead of making use of OR to push more flows out of the sender. When the bottleneck links are between the sender and relays (Figure 5.7(a)), OR is used to push more flows out the sender. This observation is expected to provide a guideline on designing distributed radio-channel assignment for OR in multiradio, multichannel systems.

Table 5.3   Channel assignment, routing, and scheduling of opportunistic forwarding strategies for Figure 5.7(a) and Figure 5.7(b)

Figure 5.7(a)		Figure 5.7(b)	
CTS	Time fraction	CTS	Time fraction
$\{(v_1^1, \langle v_2^1, v_3^1 \rangle)\}$	0.14	$\{(v_1^1, \langle v_3^1 \rangle), (v_2^2, \langle v_4^2 \rangle)\}$	0.354
$\{(v_1^1, \langle v_3^1, v_2^1 \rangle)\}$	0.14	$\{(v_1^1, \langle v_2^1 \rangle), (v_3^2, \langle v_4^2 \rangle)\}$	0.354
$\{(v_1^1, \langle v_2^1 \rangle), (v_3^2, \langle v_4^2 \rangle)\}$	0.36	$\{(v_2^1, \langle v_4^1 \rangle)\}$	0.146
$\{(v_1^1, \langle v_3^1 \rangle), (v_2^2, \langle v_4^2 \rangle)\}$	0.36	$\{(v_3^1, \langle v_4^1 \rangle)\}$	0.146

5.5.2   Simulation of random networks

In this subsection, we investigate the throughput bound of OR and TR in multiradio, multichannel systems and compare the results with that in single-radio, single-channel systems. We examine both linear topology and rectangle topology. For the linear topology, we uniformly deploy 12 nodes in a line with 300 units of length. For the rectangle topology, we randomly deploy 12 nodes in a rectangle area of 200 units $\times$ 300 units. We select node n_1 at the left end (for the linear topology) and left corner (for the rectangle topology) of the networks as the destination, then calculate the throughput bound from other nodes to the destination using the LP formulations in Figure 5.2. There are therefore 11 different source–destination pairs considered in the evaluation for each topology. In all the simulations, we assume the packet reception ratio is inversely proportional to the distance with Gaussian random variation, which simulates the log-normal fading and two-ray path loss model. The transmission range is set as 100 units, and the interference range is set as twice of the transmission range. The performance metric is the normalized end-to-end throughput bound (by assuming the transmission rate is unit one). Note that although the network size is limited at 12 nodes in our simulation due to the exponential complexity of finding all the CTS's, this small

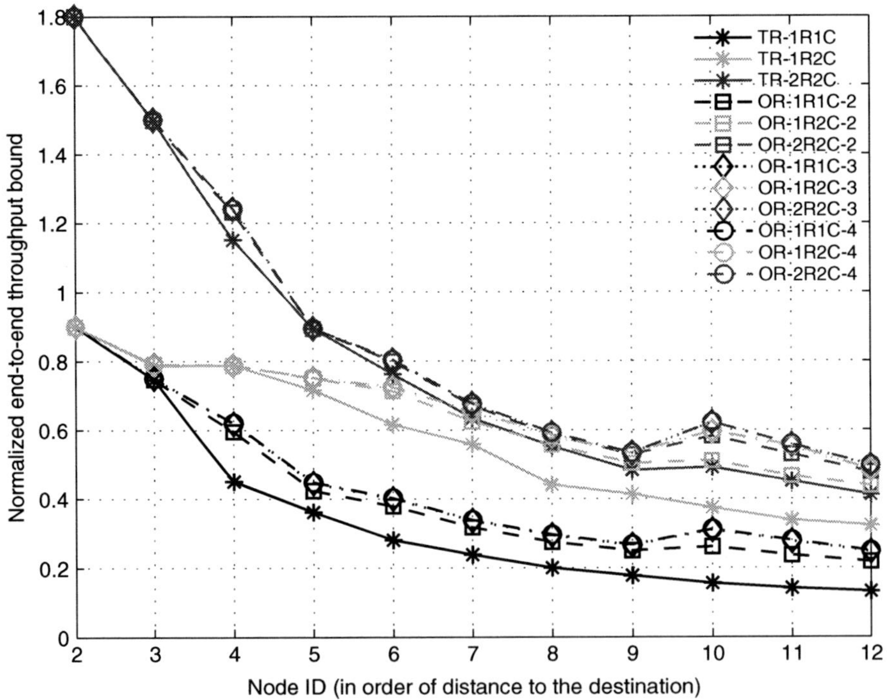

Figure 5.8 Normalized end-to-end throughput bound under different number of radios, channels and potential forwarding candidates in linear topology. Reproduced by permission of © 2010 IEEE.

network size is sufficient to allow us to gain insight of the opportunistic routing in multi-radio multi-channel networks. According to the simulation settings, the longest path between node 12 and 1 can be six hops and 11 hops in the rectangle and linear topologies, respectively. The shortest path between the node 12 and 1 can be 4 hops and 3 hops in the two topologies, respectively. So reasonable multihop network scenarios are simulated.

Figures 5.8 and 5.9 show the simulation result under linear topology and rectangle topology, respectively. In the legend, "TR" represents traditional routing, "OR" represents opportunistic routing, "$xRyC - z$" represents x radios and y channels, with z maximal number of forwarding candidates. That is, in the CTS enumeration, we only consider the opportunistic module that contains at most z number of forwarding candidates. We can see that performance shows similar trends under both topologies. With the number of radios and channels increasing, the throughput of TR and OR are both increased. Generally OR achieves higher throughput than TR, and the multiradio/channel capability has greater impact on the throughput of TR than OR. When the source is farther away from the destination, the OR presents more advantages than TR. Opportunistic forwarding using multiple forwarding candidates does help increase the throughput. An interesting result is that, for nodes 7 to 12, the throughput of 1R2C case for OR is comparable with or even

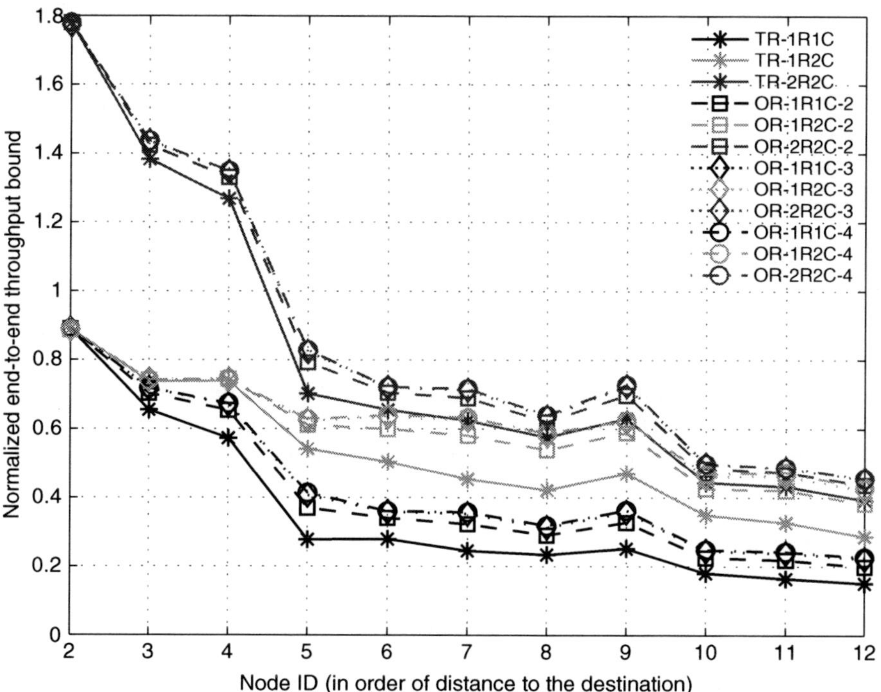

Figure 5.9 Normalized end-to-end throughput bound with different numbers of radios, channels and potential forwarding candidates in rectangle topology. Reproduced by permission of © 2010 IEEE.

greater than that of 2R2C case for TR. This result indicates that OR can achieve comparable or even better performance as TR by using less radio resource.

Another interesting observation is that the throughput gained decreases as the number of forwarding candidates increases. This result is consistent with that found in Zeng *et al.* (2007b,c). So it is not necessary to involve all the usable receivers of the transmitter into the opportunistic forwarding and selecting a few "good" forwarding candidates is enough to approach the optimal throughput. This theoretical observation may help us design practical protocols.

5.6   Conclusions and future work

In this chapter, we proposed a unified framework to compute the throughput bound between two end nodes in multiradio, multichannel, multihop wireless networks when OR is available. Our model accurately captures the unique property of OR: throughput can take place from a transmitter to any one of its forwarding candidates at any instant. We also studied the capacity region of an opportunistic module and proposed an LP approach and a heuristic algorithm to obtain an opportunistic forwarding priority scheduling that satisfies a rate vector. Numerical simulations show that the heuristic algorithm achieves desirable performance under various number of forwarding candidates. It can satisfy the rate vector with unsatisfied rate ratio below 0.7% when there are no more than five forwarding candidates. Even when there are ten forwarding candidates, the unsatisfied rate ratio is below 10%. Our methodology can be used to calculate the end-to-end throughput bound of OR and TR in multiradio, multichannel, multihop wireless networks, as well as to study the OR behaviors (such as candidate selection and prioritization). Leveraging our analytical model, we gained the following two insights: 1. OR can achieve better performance than TR under different radio/channel configurations, however, in some scenario (e.g. bottleneck links exist between the sender to relays), TR can be preferable to OR.2. OR can achieve comparable or even better performance than TR by using fewer radio resources. We also confirm that just involving a few "good" forwarding candidates is enough to approach optimal throughput. As for future work, we are interested in designing practical distributed joint radio-channel assignment and opportunistic routing protocols in multiradio, multichannel systems based on our theoretical study and observations in this chapter.

References

Aguayo D, Bicket J, Biswas S, Judd G and Morris R 2004 Link-level measurements from an 802.11b mesh network. *SIGCOMM Computer Communication Review* **34**(4), 121–132.

Alicherry M, Bhatia R and Li LE 2005 Joint channel assignment and routing for throughput optimization in multi-radio wireless mesh networks *MobiCom '05*, pp. 58–72. ACM, New York, NY.

Biswas S and Morris R 2005 Exor: Opportunistic multi-hop routing for wireless networks *SIGCOMM'05*, Philadelphia, Pennsylvania.

Chachulski S, Jennings M, Katti S and Katabi D 2007 Trading structure for randomness in wireless opportunistic routing *ACM SIGCOMM*, Kyoto, Japan.

Dubois-Ferriere H, Grossglauser M and Vetterli M 2007 Least-cost opportunistic routing. Technical Report LCAV-REPORT-2007-001, School of Computer and Communication Sciences, EPFL.

Fussler H, Widmer J, Kasemann M, Mauve M and Hartenstein H 2003 Contention-based forwarding for mobile ad-hoc networks. *Elsevier's Ad Hoc Networks* **1**(4), 351–369.

Jain K, Padhye J, Padmanabhan VN and Qiu L 2003 Impact of interference on multi-hop wireless network performance *MobiCom '03: Proceedings of the 9th Annual International Conference on Mobile Computing and Networking*, pp. 66–80. ACM Press, New York, NY.

Kim KH and Shin KG 2006 On accurate measurement of link quality in multi-hop wireless mesh networks *MobiCom '06: Proceedings of the 12th Annual International Conference on Mobile Computing and Networking*, pp. 38–49. ACM Press, New York, NY, USA.

Kodialam M and Nandagopal T 2005 Characterizing the capacity region in multi-radio multi-channel wireless mesh networks *MobiCom '05*, pp. 73–87. ACM, New York, NY.

Shah RC, Bonivento A, Petrovic D, Lin E, van Greunen J and Rabaey J 2004 Joint optimization of a protocol stack for sensor networks *IEEE Milcom*.

Tang J, Xue G and Zhang W 2007 Cross-layer design for end-to-end throughput and fairness enhancement in multi-channel wireless mesh networks. *IEEE Transactions on Wireless Communications* **6**(10), 3482–3486.

Tassiulas L and Ephremides A 1993 Dynamic server allocation to parallel queues with randomly varying connectivity. *IEEE Transactions on Information Theory* **39**(2), 466–478.

Zeng K, Lou W and Zhai H 2008 On end-to-end throughput of opportunistic routing in multirate and multihop wireless networks *Infocom*, Phoenix, AZ.

Zeng K, Lou W and Zhang Y 2007a Multi-rate geographic opportunistic routing in wireless ad hoc networks *IEEE Milcom*, Orlando, FL.

Zeng K, Lou W, Yang J and Brown DR 2007b On geographic collaborative forwarding in wireless ad hoc and sensor networks *WASA'07*, Chicago, IL.

Zeng K, Lou W, Yang J and Brown DR 2007c On throughput efficiency of geographic opportunistic routing in multihop wireless networks *QShine'07*, Vancouver.

Zhai H and Fang Y 2006 Impact of routing metrics on path capacity in multirate and multihop wireless ad hoc networks *IEEE ICNP*.

Zhang J, Wu H, Zhang Q and Li B 2005 Joint routing and scheduling in multi-radio multi-channel multi-hop wireless networks *IEEE Broadnets*.

Zhong Z, Wang J and Nelakuditi S 2006 Opportunistic any-path forwarding in multi-hop wireless mesh networks. Technical Report TR-2006-015, USC-CSE.

Zorzi M and Rao RR 2003 Geographic random forwarding (geraf) for ad hoc and sensor networks: energy and latency performance. *IEEE Transactions on Mobile Computing* **2**(4), 349–365.

6

Medium access control for opportunistic routing – candidate coordination

One important and challenging issue in OR is candidate coordination. That is, in order to avoid duplication, we should ensure that only the "best" receiver of each packet forwards it. However, it is nontrivial to achieve this goal in an efficient way. The existing candidate-coordination schemes have some inherent inefficiency such as long time delay at each one-hop transmission and potential duplicate forwarding. Improperly designed coordination schemes will aggravate these problems and even overwhelm the potential gain provided by OR.

In this chapter, we carry out a comprehensive study on candidate coordination in OR and propose a new scheme, "fast slotted acknowledgment" (FSA), to further improve the efficiency of OR. This adopts a single ACK to confirm successful reception and suppresses other candidates' attempts to forward the data packet with the help of a channel-sensing technique. We also confirm the benefit of our scheme by simulation. The result shows that FSA can reduce the average end-to-end delay by up to 50% when the traffic is relatively light and can improve the throughput by up to 20% under a heavy traffic load.

The rest of this chapter is organized as follows. Section 6.1 reviews state-of-the-art coordination schemes in detail. Section 6.2 presents FSA's design and analysis. This is followed by Section 6.3, where evaluation and analysis of FSA performance are presented. Section 6.4 concludes.

Multihop Wireless Networks: Opportunistic Routing, First Edition.
Kai Zeng, Wenjing Lou and Ming Li.
© 2011 John Wiley & Sons, Ltd. Published 2011 by John Wiley & Sons, Ltd.

6.1 Existing candidate coordination schemes

In this section, we review five state-of-the-art candidate coordination schemes: GeRaF collision avoidance MAC, ExOR batch-based MAC, contention-based forwarding (CBF), slotted acknowledgement and compressed slotted acknowledgement.

6.1.1 GeRaF collision avoidance MAC

The GeRaF collision-avoidance MAC-coordination scheme is proposed in (Zorzi and Rao 2003). This scheme applies to a sensor network scenario, where nodes wake up and go to sleep periodically. It is basically a geographic opportunistic routing, where each transmitter knows its own location information and that of its neighbors, and the destination. The transmitter node's location information as well as the destination node's location information are embedded in the ready-to-send (RTS) frame. It assumes that sensor nodes are equipped with two radios (Schurgers *et al.* 2002). Therefore, the data traffic and the wakeup signaling can be operated on two different channels simultaneously. That is, all messages exchange on the first "data" frequency while the busy tone is sent on the second frequency. We present this protocol from transmitter and receiver point of view as described in Zorzi and Rao (2003).

6.1.1.1 Transmitter behavior

When a sleeping node has a packet to send, it wakes up and enters the active state and monitors both data and busy tone frequencies for τ seconds. If either frequency is busy, the node reschedules an attempt at a later time. If both frequencies are sensed idle during this entire time interval, the node broadcasts a RTS message that contains the location of itself and the intended destination. After sending the RTS, the transmitting node listens to the subsequent slots for clear-to-send (CTS) messages from potential forwarding candidates. In each of the CTS slots following the end of the RTS message, the transmitting node acts as follows:

1. If only one CTS message is received, it starts transmission of the data packet. The initial part of this packet acts as a CTS confirmation for the forwarding candidate which issued the CTS.

2. If it receives no CTSs, it sends a CONTINUE message and waits again for CTSs. It times out after N_p empty CTS slots (which forces the transmitter to abort the handshake and to reschedule the transmission at a later time).

3. If it hears a signal but is unable to detect a meaningful message, it assumes that a CTS collision took place and will send a COLLISION message, which will trigger the start of a collision resolution algorithm and will wait again for CTSs.

Like the IEEE 802.11 protocol, an immediate ACK is expected after packet transmission. The node can go to sleep if the ACK is correctly received. If the transmitter does not correctly receive an ACK within a given time interval, it times out and considers the packet delivery failed. It will reschedule the same packet for future transmission. The transmitter will drop the packet after $N_{rtx\,limit}$ failed attempts, it will generate an error message for the higher layers. The transmitter transmits the busy tone to prevent interference from hidden terminals while listening for CTSs and ACK.

A problem with this scheme is that packet duplication cannot be completely avoided. Actually, it is a common problem of most of the existing opportunistic candidate coordination MAC protocols. After sending the ACK, the forwarding candidate becomes the actual forwarder and is in charge of the packet delivery. If the ACK is lost, the transmitter will not be aware of this fact and will retry the transmission of the same packet. This duplication might be exaggerated at the later hops if the same scenario happens. However, this duplication can be alleviated if intermediate node is able to remember the sequence numbers of the latest packets, and discard the duplicated ones. The destination node can also drop the duplicated packets. This inefficiency caused by packet duplication is expected to be limited, given the fact that losing an ACK when the packet transmission was successful is a low-probability event.

6.1.1.2   Receiver behavior

When a node wakes up and is in the listening mode, it senses the channel to see if there is any transmission. If nothing is detected during the listening time interval, the node goes back to sleep. Otherwise, if the node detects the start of a transmission, it changes into the receiving state. Meanwhile, it activates the busy tone on the second frequency for a duration of T_{RTS} (the time for RTS transmission). If no valid RTS is received, the node goes back to the listening state. On the other hand, if a valid RTS is received, the node reads the information in it (i.e. the location information of the transmitter and the destination) and determines its own priority as a forwarding candidate. This priority is based on the relative location of the node itself compared with the distance between the transmitter and the intended destination.

A relay priority assignment proposed in Zorzi and Rao (2003) is as follows. The portion of the coverage area of the transmitter that is closer to the destination than the transmitter itself is divided into N_p regions R_1, R_2, ..., and R_{N_p}, such that all points in R_i are closer to the destination than all points in $R_j \; \forall \; j > i$, $i = 1, 2, \ldots, N_p - 1$. After the RTS, all nodes in R_1 will send a CTS message in the first CTS slot, whereas all others with lower priorities will keep silent. All nodes will then listen for the message from the transmitter in the latter part of the CTS slot.

If a data packet transmission (which contains the identification of the forwarding candidate which sent the CTS) is heard, only the designated forwarding candidate will continue to receive, whereas all others will go back to sleep to save energy.

If a CONTINUE message is heard in the second part of the first CTS slot, it means that there are no forwarding candidates in R_1 and all forwarding candidates in R_2 will contend in the second CTS slot.

If an ABORT message is received, it indicates that the transmitter has reached the maximum allowed number of CTS slots and the forwarding coordination is aborted.

If a COLLISION message is received, it means that more than one CTS frame was generated in the CTS slot. All forwarding candidates who did not transmit CTS will quit from the forwarding contention because they can infer that higher priority forwarding candidates are present. The forwarding candidates involved in the collision will start the collision-resolution mechanism. Each colliding forwarding candidate will decide whether or not to continue with a probability of 0.5 for each outcome. A candidate that decides to continue will send a CTS in the next slot. There are three possible events:

1. Only one forwarding candidate sends the CTS. Then the transmitter starts the packet transmission and all other forwarding candidates quit the contention.

2. More than one CTS frame is sent in the same slot. Then the transmitter sends a COLLISION message. Those forwarding candidates that did not send a CTS drop out, while those involved in the collision decide whether to continue as before until the collision is resolved.

3. No CTS is heard. A CONTINUE message is sent by the transmitter, and all forwarding candidates that did not select the current slot decide again independently whether to continue as before.

The above procedure have a high probability of terminating within a few slots. In order to force it to be terminated, the transmitter can send an ABORT message if the collision is not resolved within a maximum allowed number of CTS slots. Finally, any node that receives a message it does not understand quits the contention.

Nodes that heard the RTS correctly will follow the sequence of steps above and they are guaranteed to either become the actual relay node or to drop out at some point. In order to avoid the hidden terminal problem, each forwarding candidate involved in the above contention procedure will keep the busy tone active until it drops out or, if it wins the contention, until the whole data packet has been received.

6.1.2 ExOR batch-based MAC

ExOR is proposed for improving throughput in static multihop wireless networks (Biswas and Morris 2005). It assumes that each node knows the global information of the network, and each node can compute the shortest traditional single path to the destination in terms of ETX but only considering forward link quality. It operates on batches of packets and all packets are broadcast.

6.1.2.1 Source behavior

The source node includes a list of forwarding candidates in the header of each packet, prioritized by the estimated ETX to the destination. The destination node ID is at the head of the list with the highest priority. In order to avoid overwhelming overheads in a network with a large number of nodes, only the neighbors whose link quality to the source is larger than 10% will be considered in the forwarder list. Then the source node broadcasts the first batch of packets. It moves to the next batch only if it infers that more than 90% of the packets in the current batch are received by higher priority forwarders or the destination.

6.1.2.2   Forwarder behavior

When the neighbors of the source node receive the broadcast packet, they will check if they are in the forwarding candidate list. If they are not in the forwarder list they will drop the packets. If they are, they will buffer the successfully received packets and wait until the end of the batch.

The highest priority forwarding candidate then broadcasts the packets in its buffer as its own fragment of the batch. Each packet includes a copy of the sender's batch map containing the forwarding candidate's best guess of the highest priority node to have received each packet in the batch. Upon receiving a packet from a higher priority candidate, a forwarder updates its batch map by replacing its own guesses of the highest priority nodes for each packet with those indicated in the batch map of the received packet. Note that the destination node is a special case; when its turn comes it sends ten packets only with header but no data. In this way, this batch map serves as an acknowledgement that propagates back to the source, which is highly reliable as it has been accumulated several times by intermediate forwarders.

In addition, each packet includes a fragment size (the number of packets in the current fragment) and its fragment number (offset of itself in the current fragment). Each remaining forwarder then transmits in order to avoid collision. A transmission timer is set at each forwarder and upon the expiration of this timer the forwarder begins to transmit its fragment. Ideally, a lower priority forwarder should wait until all the forwarders with higher priority than itself finishes transmission. However in practice it is hard to estimate the time spent by each higher priority forwarder in transmission due to packet losses. Thus, in ExOR, rather than starting the transmission after each forwarder overhears the last packet sent by the forwarder before it, the forwarder estimates the transmission start time based on the fragment size and fragment numbers piggybacked in the overheard packets from the forwarder before itself, which is more reliable. When a node does not hear any forwarded packets from higher priority nodes, it simply assumes that each of them transmits five packets. Each forwarder sends only packets which were not acknowledged in the batch maps of higher priority nodes.

The forwarders continue to cycle through the priority list until the destination has 90% of the packets. The reason is that the last few packets in a batch would take the most overheads to send, which may overwhelm the benefit of OR. The remaining packets are transferred with traditional routing.

6.1.3   Contention-based forwarding (CBF)

Contention-based forwarding belongs to geographic routing and was proposed in Fussler *et al.* (2003). Traditional geographic routing requires a forwarding node to know all its neighbors' positions in realtime which is usually done using periodic beacon messages. This does not work well under high node mobility, since nodes' location information collected in this way is often outdated, which leads either to a significant decrease in the packet delivery ratio or to a drastic increase in load on the wireless channel. To this end, CBF proposes to perform position-based unicast forwarding without the help of beacons. The idea is to select the next hop using a distributed contention mechanism based on the actual positions of all current neighbors. In order to achieve this, in CBF the nodes that received a packet from a forwarding node set a biased timer based on their own positions. And to avoid packet duplication, the first node that is chosen suppresses the other nodes.

6.1.3.1   Timer-based contention

Next we briefly introduce the decentralized timer-based contention mechanism in CBF. In order to greedily maximize the remaining distance to the destination (hop progress) of each packet, the timer's runtime is set to be inversely proportional to the hop progress P:

$$t(P) = T(1 - P), \tag{6.1}$$

where P is defined as

$$P(f, z, n) = max\left\{0, \frac{\mathrm{dist}(f, z) - \mathrm{dist}(n, z)}{r_{\mathrm{radio}}}\right\}, \tag{6.2}$$

where f is the position of the forwarding node, z is the position of the destination, and n is the position of the neighboring node whose timer is set, "dist" stands for Euclidean distance between two positions and r_{radio} is the nominal radio range.

The above timer runtime configuration inevitably incurs packet duplication in the situation where the best suited node has a hop progress of P_1, and at the same time there exists at least one node with a hop progress of P, such that $t(P) - t(P_1) < \delta$, where δ stands for the minimum time interval needed for suppression. As shown by the analysis in Fussler *et al.* (2003), the probability of packet duplication decreases as the distance of a neighbor node to the forwarder increases, when all the nodes are uniformly distributed.

6.1.3.2   Suppression techniques

There are three suppression techniques proposed in Fussler *et al.* (2003):

1. *Basic Suppression*. When a node's timer expires, it simply assumes that it is itself the next hop forwarding node and broadcasts the packet. When another node receives this broadcast and at the same time still has a timer running for the same packet, the timer is canceled and the node does not forward the packet. This method, although simple, will introduce duplicated transmissions in addition to those incurred by the time interval needed for suppression. That is, some nodes may not be able to hear the broadcast of the assumed next hop forwarder; they can be located outside of the transmission rage of the next hop forwarder. Using the basic suppression technique, it is shown that at most three packet duplications could happen in addition to those incurred by the suppression time interval.

2. *Area-based Suppression*. To solve the problem of basic suppression, area-based suppression is proposed. The basic idea is to artificially restrict the nodes that participate in the next hop forwarder contention process to those that lie within a smaller geographical area than before, such that every node will hear the transmission of any other node within this area (every node is in the transmission range of every other). That area is called a suppression area. A *Reuleaux Triangle* is adopted, which covers the area (from the forwarder to the destination and within the forwarder's transmission range) well with good forwarding progress. The Reuleaux Triangle trades off the number of nodes contained within the suppression area against the inclusion of better candidate forwarding nodes. In case that there are no nodes in the suppression area, two backup areas are used, in which a similar contention and suppression process is repeated in each of them.

3. *Active Selection*. The above two methods still cannot eliminate the duplicate transmissions caused by the suppression time interval. To this end, active selection prevents all kinds of packet duplications by introducing additional control message overhead at the forwarding node. Instead of sending the packet directly, the forwarding node first sends a request-to-forward (RTF) signal containing the forwarding node's location and the destination node's location. Each neighbor node that hears the RTF and with a positive hop progress sets a replay timer as in the basic suppression scheme. Upon the expiration of the reply timer, the node sends a clear-to-forward (CTF) packet and other nodes that received the CTF suppress themselves from forwarding. Finally, the forwarding node selects a node from the set of nodes that have sent a CTF, and transmits the packet to that node via unicast. This process resembles the request-to-send (RTS) and clear-to-send (CTS) mechanism in IEEE 802.11 MAC. In fact, the active selection method can be integrated with RTS/CTS schemes to avoid the "hidden terminal" problem. In active selection, the forwarding node has the ultimate authority over the selection of the next hop forwarder.

6.1.4   Slotted acknowledgment (SA)

Slotted acknowledgment was is proposed by Biswas and Morris (2003). It applies a similar acknowledgment scheme to the one used in traditional 802.11. However, it requires each candidate who has received the data packet to broadcast an ACK in different time slots according to their priorities. Instead of only indicating the success of reception, each ACK contains the ID of the highest priority successful recipient known to the ACK's sender. All the candidates listen to all ACKs before deciding whether to forward the data packet in case a lower prioritized candidate's ACK reports a higher prioritized candidate's ID. In order to protect all the ACKs from being interrupted by other transmissions, SA extends the Network Allocation Vector (NAV) in the MAC header of the data packet to reserve the channel for longer time. Thus the total coordination time for SA with n candidates is $n \times (T_{SIFS} + T_{ACK})$, where SIFS is short interframe space (802.11b 1999 n.d.). This scheme has a serious vulnerability that makes it fail to work properly in some scenarios. Taking one transmission with three candidates as an example in Figure 6.1. Suppose that when the sender is transmitting, another node within the sender's transmission range, which is willing to transmit, does not hear the data packet clearly (for example, received a corrupted one that cannot get the NAV value from the MAC header). At the same time, the highest priority candidate also failed to receive the data packet. In this case, the first ACK is missing and the potential transmitter that does not update its NAV accordingly senses the channel to be clear for $2 \times T_{SIFS} + T_{ACK}$, which is obviously greater than DIFS (Distributed Inter Frame Space), the idle time needed before sending packet in 802.11 protocols (802.11b 1999 n.d.). Thus it sends its own packet which will collide with the subsequent ACKs from $candidate_2$ and $candidate_3$ as illustrated in Figure 6.1. No one hears a clear ACK, the consequence is that both $candidate_2$ and $candidate_3$ will forward the packet, which results in duplication, and the sender will unnecessarily retransmit the packet. The scenario shown above is not rare, especially in networks under heavy traffic loads.

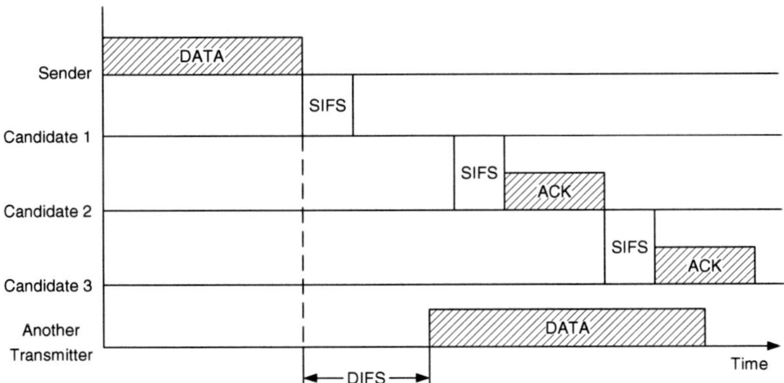

Figure 6.1   SA with first ACK missing. Reproduced by permission of © 2009 IEEE.

6.1.5   Compressed slotted acknowledgment (CSA)

Zubow *et al.* (2007) tried to alleviate the potential collision in SA by introducing the channel assessment technique and refine SA to a "compressed slotted acknowledgment". The general idea is as follows: with a delay of SIFS after receiving the data packet, the highest priority candidate sends its ACK out. From this time point all other candidates that also successfully received the data packet sense the channel by received signal strength indicator (RSSI), a parameter in PHY layer. If the RSSI value increases significantly within the predefined detecting period determined by the priority, the ACK is considered as sent and they will continue to wait for their corresponding ACK slots before sending ACKs. Otherwise, if no such increase in signal strength is observed, the other candidates conclude that the highest priority candidate did miss the data packet. In that case, the second highest priority candidate prematurely sends its ACK to compress the channel's idle space to be smaller than DIFS. All the other candidates behave in the same way as before except that all subsequent events happen earlier. Figure 6.2 depicts a case with three candidates. The use of the channel assessment technique makes the SA's fixed ACK slots mechanism more flexible and enables the CSA to alleviate the potential collision more effectively. However, this detection-based scheme still requires multiple ACKs and thus suffers from the same high coordination delay as SA.

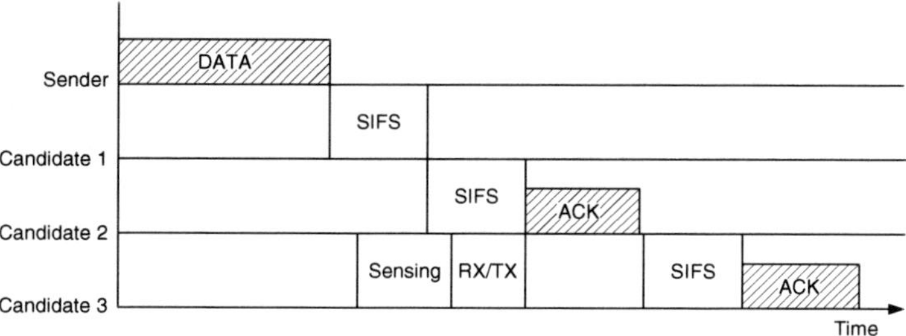

Figure 6.2   CSA with the first ACK missing where RX/TX is the turnaround time for radio to change from receive state to transmit state. Reproduced by permission of © 2009 IEEE.

6.2   Design and analysis of FSA

6.2.1   Design of FSA

The main objective of FSA is to achieve an agreement among multiple candidates with lower coordination delay than SA and CSA. At the same time, FSA must be robust enough to deal with potential collision and unnecessary retransmission. Since all the inefficiencies in SA and CSA are mainly due to the use of multiple ACKs, we adopt a single ACK in FSA, which will be sent by the highest priority candidate in the set of successful receivers. This single ACK plays two roles. On one hand, it informs the sender that reception has been successful, which is the same as SA and CSA; On the other hand, it suppresses all the other lower priority candidates' attempts to forward the data packet. This is different from the ACKs in SA and CSA schemes, which are to help candidates share the information about the reception status. Accordingly, we also choose to use the channel assessment technique to detect the appearance of ACK.

The FSA works as follows. Each candidate waits for $T_{SIFS} + (n - 1) \times T_{Sensing\_Slot}$ before deciding whether it should broadcast ACK, where n is its priority order in the candidate set. So with a time delay of SIFS after the data packet was received, the highest priority candidate sends out an ACK. From that point in time, all the other candidates detect the channel for a $T_{Sensing\_Slot}$ time to tell whether they detect this ACK. If the answer is positive, they stop detecting the channel and simultaneously suppress their own attempts to send ACK and to forward the data packet. Otherwise, if they did not detect any signal within this period, they would think that the highest priority candidate missed the data packet. In that case, the second highest priority candidate takes the responsibility of sending ACK in the beginning of the second $T_{Sensing\_Slot}$ and all the remaining lower priority candidates continue to monitor this ACK within this time. The coordination process goes on like that until some successful receiver finally sends an ACK in the channel.

An example of a transmission with three candidate nodes is illustrated in Figure 6.3. During this one-hop transmission, $candidate_1$ failed to receive the packet. Thus $candidate_2$ and $candidate_3$ detect nothing in the first $T_{Sensing\_Slot}$ time. Then $candidate_2$ thinks that it is itself the highest successful receiver and sends ACK at the beginning of the second $T_{Sensing\_Slot}$. $candidate_3$ detects this ACK and immediately suppresses itself. The total coordination time for FSA with n candidates is: $T_{SIFS} + T_{ACK} + (n - 1) \times T_{Sensing\_Slot}$ compared with SA and CSA's $T_{SIFS} + n \times T_{ACK}$. Since the $T_{Sensing\_Slot}$ is far less than T_{ACK} (for 802.11b, the former is 15 microseconds while the latter is more than 200 microseconds), FSA can significantly reduce the time cost for candidate coordination.

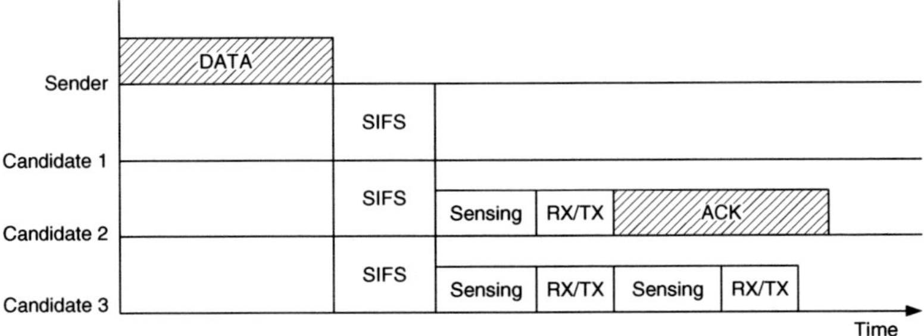

Figure 6.3   FSA with the first ACK missing. Reproduced by permission of © 2009 IEEE.

6.2.2   Analysis

The key difference of FSA from SA and CSA is that it only uses a single ACK to suppress other potential forwarders and acknowledge the sender. Fast slotted acknowledgment uses channel assessment technique to infer some raw information such as whether some packet is transmitting rather than more detailed information like the content of the packet. At first sight, it seems to be less reliable than SA and CSA. However, this is not true. On the contrary, because the information required by FSA is raw, it can be obtained more easily and reliably, which makes the whole scheme work well in a wireless environment, where the most distinct property is unreliability. Suppose one transmission with two candidates A and B where A possesses higher priority and both candidates got the data packet. If the link between A and B is not good when A is transmitting its ACK, then the ACK received by B may be corrupted. In SA and CSA coordination schemes, node B needs to get the ID of the highest priority successful recipient known to the ACK's sender (in the case where it is A itself) from the received ACK. However, with this corrupted ACK, B will fail to do that and will consider itself as forwarder, which leads to duplicate forwarding and unnecessary retransmissions. However, in FSA, B just needs to know that a transmission has happened instead of the detailed content within the received packet. Even if the ACK is corrupted, B may still be able to infer that there is an ACK transmission from higher priority node A. Thus FSA is more robust in this case.

Another apparent weakness of FSA is that a single ACK would not be reliable enough to ensure that the send would receive it correctly, because once this single ACK is lost the sender needs to retransmit the data packet unnecessarily. With multiple ACKs like SA and CSA, the sender would hear at least one clear ACK with high probability. However, it is also not true because of the following reasons. 1. If the data packet (which is generally longer than ACK and sent in higher rate) has already been received by the corresponding candidate successfully, the subsequent ACK sent along the reverse direction at a lower rate (1 mbps for 802.11b) will be received by the sender successfully with very high probability (Sang *et al.* 2007). 2. The other ACKs, except the first one in SA and CSA, are sent in relatively long intervals (several ACK slots) after receiving the data packet. The link states may have already changed at that time. Then the following ACKs may not be able to be received correctly by the sender. So the added reliability by those extra ACKs is quite limited. In other words, a single ACK is already strong enough and multiple ACKs are dispensable.

The real potential vulnerability of FSA is its dependence on the precision of the channel assessment technique. For example, if the detecting node considered other interference to be ACK sent by some higher priority candidate, it will falsely suppress itself from forwarding the packet. It is also possible that, in some situations, the detecting node fails to sense the transmission of ACK from higher priority candidate and then sends its own ACK, which will collide with the transmitting one. However, this dependence problem can be greatly alleviated through careful design. For the first case, we make the whole coordination process highly synchronized and also introduce more precise channel-assessment technique (see details in following subsection). Thus such a probability will be constrained at a rather low level. Even if this scenario indeed happens, the consequence is just that those lower priority candidates suppress themselves "overcautiously" and cause the sender to retransmit the packets unnecessarily. This will make opportunistic routing behave like traditional routing. For the second case, as we described at the beginning of this section, this possibility will be very low because the forwarding candidates are usually in each other's carrier sensing ranges. If this case really happens, the consequence will be duplicate forwarding and unnecessary retransmission, which has serious impact on the performance of OR protocols. However, we should notice that this false detection results in the same consequences in the CSA scheme, which means that FSA will not introduce extra chance for duplicate forwarding in the worst case.

6.2.3   More on channel assessment techniques

Carrier Sensing Multiple Access with Collision Avoidance (CSMA-CA) is the de facto medium access control protocol for 802.11 WLAN. It follows the LBT (Listen Before Talk) principle and works in a time-slots manner, which requires the sender to sense the channel status within one time slot before sending packets. Such a sensing mechanism is called clear channel assessment (CCA) (802.11b 1999 n.d.). Generally speaking, CCA performance could be characterized by a pair of detection and false alarm probabilities (P_d and P_{fa}) in which P_d refers to the probability of detecting the channel to be busy when the channel is indeed busy and P_{fa} refers to the probability of detecting the channel to be busy when the channel is actually idle. There is an inherent tradeoff between P_d and P_{fa} with the constraint of limited detection time (Zhen *et al.* 2008). The CCA module can be implemented in two ways:

1. *Energy detection (ED).* Energy detection based CCA is a simple noncoherent detection approach. It integrates the square of the incoming signal from the radio front end during the CCA window to get an average signal strength, then compares it with a predefined threshold level, which represents the normal background noise and makes a judgment. The main advantage of ED-based CCA is simplicity and the main disadvantage is relatively poor detection reliability (especially in 802.11b/g, which works on 2.4 GHZ, coexisting with other technologies such as microwave ovens and Bluetooth devices).

2. *Preamble detection (PD).* Preamble detection based CCA tries to use the correlation of well-known preambles with the received signal to detect the presence of a packet. It can be implemented by a crosscorrelation-based matched filter which is more complex as it can fully take advantage of the processing gain and thus has a more enhanced reliability. The main disadvantage of PD-based CCA is that it needs to run continuously and thus brings a relatively high energy cost.

Both detection methods are supported by most of the current wireless cards and can promise a P_d of no less than 99% with the CCA window specified by IEEE 802.11 standards (http://standards.ieee.org/). However, the PD-based CCA outperforms the ED-based CCA in P_{fa}, especially in noisy scenarios. Thus in FSA we choose the PD-based CCA technique. Another reason that we prefer PD-based CCA is that its main disadvantage can be avoided in the scenario of coordination in OR protocols because what we need to know is that if any ACK being sent during the CCA window and do not care the channel state that is out of this period. Thus we do not require the PD module running continuously and it can be turned on only when the MAC layer requires a CCA from the PHY layer, just as an ED module does.

6.3   Simulation results and evaluation

In this section, we evaluate and compare the performance of FSA with SA and CSA in GloMoSim (Zeng *et al.* 1998). We also introduce a perfect detection-based scheme, IDEAL, as the baseline for comparison. The IDEAL scheme is the same as FSA, except that all the ACKs in IDEAL are 100% reliable and the detection judgement is 100% precise. As our focus is on the efficiency of different coordination schemes given the same candidate set and the corresponding forwarding priorities, we use an existing candidate selection algorithm based on a node's geographic locations and adopt the local metric expected one-hop throughput (EOT) (Zeng *et al.* 2007) to select candidates. We will elaborate this local metric in Chapter 8.

Because the existing popular network simulators, such as NS-2 (McCanne and Floyd 1997), OPNET (www.opnet.com), GloMoSim (Zeng *et al.* 1998), have not implemented the PHY layer's function like energy integration or matched filter module currently, we have done some modification to the PHY layer in GloMoSim and make the detection judgment based on the following probability model. We define the CCA error floor (Zhen *et al.* 2008) at the optimal threshold, which can be achieved by equating $1 - P_d$ and P_{fa} where P_d is detection possibility and P_{fa} is the false alarm possibility. Then the CCA error floor for ED-based CCA and PD-based CCA can be expressed in terms of the Q function (Kay 1998):

$$P_{CCA\_ef\_ed} = Q\left(\sqrt{N}\frac{SNR}{1 + \sqrt{1 + 2SNR}}\right)$$

$$P_{CCA\_ef\_pd} = Q\left(\sqrt{\frac{N}{2}}SNR\right)$$

We define the following performance metrics.

- Throughput: the ratio of the number of received bits to the whole session time.

- Delay: the per-packet end-to-end time delay from the packet being sent out until it reaches the destination.

- Packet deliver ratio: the number of successfully received packets over the number of sent packets.

- Number of transmissions: the total number of data transmissions occurring during the whole simulation time.

- Duplicate delivery ratio: the number of duplicate packets received at all the destinations over the total number of received packets.

- Retransmission ratio: the transmission number needed for a successful one-hop forwarding.

The simulation results of all metrics except for the number of transmissions are averaged over 25 flows under 5 simulation runs with different seeds.

6.3.1 Simulation setup

We developed a simulation environment with Glomosim. The MAC protocol is based on 802.11b, but with some modifications. Since the source code of SA

Table 6.1   Simulation parameters

Simulation parameter	Value
Number of nodes	50
Stationary or dynamic	stationary
Data transmission rate	11 mbps
ACK transmission rate	1 mbps
Retry limit	5
Contention window	31–1023
Radio sensing threshold for data	−100 dbm
Radio receiving threshold for data	−83 dbm
Radio sensing threshold for ACK	−100 dbm
Radio receiving threshold for ACK	−91 dbm
Pathloss model	two-ray
Fading mode	rician
Rician k factor	4
Radio reception SNR	10
Hello packet interval	1s
Size of candidate set	3
CCA window	15 μs
SIFS	10 μs
Radio receive/transmit turnaround time	5 μs

Reproduced by permission of © 2009 IEEE.

and CSA schemes are not publicly available, we implemented our own version. Table 6.1 lists all the related simulation parameters.

All these 50 nodes are randomly uniformly distributed in a $d \times d\,m^2$ square region where $d = 1400, 1500, \ldots, 1800$. The corresponding average number of neighbors per node and average hop counts per flow are listed in Table 6.2.

We randomly choose 25 communication pairs in the network. The sources are CBR (constant bit rate) and each packet being 512 bytes long. User Datagram Protocol (UDP) is used at the transport layer. Each communication session lasts

Table 6.2   Average number of neighbors per node and average hops per packet under different network densities

Terrain size	Neighbors	Hops
1400	12.42	2.17
1500	10.90	2.50
1600	9.65	2.67
1700	8.60	2.96
1800	7.79	3.17

120 s. Before all the transmissions start, the simulation environment will go through a 30 s warm-up phase, during which each node sends out a "Hello" packet periodically to learn the neighbors, information and this learning process lasts through the simulation.

6.3.2   Simulation results and evaluation

6.3.2.1   Delay

Figure 6.4 shows the average per packet end-to-end time delay of SA, CSA, FSA and IDEAL. In order to make a fair comparison, we set the packet interval of all the data flows to be 120 ms, which promises all the schemes can handle the traffic demand (in this case, all the protocols achieve 100% delivery ratio and almost the same average per flow throughput of 34 kbps, thus we will not show the performance comparison of these two metrics in this setting). We see that SA has the highest delay value under all terrain side lengths and CSA performs slightly better. Fast slotted acknowledgment achieves a far lower time delay than these two schemes and very close to the performance of IDEAL, which has the lowest delay value. From this result, firstly, we can get the conclusion that the use of the channel assessment technique indeed can alleviate the potential collision problem caused by the ACK's unexpected missing. Secondly, we also notice that the time delays for CSA, FSA and IDEAL grow very slow as the terrain side length increases. This

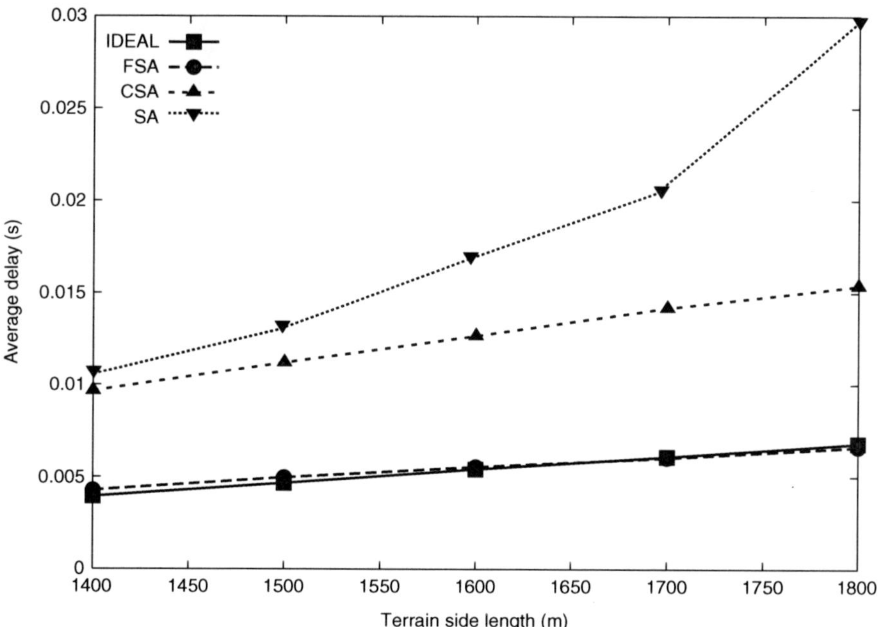

Figure 6.4   Average per-packet end-to-end time delay. Reproduced by permission of © 2009 IEEE.

demonstrates that applying the channel assessment technique can also make the delay more stable under different network densities. Finally, we observe that FSA experiences less than half the time delay of CSA under all the terrain side lengths. This reduction in time delay can be attributed mainly to the single ACK design.

6.3.2.2   Number of transmissions

Figure 6.5 shows the total number of data transmissions during the simulation time. We see that as the terrain side length increases all schemes need a greater number of transmissions to deliver these data flows. This can be explained by the simultaneous increment of average hop count shown in Table 6.2. We also can observe that FSA needs a lower number of transmissions than CSA and SA. This proves that FSA can not only greatly reduce the time cost for the coordination process but also can achieve better coordination reliability, which contributes to the reduced number of transmissions.

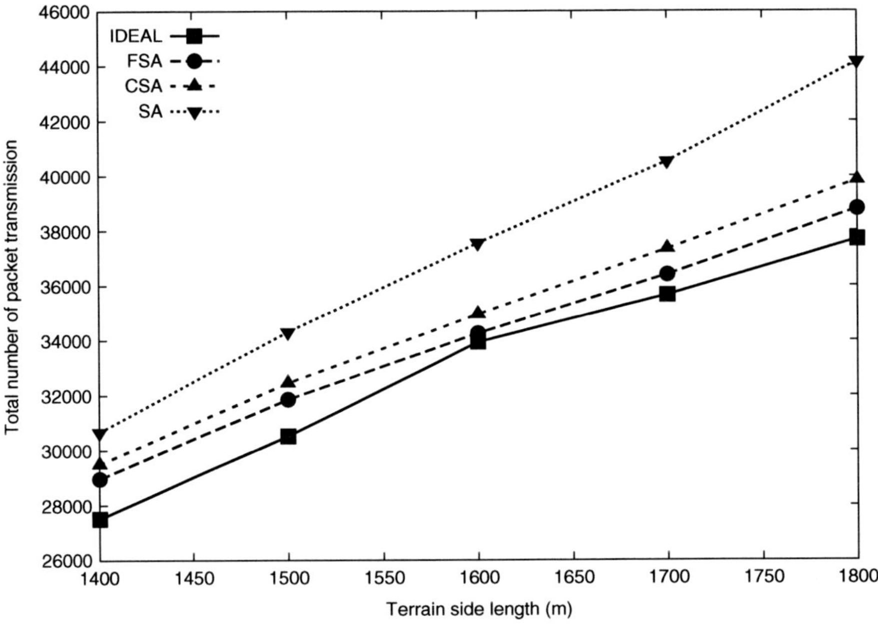

Figure 6.5   Total number of transmissions needed for delivering all the data flows.

6.3.2.3   Duplicate deliver ratio and average retransmission ratio

The per-packet duplicate ratio shown in Figure 6.6 and average retransmission ratio shown in Figure 6.7 can further demonstrate that FSA is more reliable. In Figure 6.6, we see that of all the data packets received successfully by the destinations of these data flows, there are approximately 6%−14% duplicated ones for SA

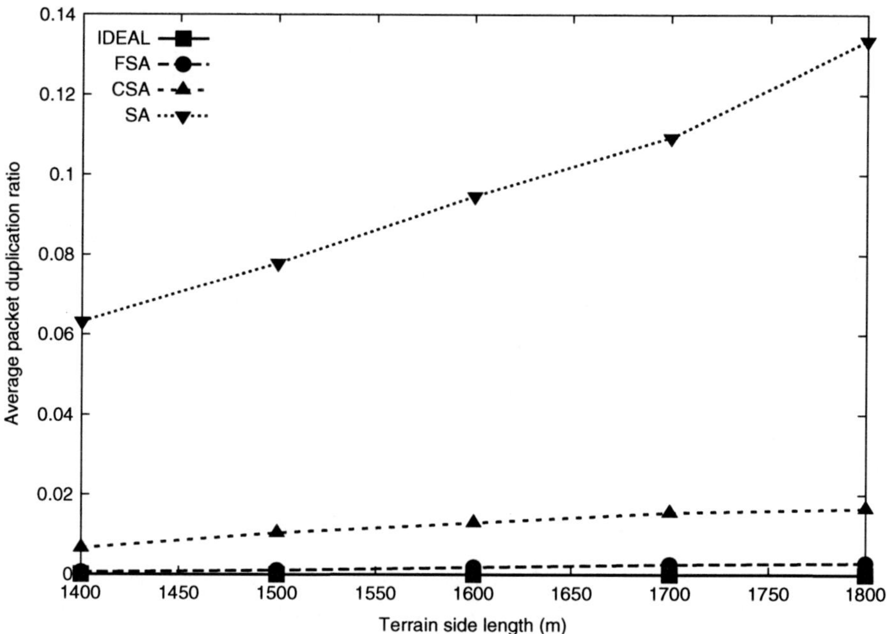

Figure 6.6   Average per-packet duplicate ratio counted in the final receivers. Reproduced by permission of © 2009 IEEE.

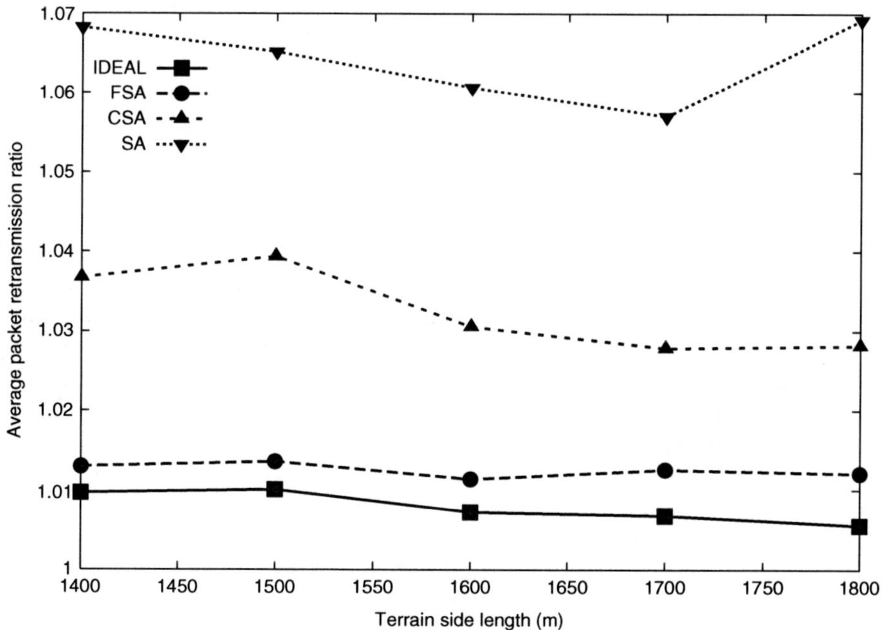

Figure 6.7   Average one-hop retransmission ratio. Reproduced by permission of © 2009 IEEE.

and 0.4%–2% for CSA under different terrain side lengths. However, the duplicate ratios for FSA are almost zero under all terrain side lengths, which are very close to the performance of IDEAL. This confirms our analysis in Section 6.2.2, which concludes that the probability of candidates in FSA missing the presence of a higher priority candidate's ACK and of this resulting in duplicate forwarding is very low, but the probability of candidates in CSA receiving corrupted ACKs from other candidates and leads to duplicate forwarding is not negligible. In Figure 6.7, we see that IDEAL achieves an average retransmission ratio of approximately 1.01 under all terrain side lengths. The ACKs in IDEAL scheme are exempt from fading or interference, so the only reason for retransmission in IDEAL is that all the candidates fail to receive the data packet. Such low retransmission ratio shows that OR schemes with multiple candidates can indeed greatly increase the forwarding reliability. We also note that FSA's performance is close to the IDEAL, with an average higher ratio of 0.5%. This shows that the use of a single ACK in FSA is sufficiently reliable to acknowledge the sender.

6.3.2.4 Packet delivery ratio and throughput

In order to evaluate the throughput performance of all the schemes, we set the packet interval of all data flows to be 70 ms, which makes a relatively heavy traffic load. From Figure 6.13 we see that SA and CSA are unable to handle the traffic demand and can only achieve packet delivery ratio of 87%–81% and 81%–74% respectively under different terrain side lengths. However, FSA still performs well and achieves 100% delivery ratio under different terrain sizes, just like IDEAL. Since our throughput metric is the ratio of the number of received bits to the whole session time and all the schemes have the same simulation time, thus the throughput is proportional to the packet delivery ratio. From Figure 6.8 we can see that FSA can achieve a throughput of approximately 54 kbps, with an average gain of 12.5%–20% compared with CSA's throughput under different terrain side lengths. The reason is that FSA takes less medium time in each hop's transmission and thus possesses a higher throughput capacity and the traffic demand in this setting is still within its capacity range. The significant higher time delay of SA and CSA shown in Figure 6.9 is also due to the fact that the traffic demand under this setting is beyond the throughput capacity of these two schemes, which cause each packet to suffer a long waiting time in the packet queue of every intermediate relay node. This long queuing delay further aggravates the duplication and retransmission problems, which can be observed in Figure 6.11 and Figure 6.12. From Figure 6.10 we observe that SA still has the highest number of transmissions during the simulation, but it can achieve the lowest throughput. This is not strange because the potential collision problem is exacerbated under a heavy traffic load. However, we also observe that the total transmission number of FSA and IDEAL is higher than CSA under terrain side lengths of 1700 m, 1800 m. This "abnormal" case also proves that nodes in FSA and IDEAL are more positive in transmitting rather than waiting in queue or backing off.

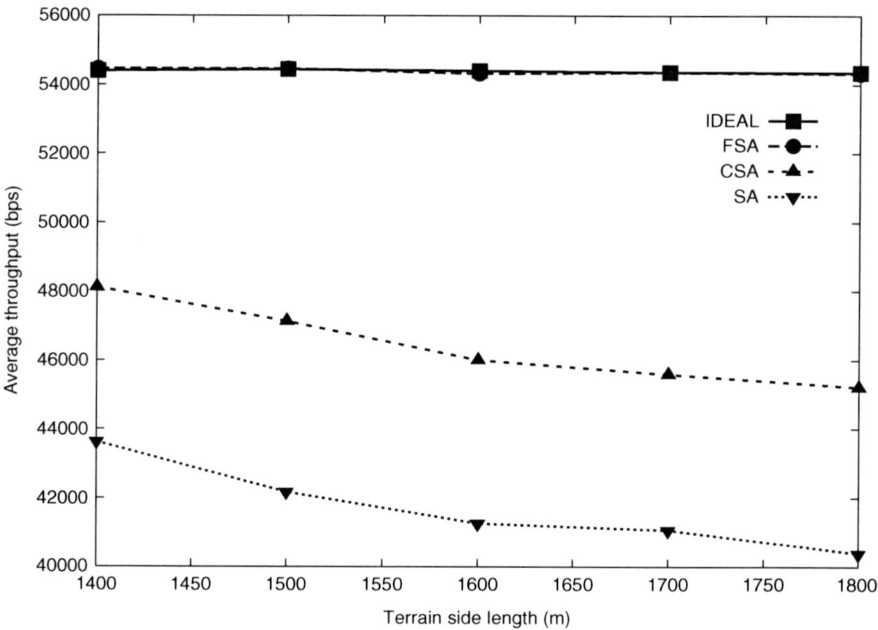

Figure 6.8   *Average per-flow throughput.*

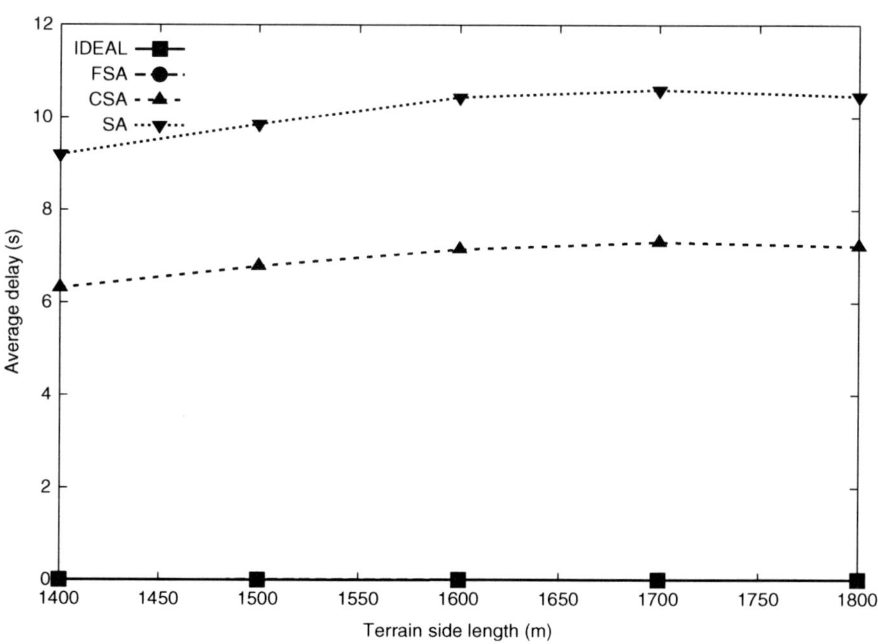

Figure 6.9   *Average per-packet end-to-end time delay.*

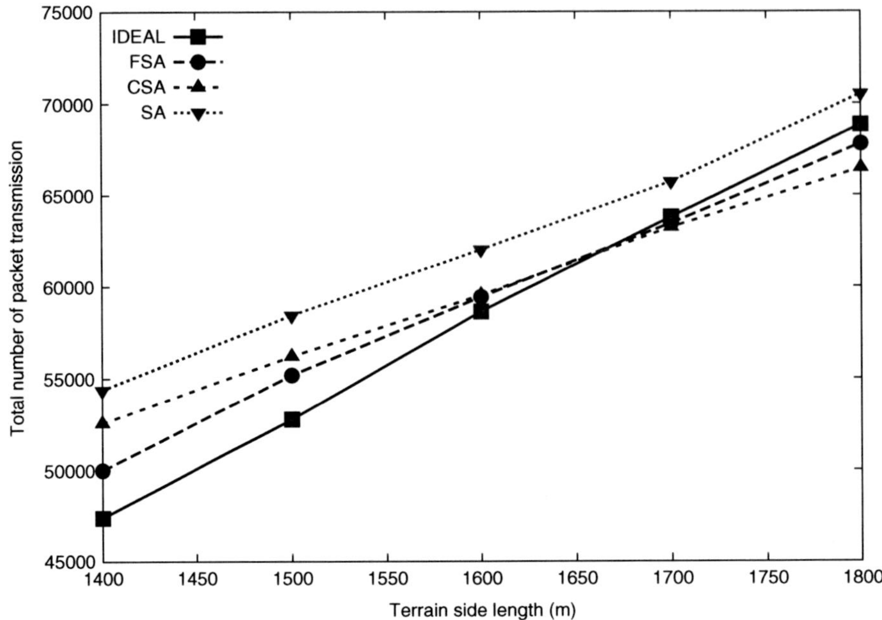

Figure 6.10   Total number of transmissions needed to deliver all the data flows.

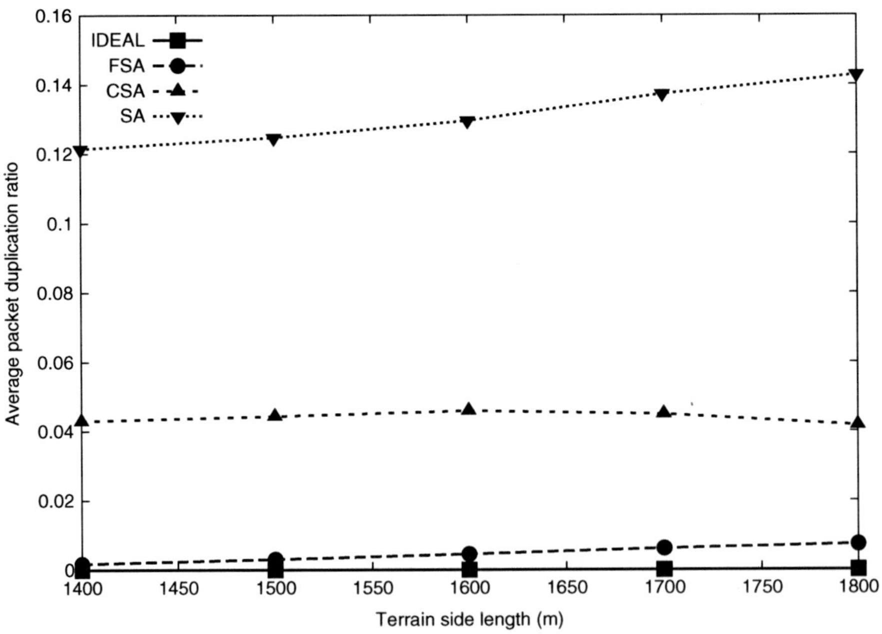

Figure 6.11   Average per-packet duplicate ratio counted in the final receivers. Reproduced by permission of © 2009 IEEE.

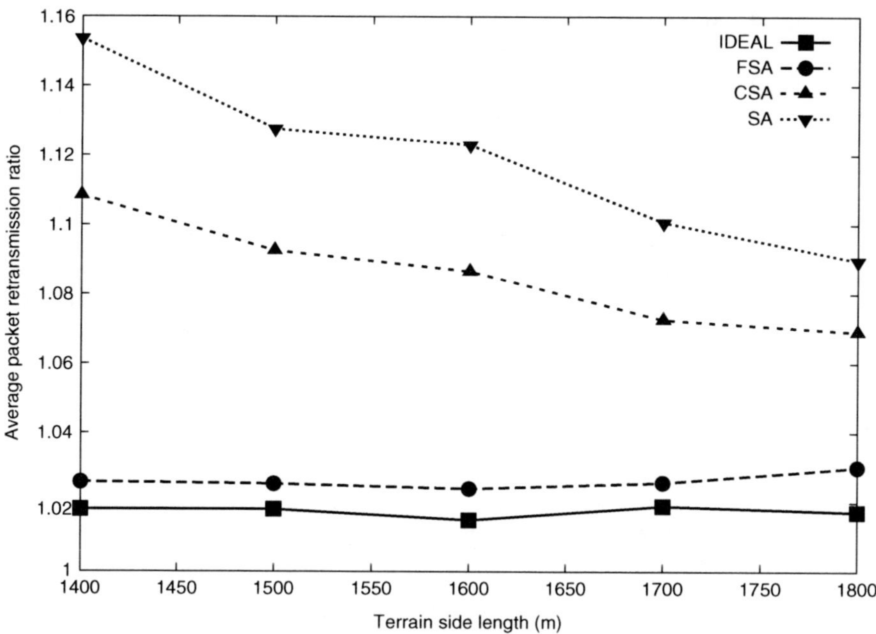

Figure 6.12   Average one-hop retransmission ratio. Reproduced by permission of © 2009 IEEE.

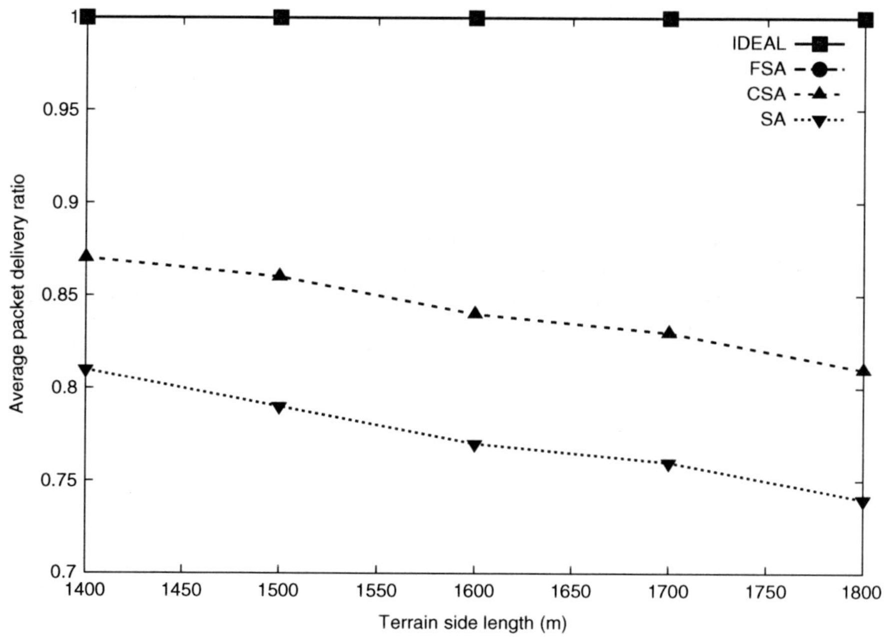

Figure 6.13   Average per-flow delivery ratio. Reproduced by permission of © 2009 IEEE.

6.4 Conclusions

In this chapter, we analyzed the coordination problem in opportunistic routing. We first reviewed and analyzed five state-of-the-art candidate coordination schemes, and based on this analysis we proposed a new coordination scheme: "fast slotted acknowledgment" (FSA), which fully takes advantage of the channel detection approach to coordinate the packet forwarding among multiple candidates. We compared FSA with those state-of-the-art schemes and simulation results show that it achieves better performance in all the metrics, especially in time delay. The simulation also validated that FSA can achieve similar performance to ideal coordination where relay priority can be ensured and duplicate packet forwarding is avoided.

References

Biswas S and Morris R 2003 Opportunistic routing in multihop wireless networks *HotNets-II*, Cambridge, MA.

Biswas S and Morris R 2005 Exor: Opportunistic multi-hop routing for wireless networks *SIG-COMM'05*, Philadelphia, PA.

Fussler H, Widmer J, Kasemann M, Mauve M and Hartenstein H 2003 Contention-based forwarding for mobile ad-hoc networks. *Elsevier's Ad Hoc Networks* 1(4), 351–369.

Kay SM 1998 *Fundamentals of Statistical Signal Processing, Volume 2: Detection Theory*. Prentice Hall Signal Processing Series, Upper Saddle River, NJ.

McCanne S and Floyd S 1997 *The LBNL Network Simulator* Lawrence Berkeley Laboratory.

Sang L, Arora A and Zhang H 2007 On exploiting asymmetric wireless links via one-way estimation *MobiHoc '07: Proceedings of the 8th ACM International Symposium on Mobile ad hoc Networking and Computing*, pp. 11–21. ACM, New York, NY, USA.

Schurgers C, Tsiatsis V, Ganeriwal S and Srivastava M 2002 Optimizing sensor networks in the energy-latency-density design space. *IEEE Transactions on Mobile Computing* 1(1), 70–80.

Zeng K, Lou W and Zhang Y 2007 Multi-rate geographic opportunistic routing in wireless ad hoc networks *IEEE Milcom*, Orlando, FL.

Zeng X, Bagrodia R and Gerla M 1998 Glomosim: a library for parallel simulation of large-scale wireless networks *Proceedings of PADS'98*, Banff, Canada.

Zhen B, Li HB, Hara S and Kohno R 2008 Clear channel assessment in integrated medical environments. *EURASIP Journal on Wireless Communications and Networking*, 1–8.

Zorzi M and Rao RR 2003 Geographic random forwarding (geraf) for ad hoc and sensor networks: energy and latency performance. *IEEE Transactions on Mobile Computing* 2(4), 349–365.

Zubow A, Kurth M and Redlich JP 2007 Multi-channel opportunistic routing *IEEE European Wireless Conference*, Paris, France.

7

Integration of opportunistic routing and network coding

All the coordination schemes or protocols discussed in the previous chapter need to modify the MAC layer or depend on physical layer information, which may not be an easy task. In order to use off-the-shelf IEEE standard compatible hardware devices, it is desirable to make minimum (or zero) modifications to the lower layers and focus on the routing layer and above.

To this end, this chapter takes a different approach to the above issue by integrating network coding into opportunistic routing. In particular, it presents how network coding can help ease the coordination in opportunistic routing. A classical work integrating opportunistic routing and network coding, MORE, will be introduced. Then, a broadcast scheme integrating symbol level network coding with opportunistic listening will be described, with applications to mobile content distribution (MCD) in vehicular ad hoc networks (VANETs).

The rest of this chapter is organized as follows. Section 7.1 gives a brief introduction to MORE, a state-of-the-art MAC-independent opportunistic routing protocol. Section 7.2 introduces the problem of mobile content distribution in VANETs. We survey the related works on mobile content distribution in VANETs in Section 7.3. Then, Section 7.4 reviews symbol level network coding (SLNC) and its major advantages compared with PLNC are shown through both theoretical analysis and simulation. After that, for two example mobile content distribution applications – popular content broadcast and live multimedia streaming, we present novel broadcast schemes based on the integration of SLNC and opportunistic listening. Section 7.5 describes CodeOn, a cooperative popular content broadcast scheme for VANETs, and Section 7.6 presents CodePlay, a cooperative live multimedia

Multihop Wireless Networks: Opportunistic Routing, First Edition.
Kai Zeng, Wenjing Lou and Ming Li.
© 2011 John Wiley & Sons, Ltd. Published 2011 by John Wiley & Sons, Ltd.

streaming protocol for VANETs. For both of them, detailed simulation results and performance evaluation are given. Finally, Section 7.7 concludes this chapter.

7.1 A brief review of MORE

In the original opportunistic routing ExOR (Biswas and Morris 2005), the MAC layer is tied with routing to provide opportunistic gain. However, due to the strict scheduling on forwarders' access to the medium, it actually prevents spacial reuse to a large extent. If multiple packets are received by the same receiver, those packets may contain duplicate information that wastes the precious network bandwidth. In the case of a large network with highly lossy wireless medium, this problem is aggravated because the strict scheduling and coordination among forwarders is hard to realize due to multipath fading and packet collisions. Thus, MORE was proposed by Chachulski *et al.* (2007) to solve this challenge. By integrating random linear network coding[1] at the intermediate forwarders, nodes forward the coded packets randomly instead of using the structured scheduling approach in ExOR.

Here we illustrate the motivation of MORE using a simple example. Consider the topology in Figure 7.1 with a single unicast flow. Traditional routing first determines the best path before transmission, which is "$A \rightarrow B \rightarrow C$." However, there is always a chance for the destination node C to receive packets from source A directly. If, in one transmission, B and C both receive packet p_1, this will cause a wasted transmission at node B using traditional routing because B does not know C's reception status. ExOR solves this problem by enforcing a strict transmission coordination for node B, which is in A's forwarder list – B waits for C's feedback signal and only if C has not received a packet in B's queue will B transmit that packet. Such coordination reserves the wireless medium for a single forwarder at

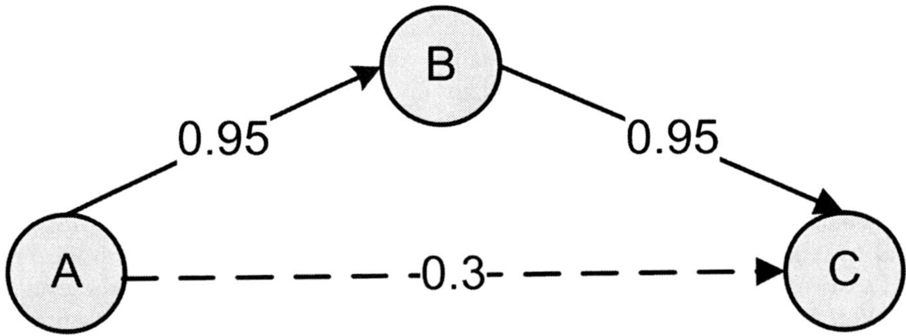

Figure 7.1   Motivating example for MORE. A unicast flow is from source node A to destination node C. The reception success probability of each link is denoted on that link.

---

[1] Unless otherwise stated network coding refers to packet-level network coding in the rest of the chapter.

any time, as each node in a forwarder list needs to wait for the transmission by all the higher priority forwarders, which actually limits spacial reuse. On the other hand, to enforce such stringent coordination requires modifications at the MAC layer or even the PHY layer, as we can see from the previous chapters. A similar problem exists for multicast flows (Chachulski *et al.* 2007).

Therefore, MORE adopts network coding (Ahlswede *et al.* 2000) as a remedy. Network coding basically breaks the store-and-forward packet forwarding paradigm. It allows intermediate nodes to combine received packets. Bandwidth efficiency can be improved because each coded packet could benefit multiple receivers simultaneously. Meanwhile, because each coded packet could be equally useful to the receiving node, the complicated coordination of retransmission can be eliminated, which in turn simplifies the protocol design.

In the above example, assume packet p_1 is received by both node B and C while p_2 is only received by B. As B is unaware of C's fortunate reception, B can simply forward the sum $p_1 + p_2$ to C, which can recover p_2 since it has p_1.

In fact, it is sufficient to generalize the above approach to random linear network coding (RLNC) (Ho *et al.* 2006). Using RLNC, the source broadcasts native packets, intermediate nodes perform random linear combinations of packets they receive $(c_1 p_1 + \cdots + c_m p_m)$, where $c_1, \cdots, c_m$ are random coefficients from a Galois field $\mathbb{F}_q$. The destination sends an acknowledgement (ACK) along the reverse path to the source once it receives all the native packets. In this way, node coordination is not required and spacial reuse is preserved, which essentially "trades off structure for randomness".

Next we describe the operation of MORE in more detail.

- **Source**. The source generates K coded packets as a batch, from uncoded packets $p_1, \cdots, p_K$, which are called *native packets*. A *coded packet* is generated as follows:

$$p_j = \sum_{i=1}^{K} c_{ji} p_i,$$

  where c_{ji} are random coefficients from a Galois field $\mathbb{F}_q$, and $\vec{c} = (c_{j1}, \cdots, c_{jK})$ is called a *coding vector*. The sender attaches the following information in the packet header: the packet's coding vector (which will be used in decoding), the ID of the current batch, the IP addresses of the source and destination, and the forwarder node list, which is computed from the ETX metric (De Couto *et al.* 2003).

- **Forwarders**. Each forwarding node listens to all transmissions opportunistically. Whenever a new packet arrives, and if the node is in the forwarder list specified by the packet, it checks whether this packet is *innovative*. A received coded packet is said to be "*innovative*" if the packet is linearly independent from the packets of the same batch that have already been received by the node, which could be done using Gaussian elimination (Koetter and

Médard 2003). If so, the packet's arrival will trigger a broadcast of a new coded packet by the node. To generate the new coded packet, the node computes a random linear combination of all the received coded packets in its buffer (from the current batch), which can be easily shown to be a linear combination of the K native packets. Assume that the node has already received packets of the form p'_j. A a new coded packet is created as $p'' = \sum_j \alpha_j p'_j$, where α_j are random numbers. p'' can be expressed as:

$$p'' = \sum_j \left(\alpha_j \sum_{i=1}^{K} c_{ji} p_i \right) = \sum_i \left(\sum_j \alpha_j c_{ji} \right) p_i,$$

where $(\sum_j \alpha_j c_{j1}, \cdots, \sum_j \alpha_j c_{ji}, \cdots, \sum_j \alpha_j c_{jK})$ is the coding vector of p''.

- **Destination**. The destination checks each received a packet to see whether it is innovative. If K innovative packets are received, it can decode the native packets by solving a linear equation:

$$\begin{pmatrix} p_1 \\ \vdots \\ p_K \end{pmatrix} = \begin{pmatrix} \vec{c}_1 \\ \vdots \\ \vec{c}_K \end{pmatrix}^{-1} \cdot \begin{pmatrix} p'_1 \\ \vdots \\ p'_K \end{pmatrix},$$

  where $\vec{c}_i$ is the coding vector of p'_i, the ith received packet. As soon as all the K native packets are decoded, the destination sends an ACK message to the source along the reverse path to inform the source to move to the next batch.

The design of MORE considers several practical issues, such as how many packets a node should forward, when a sender should stop and purge and how a node codes packets efficiently. Interested readers are referred to Chachulski *et al.* (2007) for more details.

Note that, MORE also includes an extension from unicast to multicast. The idea is to find a multicast tree that is taken as the union of the shortest paths to each destination, using the ETX metric. The forwarder list of the multicast flow is the union of all the forwarder lists of the unicast flows. Experimental results show that, for multicast, MORE has a higher average gain in throughput compared with ExOR and traditional routing. Nevertheless, MORE is considered as inefficient when applied to multicast or broadcast because almost every node in the network may become a forwarding node, which may cause heavy congestion and packet collision (Koutsonikolas *et al.* 2009). Moreover, in mobile MWNs such as VANETs, tree-based multicast schemes fall short in that they incur large overheads in maintenance of the tree structure as the topology changes very fast. This calls for new design rationales for broadcast protocols in mobile MWNs based on network coding and opportunistic listening. In what follows, we will first introduce the mobile content distribution problem in VANETs and then present CodeOn, a content broadcast scheme for VANETs, which solves the above challenge by trading off structure for even more randomness than MORE.

7.2   Mobile content distribution in VANETs

Vehicular communication has been a topic of great interest in recent years. Typically, a vehicular network consists of roadside units (RSUs), which are access points along the road, and on-board units (OBUs) mounted on vehicles, which can either communicate with the RSUs or with other OBUs in an ad hoc manner. Since the advent of dedicated short-range communications (DSRC) (Jiang *et al.* 2006); see also www.standards.its.dot.gov/Documents/advisories/dsrc advisory.htm., IEEE 802.11p and IEEE 1609 standards (WAVE 2006), people have envisioned and designed numerous tempting applications of vehicular networks, ranging from safety warnings (Li *et al.* 2009) and intelligent navigation, to mobile infotainment (Lee *et al.* 2006). Among them, content distribution, especially "popular" multimedia content distribution to vehicles inside a geographical area of interest, is particularly attractive. Examples of such mobile content distribution (MCD) include live video broadcast of road traffic and conditions to vehicles driving towards it for intelligent navigation, which is especially useful during inclement weather; periodical broadcasts of multimedia advertisements for local businesses in a city to vehicles driving through a segment of suburban highway (like a digital billboard) and the dissemination of an accurate update of the GPS map about a city or a scenic area.

The challenges of providing a MCD service in VANETs are threefold. On the one hand, content distribution, especially multimedia content consisting of audio and video, requires *high distribution rate* and *short delay*. For live multimedia streaming, further quality of service requirement, such as the deadline for receiving a packet, which translates to smooth playback or *stable downloading rate*, is also important. On the other hand, wireless is the well-known shared and lossy medium with very limited bandwidth and throughput drops dramatically after a few hops in multihop wireless networks (Jain *et al.* 2005). Moreover, the high mobility of VANETs, leading to fast and unpredictable topological changes will further exacerbate the frequent packet losses and collisions.

Network coding is a common technique adopted in content distribution as an effective approach to improving the bandwidth efficiency and simplifying the protocol design. However, traditional network coding combines information at the packet level. Generally speaking, if a packet is received in error, then it is discarded by an intermediate node, which involves more transmissions and wastes the channel bandwidth. Under adverse channel environments such as the mobile VANET, the link-loss problem is worse than in static networks like wireless mesh networks (Torrent-Moreno *et al.* 2004; 2005; 2006). Because of the low packet reception success rates, using packet-level network coding (PLNC) is no longer sufficient to meet the requirements of some MCD services (e.g, multimedia streaming) that require high and stable data rate and short delay in VANETs (Park *et al.* 2006; Yang *et al.* 2010). In addition, protocols adopting PLNC also face the well-known hidden-terminal problem in MWNs, which is notoriously difficult to solve in multicast/broadcast (Dutta *et al.* 2009; Ni *et al.* 1999).

On the other hand, the benefit of network coding often tends to be offset by severe packet collisions due to lack of proper transmission coordination mechanisms among vehicles (Li *et al.* 2011). However, too much coordination again incurs problems as we have mentioned before: the overhead required by strict coordination per se would in turn negatively affect the protocol performance. Obviously, adopting a broadcast-tree-based method is impractical because it needs to collect the global topology information in real-time. There is therefore a choice about the level of node coordination that is necessary. We contend that it is always beneficial to reduce the effort spent in coordination to the minimum amount.

Novel techniques are needed to address all of the above challenges. The proposed schemes in this chapter are based on symbol-level network coding (SLNC) (Katti *et al.* 2008). In contrast to traditional *packet level network coding*, SLNC allows intermediate nodes to combine packets at symbol level, where a *symbol* is typically composed of several physical layer symbols of a modulation scheme. Symbol-level network coding allows a node to recover correctly received symbols from erroneous packets. As symbol error rate is smaller than the packet error rate, in addition to the benefits one can gain from PLNC, SLNC provides better error tolerance and thus increased successful packet reception rate. A further study of SLNC also shows that SLNC in fact enables higher spacial reusability by allowing concurrent transmissions within shorter distances (Li *et al.* 2011; Yang *et al.* 2010), which enables much simpler node coordination than using PLNC.

In addition, by fully exploiting SLNC and opportunistic listening, a new push-based protocol design concept is proposed in this chapter. Specifically, the content source actively "pushes" information to nearby vehicles, while a dynamic set of temporary relay nodes is chosen to help further broadcast this information to all the other vehicles in the VANET. The source's responsibility to guarantee reception and quality of service for the whole network is distributed to each relay node, which only needs to ensure the reception of its neighboring vehicles (Yang *et al.* 2009; 2011). The node coordination is kept to the minimum, since whoever receives the largest amount of useful contents within a local range will be chosen as a relay node, and the relay nodes simply compete to access the wireless channel randomly based on carrier sensing, as that in IEEE 802.11.

In this chapter, we will demonstrate the effectiveness of our proposed methods using two example MCD applications: popular content broadcast and live multimedia streaming. Specific schemes designed for each of these two applications are presented, where the concept of opportunistic listening are exploited in different ways.

7.2.1   Model and assumptions

In this chapter, we consider the basic network architecture for the MCD service, which is illustrated in Figure 7.2. We assume there are one or more access points (APs, or road side units) deployed in the *area of interest* (AoI), which can be either a highway segment or an urban area. The content provider (e.g. a citywide traffic administration bureau) would like to distribute some content/multimedia content through the APs to vehicles inside the AoI. However, due to the deployment cost and limited communication range of APs, the entire AoI is not fully covered by APs. For vehicles outside the direct communication range of an AP, they form a VANET and cooperatively collect/distribute the content.

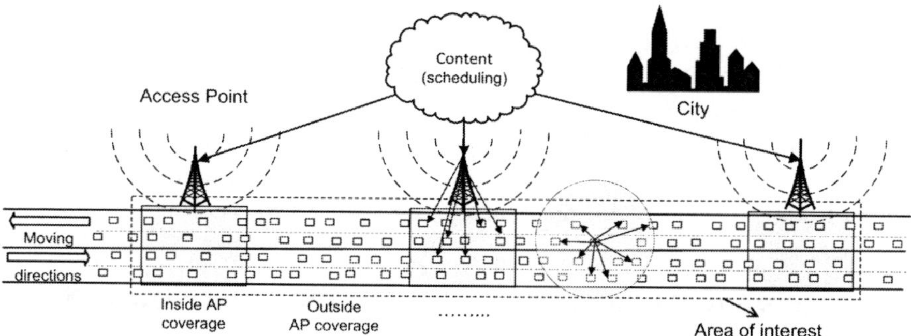

Figure 7.2   The architecture for MCD. Inside the AP coverage, AP broadcasts and vehicles receive; outside the AP coverage, vehicles distribute their received contents cooperatively. Reproduced by permission of © IEEE 2011.

Each vehicle is equipped with an on-board unit including a wireless transceiver (single radio). The wireless interface operates on multiple channels (www.standards.its.dot.gov/Documents/advisories/dsrc advisory.htm; Jiang *et al.* 2006). To model the coexistence of safety and commercial applications, we consider two representative channels. The control channel is used to broadcast safety messages, which may contain vehicles' locations, speeds etc.; one service channel is dedicated for MCD. In order to guarantee the quality of service of safety messages (the interval between two consecutive safety messages should be smaller than 100 ms – *et al.* 2009), time is divided into periodical, 100 ms slots and all vehicles and APs are synchronized to switch simultaneously between the control channel and service channel. The utilization of time and channels is depicted in Figure 7.3. Although there are advanced MAC protocols that dynamically adjust the time shares of control channel and service channel for better service (Mak *et al.* 2009), we fix it to $1/2 : 1/2$ for simplicity.

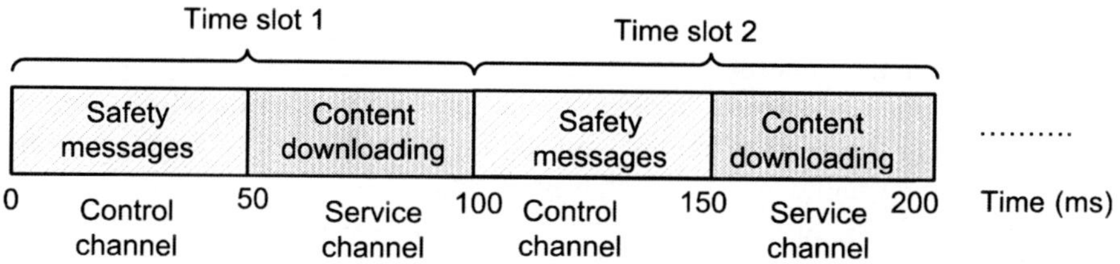

Figure 7.3   The time and channel utilization of each vehicle and each AP. Reproduced by permission of © IEEE 2011.

In the control channel, each AP and each vehicle broadcasts one beacon message in each slot. When a vehicle is in the range of an AP, it merely listens to the AP's content broadcast in the service channel; otherwise, it may share its received content with neighboring vehicles cooperatively. Vehicles outside the AoI do not involve in content distribution.

In addition, we assume all vehicles are equipped with Global Positioning System (GPS) devices, from which vehicles obtain their real-time locations and synchronize their clocks (error smaller than 100 ns). GPS devices have a low cost and are available to most drivers nowadays. When vehicles are temporarily out of satellite coverage, they can use auxiliary techniques to determine their location, and rely on their own hardware clocks. Note that GPS time synchronization is required by the IEEE 1609.4 standard for multichannel operations (WAVE 2006).

7.3 Related works on mobile content distribution in VANETs

7.3.1 Cooperative downloading of general contents in VANETs

In (Nandan *et al.* 2005), Nandan *et al.* first studied cooperative downloading in VANETs. They proposed SPAWN, a pull-based, peer-to-peer content downloading protocol for VANETs that extends BitTorrent. Later, they proposed "AdTorrent" (Nandan *et al.* 2006), which is a semi-push-based peer-to-peer protocol for vehicles to download advertisements they are interested in. In both SPAWN and AdTorrent, the peer and content selection mechanisms have a high overhead and are not scalable, especially when most of the vehicles are interested in downloading popular contents. Also, they suffer from the "coupon collector problem," which enlarges downloading delay. Moreover, they use Transport Control Protocol (TCP) for content delivery, which performs poorly over multihop lossy wireless links in highly mobile VANETs.

7.3.1.1 Network coding for content downloading

To avoid such problems, many researchers resort to network coding (Ahlswede *et al.* 2000; Ho *et al.* 2006). Lee *et al.* proposed CodeTorrent (Lee *et al.* 2006), a pull-based content distribution scheme using network coding, where vehicles need to explicitly initiate requests to download a piece of content. CodeTorrent restricts the peer selection and content delivery to the one-hop neighborhood of a vehicle, thus eliminating the need of multi-hop routing. Also, the use of network coding mitigates the peer and content selection problems.

Lee *et al.* (2008) further studied the practical effects of content distribution in VANETs using network coding based on a variation of CodeTorrent. It is shown that the resource constraints such as disk access, computation and buffer have significant impacts on the performance. They discussed approaches to reduce the communication and computation overhead of network coding while maintaining the gain of it.

The above schemes are all pull based in essence. They could suffer from large downloading delay, because nodes passively respond to their neighbors' requests and the bandwidth is wasted (being idle much of the time). For example, in CodeTorrent it takes 200 seconds to download a 1 MB file in an urban scenario (Lee *et al.* 2006). If a node wants to receive new information continuously, it must send out requests frequently. The transmissions from multiple responders tend to collide with each other, leading to low efficiency in turn. Park *et al.* proposed a push-based content delivery scheme for emergency related video streaming using network coding (Park *et al.* 2010; 2006). However their "push" protocol design essentially reduces to controlled flooding, which tends to be inefficient.

In fact, with packet-level network coding (PLNC), it is difficult to achieve high downloading performance especially under lossy wireless links in VANETs, whether or not a push-based protocol design is adopted. The wireless medium in VANET has been shown to be lossy by empirical analysis (Ramachandran *et al.* 2007; Taliwal *et al.* 2004; Torrent-Moreno *et al.* 2006). In practice, network coding for a large file is usually done within each block of the file, namely a *generation* (Ahmed and Kanhere 2006; Lee *et al.* 2006; 2008). In order to maintain reasonable coding/decoding complexity while reducing the protocol overhead, the basic coding unit (coded piece) should be larger than a usual packet. During the transmission of such a coded piece, any error to the coding vector or message body will render the whole piece useless, leading to degraded downloading performance. This is part of the reason that we resort to SLNC (Katti *et al.* 2008), which has much better resiliency to transmission errors due to symbol-level diversity.

7.3.1.2 Transmission coordination in content downloading

Transmission coordination is an important issue for MCD in VANETs. Bad coordination could result in severe packet collisions that affect downloading performance. However, this issue has not been well addressed in previous works. In Park *et al.* (2006), a simple time-out mechanism is used for each vehicle to decide when to transmit a coded packet. However, this mechanism does not take into account vehicles' content reception status, which leads to a non-negligible chance of duplicate information. Packet collisions are severe when the network is dense.

Zhang *et al.* (2009a) studied this problem from a link-layer perspective, and proposed VC-MAC, a cooperative medium access control (MAC) protocol for gateway downloading scenarios in vehicular networks. In order to avoid possible interference among multiple transmissions and to maximize the "broadcast throughput," a heuristic relay selection algorithm with a backoff mechanism is proposed. However, the "broadcast throughput" is purely based on link quality, which is not content aware. The relay chosen by VC-MAC may have nothing innovative to transmit to its neighbors.

In CodeOn in this chapter, we explicitly consider the content usefulness of nodes for higher rate content downloading. A dynamic set of relay nodes, which are selected based on their content availability and usefulness, actively broadcast (push) useful contents to neighboring nodes and make medium access decisions based on both their content usefulness and local channel status.

7.3.1.3 Multichannel compatibility

There has been little consideration of the compatibility of content downloading with other channels. In Mak *et al.* (2009), the authors propose mechanisms to adjust the time share of the service channel to enhance the performance of content downloading while guaranteeing the quality of service (QoS) of safety messages. In this chapter, we consider the coexistence of a service channel with the control channel, with the difference that we design a better PCD protocol given a fixed time share of service channel. We also use the control channel for better content downloading.

7.3.1.4 Other related works

Zhao *et al.* (2007) proposed data pouring, a push-based data dissemination protocol for VANETs. They focus on broadcasting small data items to all vehicles inside an area, while we aim at disseminating large popular files. Zhao *et al.* (2008) also studied the problem of drive-thru access to roadside APs, and proposed a vehicle-to-vehicle relay strategy to extend the coverage of APs. Yang *et al.* (2009), proposed a push-based, reliable broadcast protocol for wireless mesh networks using network coding.

In addition, Fiore and Barcelo-Ordinas (2009) focused on cooperative downloading in urban VANETs. The Roadcast (Zhang *et al.* 2009b) is a popularity-aware content-sharing protocol in VANETs. These protocols are mainly suitable for applications where each vehicle may be interested in downloading different files, while here we consider the broadcast of popular content.

7.3.2 Streaming of multimedia content in VANETs

Currently, streaming services such as PPLive and PPStream are widely used on the Internet. In particular, network coding has been shown to be an effective technique that can improve the user experience of video streaming service for large-scale systems. For example, Wang and Li (2007) proposed R^2, a random push-based P2P scheme using network coding. Also, Zimu Liu (2010) deployed a NC-based on-demand streaming scheme in a large-scaled commercial system which showed the benefits of NC for multimedia streaming in a real P2P network. In wireless mesh networks, Seferoglu and Markopoulou (2009) proposed a video-aware opportunistic network coding scheme across different flows. However, all these schemes are for traditional wired or wireless networks and are not suitable for VANETs, due to VANETs' unique characteristics described previously.

For VANETs, Bonuccelli *et al.* (2007), Bucciol *et al.* (2005), Guo *et al.* (2005), Park *et al.* (2006), Qadri *et al.* (2009) and Soldo *et al.* (2008) proposed several schemes for supporting various kinds of streaming services, which can be divided into two categories:

1. Schemes focusing on the application layer. Bonuccelli *et al.* 2007 proposed a real-time video transmission scheme in vehicular networks. This scheme only considers unicast sessions and relies heavily on fast and reliable feedback from the receiver side. Bucciol *et al.* (2005) carried out a series of

experiments using two vehicles under different scenarios, which proved the feasibility of video streaming between moving vehicles. Qadri *et al.* (2009) showed that by adopting error resilience coding, state-of-the-art routing protocols can support multicast video streaming in city VANETs when the network is not dense. These works mainly showed the feasibility of video streaming in VANETs and have not considered more practical issues such as dealing with dynamically changing network density and minimizing bandwidth cost, which are taken into considered in this chapter.

2. Schemes focusing on network and MAC layer. Park *et al.* (2010) proposed NCDD for emergency related video streaming in VANETs using NC. In this scheme, the transmission of each vehicle is triggered by a timer set upon the reception of every new packet. Since neighbors' current reception status is not considered, the broadcast packets are not always useful for nodes' neighbors, which decreases the bandwidth efficiency. Due to lack of coordination between concurrent transmitting vehicles, the scheme also tends to suffer from severe collisions, especially under dense vehicular traffic.

Soldo *et al.* (2008) introduced SMUG, a TDMA-based scheme to support streaming media dissemination in city VANETs. A tree structure is established for broadcasting streaming video content. However, it is hard to maintain a stable and up-to-date communication structure for dynamic VANETs, thus stable streaming rate is difficult to achieve. Guo *et al.* (2005) proposed *V3*, a live video architecture for VANET, where directed broadcast is adopted for remote video request scenarios, which are different from the application in this chapter.

7.4   Background on symbol-level network coding

In this section, we first describe the symbol-level network coding technique. Then, we give a motivating example to show the potential advantage of exploiting symbol-level diversity in MCD in VANETs.

7.4.1   A brief review of SLNC

Symbol-level Network Coding was recently introduced by Katti *et al.* (2008) to improve the unicast throughput in wireless mesh networks. It arises from the observation that in wireless networks, even if a packet is received erroneously, some small groups of bits ("symbols") within that packet are likely to be received correctly. SLNC gathers these correctly received (i.e., "clean") symbols aggressively, and performs network coding on the granularity of symbols. In contrast to PLNC, SLNC gains from both symbol-level diversity and network coding. In addition, since more bit errors are tolerated than PLNC, SLNC can also gain higher throughput by encouraging more aggressive concurrent transmissions.

In general, SLNC works as follows. A symbol is defined as a group of consecutive bits in a packet, which may correspond to multiple PHY symbols of a modulation scheme. Assume the source has K packets to send, each of them expressed as a vector with elements from a Galois field $\mathbb{F}_{2^q}$. The jth symbol $\mathbb{s}'_j$ in a coded packet at the source is a random linear combination of the jth symbol in all K source packets:

$$\mathbb{s}'_j = \sum_{i=1}^{K} v_i \mathbb{s}_{ji}. \tag{7.1}$$

where $\mathbb{s}_{ji}$ is the jth symbol (at jth position) in the ith original packet, coefficient v_i is randomly chosen from $\mathbb{F}_{2^q}$, and $\mathbb{v} = (v_1, \ldots, v_K)$ is the coding vector of the coded packet, which is also the coding vector for each symbol. Each receiver node v maintains a decoding matrix for every symbol position. A newly received coded symbol for position j is called *innovative* to v, if that symbol increases the *rank* of the decoding matrix of the jth symbol position, referred to as *symbol rank*. Only innovative clean symbols are buffered.

Each coded packet transmitted by a relay node consists of random linear combinations of buffered clean symbols. For a source, every symbol in a packet is clean and shares the same coding vector. However, at a relay node, coding vectors may be different across symbols. For a coded packet to be sent by relay u, the jth coded symbol is expressed as

$$\mathbb{s}''_j = \sum_{i=1}^{R} v'_i \mathbb{s}'_{ji} = \sum_{i=1}^{R} \left(v'_i \sum_{l=1}^{K} v_{li} \mathbb{s}_{jl} \right) = \sum_{l=1}^{K} \left(\sum_{i=1}^{R} v'_i v_{li} \right) \mathbb{s}_{jl}, \tag{7.2}$$

where R is the number of buffered clean symbols at position j, $\mathbb{s}'_{ji}$ is the ith buffered clean symbol (row) at position j (column), and $\mathbb{v}_i = \{v_{1i}, \ldots, v_{Ki}\}$ is the coding vector for that symbol. $\mathbb{s}_{jl}$ is the jth symbol of the lth source packet. From Equation (7.2), $\mathbb{s}''_j$ is still a random linear combination of source symbols and its new coding vectors are $\mathbb{v}' = (\sum_{i=1}^{R} v'_i v_{1i}, \ldots, \sum_{i=1}^{R} v'_i v_{Ki})$.

In the extreme case, every symbol's coding vector is different and needs to be sent along with a packet, which incurs high overheads. To minimize these overheads the optimized run-length coding method can be adopted (Katti *et al.* 2008), where consecutive clean symbols are combined into a "run".

We use a three-node toy example to demonstrate how SLNC works (Figure 7.4 and Figure 7.5). The corresponding topology is shown in Figure 7.4. Assume source

Figure 7.4   The topology for the example in Figure 7.5. Left: numbers on the edges (links) show the symbol error probabilities; right: corresponding packet error probabilities. Reproduced by permission of © IEEE 2011.

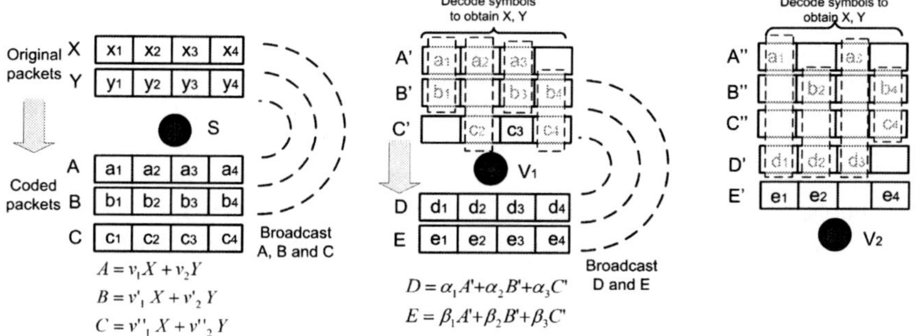

Figure 7.5 Symbol level network coding in VANET content distribution. S: source node; V_1 and V_2: downloading vehicles and relays. Reproduced by permission of © IEEE 2011.

S has two original packets X and Y to broadcast. Assume a simple scheduling: S broadcasts coded packets until V_1 can decode the original packets, and then V_1 broadcasts until V_2 decodes all original packets.

Suppose S generates and broadcasts three coded packets A, B and C, each of them divided into four symbols. Let the symbol error probability from S to V_1 be $P_{se}(S, V_1) = \frac{1}{4}$ and it happens that each packet received by V_1 contains an erroneous symbol (Figure 7.5). Luckily, at least two clean symbols are received for each symbol position. Since any two coding vectors among v, v', v'' of A, B and C are independent[2], V_1 can decode X and Y by solving four linear equations. When V_1 broadcasts two packets (say, D and E), it generates two new coded symbols at each position, and packs the eight coded symbols into D and E. Each new coded symbol is also a random linear combination of original symbols. Thus, V_2 can recover all original symbols after collecting two innovative coded symbols at each position, which may come from both S and V_1.

Next, we show the potential performance gain of SLNC compared with PLNC from two aspects: higher throughput and higher spacial reuse. We use both theoretical analysis and simulation to confirm our intuitions.

7.4.2   Motivation: why VANET content distribution benefits from SLNC

7.4.2.1   Symbol-level diversity that leads to higher throughput

The key insight of SLNC is that, for each symbol position, every correctly received coded symbol is equally useful for decoding, or it does not matter which symbol is received. For PLNC, the reception granularity is a whole packet. The symbol error rate will be much less than the packet error rate, so it is not hard to imagine that SLNC will take less transmissions to collect the information needed for decoding the same amount of content.

---

[2] This happens with high probability when the size of $\mathbb{F}_{2^q}$ is large.

To confirm the above intuition, we reuse the above toy example. We would like to compute the expected number of packets ($\mathbb{E}[Z]$) transmitted by S for node V_1 to decode using the simple example. Without loss of generality, we assume S has K source packets to broadcast, and each packet is divided into M symbols. We are interested in when V_1 is able to decode all the source symbols from S (receive at least M correct and innovative symbols in all the positions). Assume all symbols in the same position are independently received according to an i.i.d. Bernoulli process[3], where the probability of receiving a symbol correctly in one trial is $1 - P_{se}$, ($P_{se} = P_{se}(S, V_1)$). Let Z_i denote the number of packets sent (trials) for V_1 to receive exactly K correct symbols in the ith position. Then

$$P(Z_i = k) = \begin{cases} \binom{k-1}{K-1} P_{se}^{k-K} (1 - P_{se})^K & \text{if } k \geq K \\ 0 & 0 \leq k < K \end{cases} \tag{7.3}$$

and we have $P(Z_i \leq k) = \sum_{m=1}^{k} P(Z_i = m)$. Define r.v. Z as the smallest number of packets sent for V_1 to receive at least K correct symbols at all positions, then

$$Z = \max\{Z_i\}, i = 1, \ldots, M$$

We have

$$P(Z \leq k) = [P(Z_i \leq k)]^M \tag{7.4}$$

Therefore, the expected number of packets transmitted by S for V_i to decode is:

$$\mathbb{E}[Z] = \sum_{k=0}^{\infty} P(Z > k) = \sum_{k=0}^{\infty} [1 - P(Z \leq k)] \tag{7.5}$$

Plugging in the parameters in the example ($K = 2, M = 4, P_{se} = 1/4$), we obtain $\mathbb{E}[Z] = 3.67$. That is, 3.67 coded packets should be sent by S on average for V_1 to decode X and Y.

Next we compare SLNC to using PLNC for the same case. We compute the expected number of packets $\mathbb{E}[Z']$ sent by S for V_1 to receive K source packets. Since PLNC discards a packet with any erroneous symbol in it, the error probability from the packet level could be much larger than that of symbol level. For simplicity, we assume independent symbol error in one packet[4], so

$$P_{pe} = 1 - (1 - P_{se})^M \tag{7.6}$$

---

[3] This assumption is valid in VANETs. The channel coherence time: $T_c \approx \frac{0.42}{\Delta f}$, where $\Delta f = \frac{v f_0}{c}$ is the doppler spread. With average relative speed $v = 30$ m/s, central frequency $f_0 = 5.9$ GHz, $T_c = 0.72$ ms. Using the data rate 12 mbps in IEEE 802.11p, the time to send a 1 KB packet is $T_{tx} = 0.68$ ms. As $T_c \approx T_{tx}$, consecutive received packets can be regarded as independent, so are the symbols in the same positions.

[4] There exist error correction coding (ECC) techniques to enhance the error resiliency of packet transmission but they do not change the nature of the following derivation as they are still limited in error-correcting capabilities. On the other hand, ECC can also be added to SLNC (Katti *et al.* 2008).

The resulting error rates are shown in the right of Figure 7.4 where $P_{pe}(S, V_1) = 0.68$. Assuming independent packet reception, $\mathbb{E}[Z'] = K/(1 - P_{pe})$. For the simple example, S must transmit $\frac{2}{1 - P_{pe}(S, V_1)} = 6.26$ packets on average for V_1 to decode. Thus, the number of transmissions (proportional to downloading delay) of V_1 has been reduced by $\frac{6.26 - 3.67}{6.26} = 41\%$ due to the use of SLNC.

For node V_2, although it can overhear useful information from both S and V_1, as we can see from Figure 7.4, $P_{pe}(S, V_2) = 0.94 \approx 1$, while $P_{se}(S, V_2) = 1/2$. In this case, the (S, V_2) link can almost be neglected for PLNC, and SLNC is expected to achieve higher gain for V_2 than V_1.

From the above, the advantage of using SLNC than PLNC is evident for content distribution in VANET, i.e., it leads to higher downloading rate and incurs fewer transmissions[5].

7.4.2.2 Higher spacial reusability that leads to higher throughput

In multihop wireless communications, the distances between concurrent transmitting nodes play an important role in achievable network capacity. Another property that has been observed by Katti *et al.* (2008) is the higher spacial reusability offered by SLNC. However, Katti *et al.* (2008) only investigated this issue in a static wireless mesh network setting, while we further study this property in the mobile VANETs.

Here the fundamental question is, with SLNC, what is the optimal distance between two nearby relay nodes so that the concurrent transmissions of all relays achieve highest average downloading rate? That is, what is the maximum spacial reusability that can be achieved? Intuitively, if the relays are far away from each other, there is no interference but the space is not fully utilized; but if they are too close, severe interference will in turn decrease the downloading rate. First we define a quantity that reflects the average downloading rate in the network (assume all contents are useful for simplicity):

Definition 7.1 (*Average Symbol Reception Probability*) ($\chi(v_1, v_2, \ldots, v_n)$) *For n relay nodes* $v_1, v_2, \ldots, v_n$ *in the network, the average probability that every vehicle receives one symbol correctly from any of them during unit time (e.g., the period of one symbol's transmission).*

A simple case. To derive χ, let us first consider a toy scenario where there are only two relays v_1, v_2 in the network. We are interested in the relationship of average symbol reception probability with the inter-relay distance, and when concurrent transmission can gain advantage over nonconcurrent transmission (i.e., two relays transmit separately and alternatively). The following characterizes the condition when concurrent transmission is better than separate transmission:

$$\alpha_c = \chi(v_1, v_2) > [\chi(v_1) + \chi(v_2)]/2, \tag{7.7}$$

---

[5] Note that, in reality, the symbol errors may be correlated, which is related to the channel coherence time T_c. Then the actual difference between P_{pe} and P_{se} is smaller. Therefore, the gain we derived can be regarded as an upper bound.

where α_c is denoted as "concurrency gain." χ can be derived from symbol error probability (P_{se}) at each receiving node. However, it is hard to obtain the closed-form solution of P_{se} under concurrent transmissions (see Li *et al.* 2010). Therefore, we approximate χ by the *average symbol reception ratio*:

$$\chi \approx \frac{\text{Total \# of symbols correctly received by all nodes in unit time}}{\text{Total number of receivers in the network}} \quad (7.8)$$

which is obtained through Monte Carlo simulations. Given relays v_1 and v_2, for each of their neighbors w, the received signal to noise ratios (SNRs) are randomly sampled from Nakagami fading model (Li *et al.* 2010), based on which the signal-to-interference-and-noise ratio (SINR) γ is computed. Successful symbol reception has the probability $1 - P_{se|\gamma}$, where $P_{se|\gamma}$ is the symbol error rate given SINR γ. To simulate packet capture effect in reality[6], we let w receive the clean symbols in a packet from v_1 if its average SINR $\overline{\gamma_1} > \overline{\gamma_2}$, and vice versa. Note that this estimation is done in the worst case, i.e. the relays transmit simultaneously so that every symbol is possibly subject to interference.

For convenience of illustration, we define the following ranges under free space propagation model (Friis): 1. "energy detection range" (ER) (or carrier sensing range) for a transmitter, within which nearby nodes can detect its signal energy. We have $ER = \sqrt{\frac{T_p G}{Th_{ER}}}$, where Th_{ER} is the carrier sensing threshold, T_p is the transmission power and G is the antenna gain. 2. The "data communication range" (CR), in which nearby nodes can receive a data packet correctly. $CR = \sqrt{\frac{T_p G}{Th_{CR}}}$, where Th_{CR} is the data reception threshold. Normally $ER \geq CR$. These ranges imply statistical transition points across which nodes have different reception results.

We generate ten random topologies for VANET on a highway with traffic density 100/km, fix relay v_1 and change v_2's position. Each of v_1 and v_2 transmits ten packets (each having 30 symbols). The number of received symbols is recorded for every node in the network. We also compare with PLNC under the same setting. $\chi(v_1, v_2)$ under the concurrent case is compared with $[\chi(v_1) + \chi(v_1)]/2$ under the nonconcurrent case, against a changing interrelay distance.

The results are given in Figure 7.6. It can be seen that, for SLNC when $d(v_1, v_2) = 2250$ m, the concurrency gain $\alpha_c \approx 2$; when $d(v_1, v_2)$ decreases, α_c monotonically decreases until it becomes smaller than 1, at a small cross-distance d_c. While for PLNC, the average packet reception ratio is much smaller, and its cross distance is larger, which shows PLNC's inferior tolerance with concurrent transmission than SLNC.

The n-relay case. The results of multiple $(n > 2)$ relay nodes transmitting concurrently are based on that of the simple case. Without loss of generality, we assume that vehicles are uniformly distributed and are not too sparse so that neighbor conditions are similar. And n relays are assumed to lie on a straight highway (of length $\mathscr{L}$) with equal interdistance d.

---

[6] We assume no node can receive more than one symbol or packet from different transmitters at the same time.

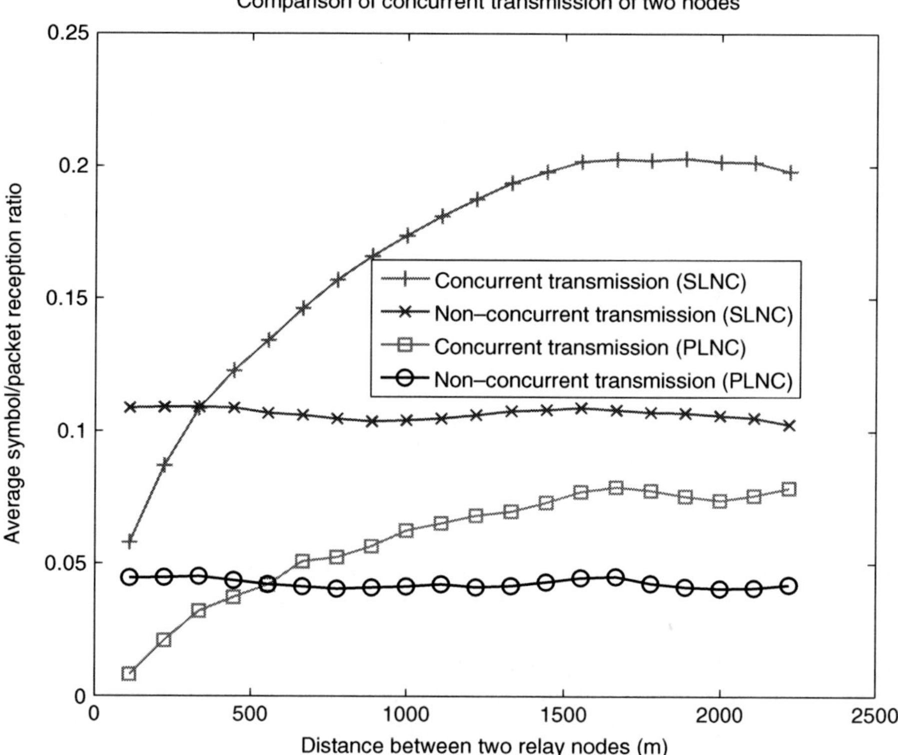

Figure 7.6   The effect of concurrent transmission of two relays. $CR = 250\,m$, $ER = 700\,m$, data rate: $12\,mbps$.

Now we derive the relationship of $\chi(v_1, v_2, \ldots, v_n)$ with $\chi(v_1, v_2)$. Let $d_c = d(v_1, v_2)$ when $\alpha_c = 1$ in the simple case. This point can be interpreted equivalently as half of the nodes around each relay are heavily subject to interference while the rest are not. Assuming uniform vehicle distribution, since d_c is larger than CR (Figure 7.6, most of these interfered nodes locate in the region between v_1 and v_2. We are interested in $d(v_1, v_2) > d_c$, when approximately only the nodes within that region experience a decrease in their symbol reception probabilities. Therefore, a third relay v_3 adds little interference to the region between v_1 and v_2, which is illustrated in Figure 7.7.

By assumption, $\chi(v_i, v_{i+1}) \approx \chi(v_j, v_{j+1})$, $\forall i, j$, so

$$\chi(v_1, v_2, \ldots, v_n) = \chi(v_1, v_2)[1/2 + (n-1)(\alpha_c - 1) + 1/2]$$

$$= \chi(v_1, v_2)[(n-1)(\alpha_c - 1) + 1], \tag{7.9}$$

where $n = \lfloor \frac{\mathscr{L}}{d} \rfloor$, and α_c is a function of d and CR. For each CR, α_c is obtained by simulation and curve fitting. Since α_c is increasing w.r.t. d, $\chi(v_1, v_2, \ldots, v_n)$ has a maximal point.

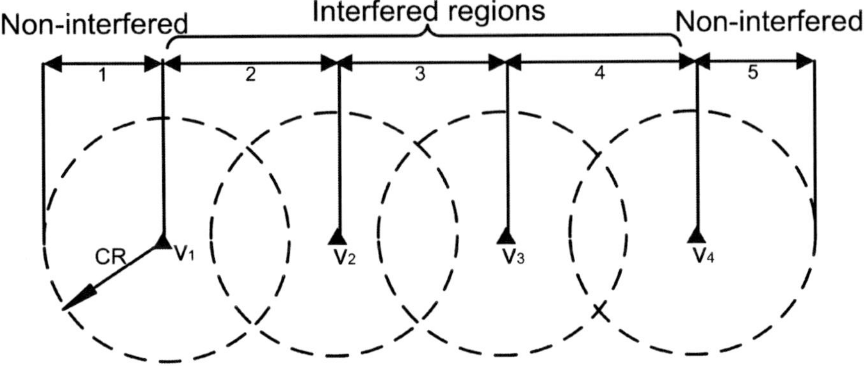

Figure 7.7   Conceptual illustration of the n-relay concurrent transmission and cal-culation of average symbol reception probability.

To see the throughput gain of SLNC versus PLNC, Figure 7.8(a) shows the average symbol reception probability $\chi(v_1, v_2, \ldots, v_n)$ as a function of D. We can see that with SLNC, for the given example communication range $CR = 277m$, the optimal distance $d_{opt} \approx 900$ m when Pr_{avg} is maximized, and Pr_{avg} is above 0.5. While with PLNC, $d_{opt} \approx 1200$ m, under which $Pr_{avg} < 0.2$. This observation confirms that SLNC tolerates transmission errors better than PLNC. In addition, SLNC allows better spacial reusability and encourages more aggressive concurrent transmissions, thus achieving higher bandwidth efficiency. We find similar conclusions under other communication range CR values, as shown in Figure 7.8(b).

The observations from the above are: 1. the optimal d_{opt} when using SLNC is always smaller than that when using PLNC, and 2. it is interesting to see that d_{opt} for SLNC is very close to energy detection range ER (or carrier sensing range). As we know, the hidden-terminal problem is a major cause of packet collisions for broadcast in a multihop wireless network (Ni *et al.* 1999; Talucci *et al.* 1997; Ye *et al.* 2002). The fundamental reason for the existence of the hidden-terminal problem is the fact that, for wireless packet transmission, the interference range is typically close to or larger than the carrier sensing range. Therefore, it is hard for a localized protocol, especially broadcast/multicast protocols, to eliminate hidden terminals based on local carrier sensing alone. With SLNC, we observed that the concurrent transmission can be successful when nodes are only ER distance apart.

The implication of this observation is significant–with SLNC, we can base the channel access decisions largely on carrier sensing, which is not the case for packet based transmissions that must consider hidden terminals. This property of SLNC was also noticed by Brodsky and Morris (2009) but was not thoroughly investigated. Its great potential to alter the content distribution protocol design, simplify the MAC protocol design and at the same time achieve higher throughput are explored by both of our proposed schemes in this chapter.

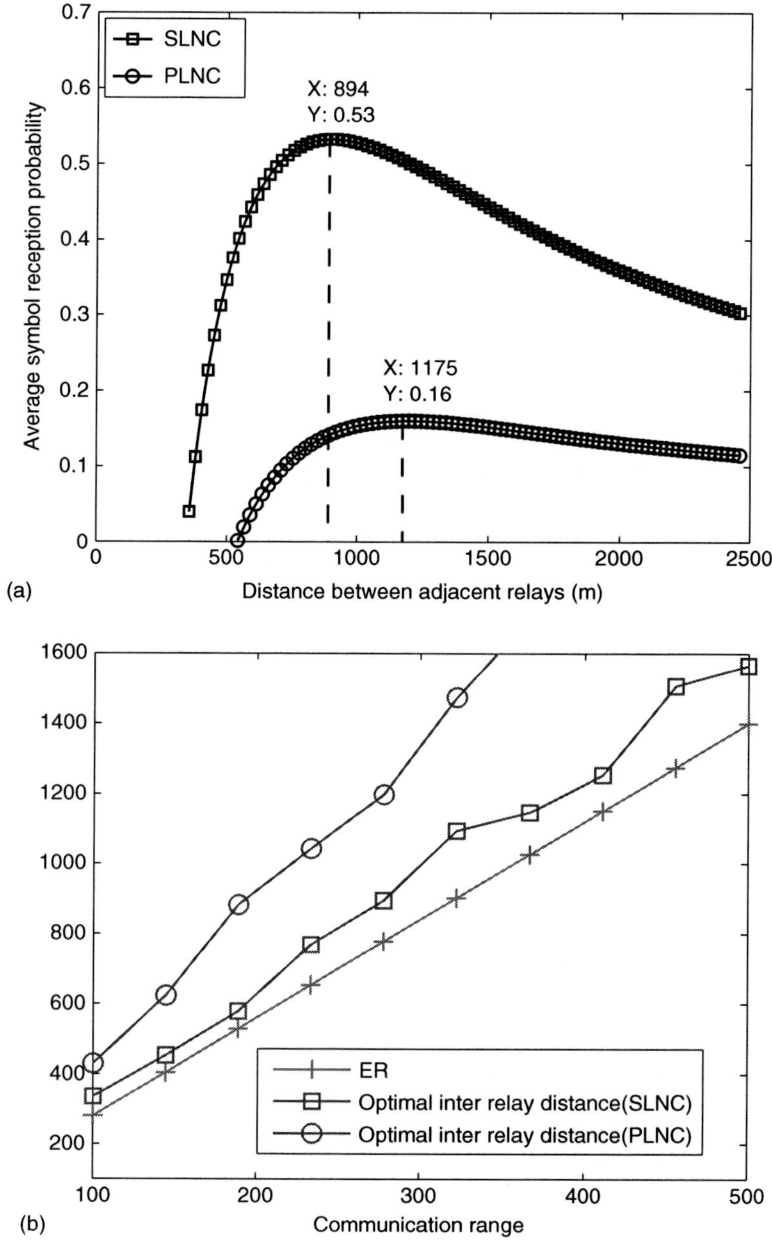

Figure 7.8   Comparison of spacial reusability of SLNC and PLNC. (a) The expected symbol/packet reception rate, for CR = 277 m, ER = 700 m, Data rate = 12 mbps. Nakagami fading model. (b) The optimal distance between two adjacent relays under various data communication ranges. Reproduced by permission of © 2010 IEEE.

7.5 CodeOn: a cooperative popular content broadcast scheme for VANETs based on SLNC

In this section, we study the cooperative broadcast of popular contents in VANETs. Next we introduce specific design requirements of such MCD service. After that, the main design will be given, followed by detailed simulation and performance evaluation results.

7.5.1 Design objectives

For any popular content distributed by the MCD service, the primary objective is to achieve low average downloading delay, which is equivalent to high average downloading rate. For each vehicle in an AoI, its *downloading delay* is defined as the elapsed time from downloading start to 100% completion. Meanwhile, it is desirable to achieve a high degree of fairness, i.e., the variation of downloading delays among different vehicles should be small. Finally, high-rate content distribution cannot come at the cost of incurring too much protocol overhead and data traffic, otherwise the MCD service would be less compatible with other possible services in the service channel. Thus it is also important to maintain high protocol efficiency.

We first define the main notations used in this chapter in Table. 7.1.

Table 7.1   Frequently used notations

Notation	Definition
F	The file to be distributed
N	Data packet size (bytes)
L	File length (number of generations)
K	Generation size (number of pieces)
J	Piece size (bytes)
M	Number of symbols in a packet
G_i	Generation i
$\mathbb{F}_{2^q}$	The Galois field used in network coding
$U(v)$	The utility of a node v
$\mathcal{N}(u)$	The neighbor set of node u
$\bar{r}_{v,i}$	Average symbol rank of G_i in vehicle v
γ	Average received SNR or SINR for a symbol

7.5.2 Design overview

CodeOn is a push-based cooperative MCD protocol, where a large file F is actively distributed from the APs to the vehicles inside the AoI through the help of a dynamic set of relay nodes. Each AP is a source for F, and F is divided into equal-sized generations (chunks), and the SLNC is performed within each

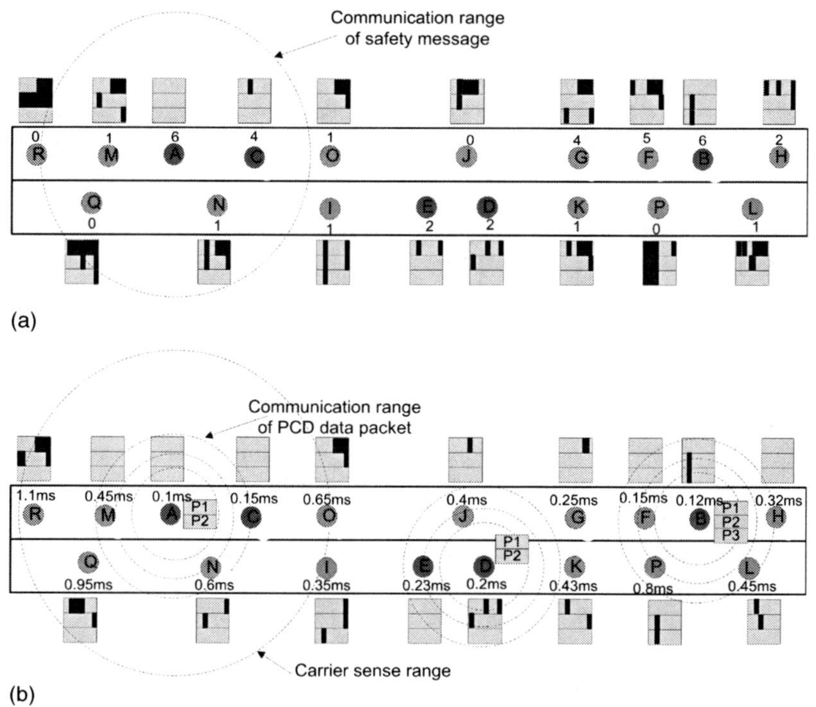

Figure 7.9   Overview of cooperative MCD in CodeOn. (a) Exchange of neighbor information and utility calculation based on nodes' reception status. All nodes' reception status are depicted. Black parts in a piece indicate corrupted symbols in a nodes' buffer. (b) Transmission coordination among potential relays, based on both node priority and carrier sense. Backoff delays are inversely related to nodes' utilities (nodes A-F have the least delays, but only A, B, D become relays). Reproduced by permission of © 2011 IEEE.

generation. In Figure 7.9, we illustrate the general process of content distribution in CodeOn, assuming F has only one generation consisting of three pieces.

Each AP/source broadcasts the source file to vehicles in its range based on vehicles' reception status, which is not shown in Figure 7.9. Outside the ranges of APs, vehicles distribute the file cooperatively by choosing a set of relay nodes distributively. This is the core to CodeOn, which consists of three steps.

1. Efficient exchange of neighbor information. This is done in each control time slot, where every vehicle broadcasts a safety message that piggybacks a sketch of its content reception status, which will be used as an implicit content request for step. In this way, zero overheads are incurred in the service time slots. To limit the impact of piggyback overheads on control time slots we will introduce a fuzzy representation of nodes' reception status later.

2. Node utility calculation. This is the first step of distributed relay selection. In the beginning of each service time slot, every node computes its own

utility based on neighbors' reception status information collected from step 1. The utility reflects each node's priority in relay selection, i.e., the total amount of useful content that this node can provide to all of its neighbors. Under such a priority assignment, the usefulness of each relay's transmission will be maximized, which enhances both the downloading rate and protocol efficiency. The utility of every node is shown in Figure 7.9 (a).

3. Transmission coordination among potential relays. As the last step of relay selection, we need to determine which nodes should actually access the channel, based on both node priority and the channel status. Each node computes a backoff delay that is inversely related to its utility, and upon the expiration of the delay it will sense the channel. If it cannot detect signal energy, it will broadcast coded contents without delay. Otherwise, it remains silent throughout the time slot. This process is captured by Figure 7.9 (b). Thanks to SLNC's better error tolerance, this aggressive way of channel access, although simple, will be shown to achieve close to maximum overall downloading rate. During the relays' transmissions, the vehicles exploit opportunistic listening and buffer overheard useful symbols. Therefore, the content reception status of vehicles may change from time slot to time slot, which may yield a different set of relay nodes at each time slot.

7.5.3 Network coding method

We now describe the way that SLNC actually operates in CodeOn. Assume F with size $|F|$ is divided into L generations $G_1, G_2, \ldots, G_L$, where each generation contains K pieces. A piece has size J and contains $\lceil J/N \rceil$ packets. Then, $|F| = L \cdot K \cdot J$. In order to reduce the overhead brought by SLNC, we adopt "*piece-division, run-length SLNC*."

The reasons are twofold. On the one hand, if a generation is divided into packets (packet-division), in order to keep low computational overheads we must use relatively small K (the computation complexity of decoding is usually $O(K^3)$), thus a large number of generations is required for large F. This reduces the gain of NC due to the "coupon collector's problem" (Lee *et al.* 2008), and increases the communication overhead for exchanging the content availability. On the other hand, using multipacket pieces (piece-division), K can be maintained at a reasonable value by scaling the piece length linearly with file size. However, the number of symbols in a piece ($\frac{J \cdot M}{N}$) increases with the piece length. In the extreme case that every symbol in a piece has a different coding vector, the communication overhead is at least $\frac{J \cdot M \cdot K \cdot q}{N}$ bits, which equals to 10 KB if $J = 20$ KB, $N = 1$ KB, $K = 32$, $M = 32$, $q = 8$. This is clearly unacceptable. Fortunately, run-length coding method (Katti *et al.* 2008) can be used to reduce the communication overhead of SLNC, in which one coding vector is used for each sequence of consecutive clean symbols (*run*). Dynamic programming is used to choose appropriate combination of runs to minimize the overhead (Katti *et al.* 2008). Therefore, in CodeOn, we combine run-length SLNC with the piece division to achieve higher network coding gain and reduce the communication overhead, which we call *piece-division run-length SLNC*. When a coded piece is transmitted, it is separated into several packets; only the header of the first packet contains the coding vectors of runs that composing the piece, while subsequent packets only have normal small headers. Thus, a piece can be regarded as a "big packet."

Compared with PLNC, the gain from symbol-level diversity can be easily seen from the analysis in Section 7.4. Meanwhile, the overhead of our method is always smaller than run-length SLNC combined with packet division. Generally, the number of coding vectors in a piece equals to the number of runs. However, using packet division a run may be fragmented into more than one runs, which needs more coding vectors in total. In the worst case, each symbol is a run and the overheads are equal. This is illustrated in Figure 7.10. In reality, the symbols errors are often bursty (due to packet collisions) so the number of runs is usually much smaller compared with the number of symbols. For example, if there are 20 runs in a 20 KB piece the overhead is about 640 B, which is 3.2% of piece size.

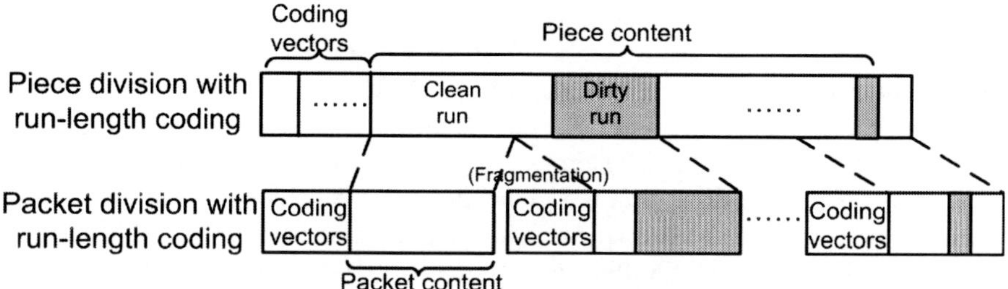

Figure 7.10   Comparison between the overhead of piece division and packet division, when both uses run-length SLNC. Reproduced by permission of © 2011 IEEE.

In order to balance the gain and overhead of SLNC in CodeOn, we fix the number of pieces in a generation (K) and the number of generations (L) (e.g. 32 and 50, respectively). Although the piece size J scales linearly with the file size because SLNC tolerates symbol errors, the size of a piece has small impact on the protocol performance.

7.5.4   Efficient exchange of content reception status

An important piece of information exchanged in CodeOn is every node's *content reception status* (i.e., how much content is downloaded for each generation), which is essential to enabling optimized, distributed transmission decisions. It could be obtained by sending gossip messages in each service time slot but this consumes a large portion of a service time slot. In CodeOn, we choose to piggyback the reception status in safety messages, thus adding zero overhead in the service channel.

However, for SLNC, representing the exact reception status of each generation will incur large overheads. The decoding matrix can be represented by a single null-space vector (Lee *et al.* 2006). However, the size of the reception status information adds up to $\frac{L \cdot J \cdot M \cdot K \cdot q}{N}$ bits, where Kq is the maximum size of one null-space vector. For $L = 50, J/N = 20, M = 32, K = 32, q = 8$, this amounts to 1 MB which is too large.

Therefore, in CodeOn we propose a fuzzy *average rank* method to represent the reception status in an efficient way. An important property of network coding is that the rank of the decoding matrix determines the amount of received information. For two nodes u and v with symbol ranks $r_{u,i,j}$ and $r_{v,i,j}$ for position j in G_i, respectively, if $r_{u,i,j} > r_{v,i,j}$, then a recoded symbol $\mathbb{s}'_j$ sent from u is innovative to v with high probability (Ahlswede *et al.* 2000). Otherwise, this does not hold[7]. We can therefore substitute each null-space vector with a rank, which has $log_2 K$ bits. For a generation G_i received by node u, there are many symbol positions with different rank values. But because the size of a piece is relatively small (e.g., $J = 20$ KB) compared to what can be transmitted in a 50 ms slot using DSRC (55 KB when data rate is 11 mbps), the ranks of various symbol positions are expected to increase at similar rates thus are similar to each other.

Therefore, we use the average rank $\lfloor \bar{r}_i \rfloor$ across all symbol positions in G_i to represent how much information is received for G_i. It is rounded to an integer, because it is more meaningful to interpret the average rank as how many "pieces" are received. It does not make much difference when the variation of $\bar{r}_i$ is smaller than 1. The range of the rank is in $[0, K]$; if $\lfloor \bar{r}_{u,i} \rfloor < K$, this means "some information in G_i is received"; and $\lfloor \bar{r}_{u,i} \rfloor = K$ means "G_i is received completely." Therefore, the total overhead becomes $L \cdot (log_2 K)$ bits, which equals 31B when $L = 50, K = 32$. Note that, this is independent of the piece size and also the file size. Now, the overhead takes an acceptable percentage ($\approx 10\%$) of the typical size of a safety message (300B) and is small enough to be piggybacked without affecting the QoS of safety applications (Xu *et al.* 2004). The average rank representation is illustrated in Figure 7.11.

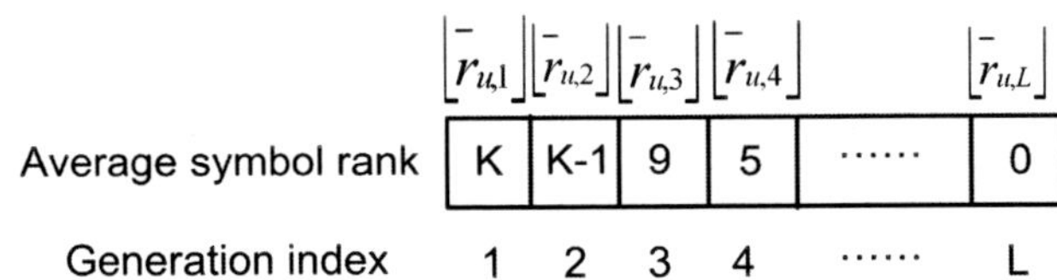

Figure 7.11   The average rank representation of a file's reception status at node u. Reproduced by permission of © 2011 IEEE.

7.5.5   Distributed relay selection in cooperative PCD

Once vehicles are out of the range of an AP, they begin distributing the content cooperatively through the VANET. Due to the mobile nature of the VANET, the very notion of "cooperative" is captured in that vehicles distributively agree on a set of relay nodes, based only on local information.

---

[7] The property was original proved under random linear packet level NC, assuming $|\mathbb{F}_{2^q}|$ is large. The same applies to SLNC, which is also based on random linear coding.

7.5.5.1   Node utility calculation

In order to determine a set of relay nodes that can bring the largest useful amount of content to the others, each node needs to calculate its own "utility" based on neighbors' content reception status collected from the safety messages in the control channel. The *utility of a generation* at node u is defined as:

$$U(G_i, u) = \sum_{v \in \mathcal{N}(u)} Step(\lfloor \bar{r}_{u,i} \rfloor - \lfloor \bar{r}_{v,i} \rfloor) \qquad (7.10)$$

where $Step(x) = x$, if $x > 0$, otherwise, $Step(x) = 0$. This quantity measures how much innovative information G_i of node u can provide to its neighbors in total.

The *utility* $U(v)$ of node v is defined as the maximum value among all generations' utilities of v. This estimates the maximum additional amount of innovative information v can provide to all neighbors, and reflects v's priority in accessing the wireless medium. We do not look at the aggregate utility of multiple generations because to transmit many generations takes a long time while the VANET topology could change dramatically.

7.5.5.2   Transmission coordination

After nodes' priorities are determined, only a subset of the high-priority nodes (relays) will become the ones who actually broadcast their contents, in order to achieve high downloading rate and prevent from severe interference. Those relays are decided via a contention process, in a local and opportunistic way. In particular, the vehicles with the highest priorities in their locality should access the channel first, and suppress the others to avoid unnecessary packet collisions.

To this end, at the beginning of each service time slot, each vehicle v sets a backoff delay Δt which is inversely proportional to its utility before it makes the, channel access decision. When the timer expires, v senses the channel; if it is clear v will broadcast a short control message that is sent immediately by the MAC layer, even without additional random backoff in 802.11. Note that an AP always has the highest utility, so it will be a relay every time if there are vehicles still in need of the file in its local range.

Backoff delay function. A straightforward one is as follows:

$$\Delta t(v) = \left(1 - \frac{U(v)}{K \cdot |\mathcal{N}(v)|}\right) \cdot \Delta t_{max} \qquad (7.11)$$

where parameter Δt_{max} is the maximum allowable backoff delay (e.g., 2 ms). However, Equation (7.11) suffers from a major problem. That is, each node v has different neighborhood and $\mathcal{N}(v)$. If v merely has one neighbor but its generation utility for G_i is K, it will have the highest priority and $\Delta t(v) = 0$. However, compared with another node w that has ten neighbors and utility $5K$, v is obviously not as beneficial to the whole network as w. Ideally, the $|\mathcal{N}(v)|$ should be a maximum possible neighborhood size ($|\mathcal{N}|_{max}$) and be the same for all vehicles, so that they have a common basis of priority comparison. However, setting it to be a fixed value is undesirable because the vehicle density will change.

Therefore, we estimate the maximum local neighborhood size. To do so, each node broadcasts its neighborhood size to others, and propagates its own estimation about the maximum neighborhood size. After several rounds, all nodes can obtain the same $|\mathcal{N}|_{max}$. Although the VANET topology may change every second so that $|\mathcal{N}|_{max}$ varies over that time, we actually need not to maintain the same $|\mathcal{N}|_{max}$ for all nodes in the network. Rather, it is sufficient for vehicles in a local one-hop range to agree on the same estimated $\widehat{|\mathcal{N}|_{max}}$, while the local propagation requires only very few rounds to converge. To achieve this, each vehicle will attach its local estimate of $\widehat{|\mathcal{N}|_{max}}$ in the safety message, and update it in a way similar to distance updates in distance vector routing.

In addition, to resolve ties, a random jitter is added to the backoff delay of each vehicle. Thus, in CodeOn, each vehicle sets its backoff delay according to the following:

$$\Delta t(v) = \left(1 - \frac{U(v)}{K \cdot \widehat{|\mathcal{N}|_{max}}}\right) \cdot \Delta t_{max} + Rand(0, T_J) \qquad (7.12)$$

where T_J is the maximum jitter.

Discussion of parameter selection. First, Δt_{max} must be large enough to distinguish two vehicles with adjacent utility rankings. For a common neighbor v_c of two vehicles v_1 and v_2, the minimum difference between $U(v_1)$ and $U(v_2)$ is 1. Therefore, the minimum difference between v_1 and v_2's backoff delays is $\min\{|\Delta t(v_1) - \Delta t(v_2)|\} = \frac{1}{K|\mathcal{N}|_{max}} \cdot \Delta t_{max}$, which should be larger than the signal propagation delay. When their distance $d(v_1, v_2) = 300$ m the propagation delay is $\frac{300}{3 \times 10^8} = 1\,\mu s$. Therefore, we can choose $\Delta t_{max} > 2$ ms, i.e., when $\widehat{|\mathcal{N}|_{max}} = 50$, $K = 32$, $\min\{|\Delta t(v_1) - \Delta t(v_2)|\} > 1.2\,\mu s$. On the other hand, Δt_{max} shall not be too large since it will waste bandwidth. For $\Delta t_{max} = 2$ ms, if transmission of one generation spans 10 service time slots (500 ms), the percentage of wasted time can be as low as $2/500 = 0.4\%$.

Second, T_J should be both large enough to distinguish two contending nodes v_1 and v_2 with the same utility, and small enough to preserve the priorities between nodes with different utilities. Assume all the contending nodes have the same neighbor set. Node utility is an integer, for node v_1, so the utility of the node v_3 with priority next to v_1 is at most $U(v_1) - |\mathcal{N}(v_1)|$ (since $\lfloor \bar{r}(v_3, i) \rfloor = \lfloor \bar{r}(v_1, j) \rfloor - 1$ for some G_i, G_j and every neighbor is counted once). Thus, the utility difference is at least $|\mathcal{N}(v_1)|$. Therefore, we have $T_J \approx \frac{\Delta t_{max}}{K}$ (e.g. 0.1 ms). Note that we do not consider $U(v_1) - U(v_3) \ll |\mathcal{N}(v_1)|$ since this is rare in reality, i.e., contending nodes always share a large proportion of neighbors.

7.5.5.3 The merit of carrier sense under SLNC

We have used *carrier sense* in the contention process for transmission opportunities by potential relay nodes. That is, *a node quits the contention for channel access whenever it detects the energy of an ongoing transmission, otherwise it is allowed to transmit concurrently with others.* Traditionally for packet-level broadcast (with/without NC), this leads to the well-known "hidden terminal" problem, because such concurrent transmissions may cause interference at their neighbors[8]. Various mechanisms have been proposed to solve this problem, such as clearing the channel within a range larger than carrier sensing range (Li *et al.* 2009). However, due to SLNC's better tolerance in transmission errors and interference, more aggressive concurrent transmission is possible. In Section 7.4, we demonstrated that the simple carrier-sensing rule actually provides near-optimal performance in terms of average downloading rate, as the impact of hidden terminals is greatly alleviated by SLNC.

7.5.6 Broadcast content scheduling

Finally, we briefly highlight the way that broadcast content scheduling is dealt with in CodeOn.

7.5.6.1 Content scheduling at APs

In CodeOn, the APs broadcast the contents in a round-robin way to maintain the "information difference" between vehicles moving out of the AP range at different times. In order to make more efficient use of the VANET bandwidth, the content scheduling should also be aware of local vehicles' reception status. Therefore an AP will sort its file generations according to their utilities; in addition to round robin, it transmits the one with both larger ID and the highest utility that hasn't been transmitted in the last "batch".

7.5.6.2 Content scheduling at vehicles

After a vehicle becomes a relay node, it broadcasts the generation with the maximum utility. To avoid from transmitting duplicate information, it is important for vehicles to decide when to stop the transmission.

To this end, we estimate the number of pieces that each relay should send in one batch. The intended number of (innovative) pieces that v sends to a neighbor w for G_i is estimated as $K_{v,w} = Step(\lfloor \bar{r}_{v,i} \rfloor - \lfloor \bar{r}_{w,i} \rfloor)$. Then, the number of pieces that v should send to all neighbors for G_i is computed as

$$Z_v(G_i) = \lceil \frac{1}{|\mathcal{N}(v)|} \sum_{w \in \mathcal{N}(v)} K_{v,w} \rceil \tag{7.13}$$

which is also the size of a batch. When the average rank $\bar{r}_{v,i}$ and those of all of its neighbors are equal to K (full rank), we set $Z_v(G_i) = 0$. Note that the above is a conservative estimation, which treats the link qualities as perfect.

In addition, we need to deal with two situations. 1. If a batch spans multiple service time slots, relay v accesses the channel deterministically by setting its $\Delta t(v) = 0$ during the following time slots in order to finish transmitting its batch. 2. If the transmission of a batch terminates before the end of some service time slot k, to avoid waste of VANET bandwidth, v will fill the rest of the channel by transmitting additional coded pieces from the same G_i until time slot k is used up.

---

[8] With packet-level broadcast, carrier sense is shown to work well under a two-transmitter setting in Brodsky and Morris (2009). Here we focus on a multi-transmitter setting instead, using SLNC.

7.5.7 Performance evaluation

7.5.7.1 Methodology

In this section, we evaluate the performance of CodeOn by simulations. We compare CodeOn with an enhanced version of CodeTorrent (Lee *et al.* 2008), which is pull based and uses PLNC. The AP is treated as a normal node. Each node periodically broadcasts a gossip message to tell others about its content availability. Based on this, a node v periodically broadcasts a downloading request, asking for the index of the rarest generation G_i among its neighbors, and attaches a null-space vector of G_i computed from v's corresponding decoding matrix. Each neighbor w, upon receiving the request, checks if it has G_i. If yes, and if the null-space vector is not orthogonal to the subspace spanned by w's coding vectors of G_i, w responds v with one coded piece from G_i via unicast, after waiting for a random backoff delay to reduce collisions. Only the first packet in a piece contains the coding vector; if that packet is lost then the whole piece is lost. Upon successful reception of a piece, node v continues sending another downloading request. Otherwise, v waits till the next period to broadcast its request. Nodes other than v exploit opportunistic overhearing, i.e., buffer a piece sent to v if that piece is useful and received correctly.

We made the following additional modifications to CodeTorrent. We equip it with multichannel capability as in CodeOn. To ensure a fair comparison, we apply the same channel switching mechanism in CodeTorrent, which results in a 1/2 reduction in the downloading rate. Also, in order to increase the success probability of overhearing, each node is allowed to overhear multiple different pieces during the same period and there is no reserved time for receiving one piece. Moreover, the packets in a piece do not have to arrive in order; a node flushes an incomplete piece after a certain time from its first reception, say 0.5 s.

In addition, we introduce a variation of CodeOn, *CodeOnBasic*, which is also push based with piece division but based on PLNC. A piece is used as a whole for encoding and decoding. A node buffers any overheard piece as long as it receives the coding vector in the first packet of that piece, and the same buffer flushing mechanism as in CodeTorrent is adopted. Moreover, in content scheduling a relay node pads a service slot with whole pieces. If the remaining service slot time is not enough for sending a whole piece, it terminates the current batch, rather than filling with individual packets. Other than that, CodeOnBasic is the same with CodeOn.

We implemented CodeOn, CodeOnBasic and CodeTorrent in NS-2.34 (www.isi .edu/nsnam/ns). For CodeOn, we implemented the run-length coding with dynamic programming algorithm to minimize the communication overhead in sending each coded piece (Katti *et al.* 2008). We simulate independent symbol errors in a packet, and in simulation the number of runs seldom exceeds 20 for 10 KB pieces. Packet capture effect is enabled and, when two packets collide, if no packet can be captured, the symbols from the point of collision are all discarded. Otherwise, the captured packet is received as usual. We do not consider vehicular buffer constraints.

We have a few notes on broadcast data rate selection. First, the safety message's communication range shall be larger than that of PCD data packets, so that the neighbor set used in relay selection can cover the set of nodes that can receive a data packet. Otherwise, the utility cannot truthfully reflect a node's total content usefulness. Considering the reliability of safety messages, we chose the base rate (3 mbps) for broadcasting safety messages. Second, we want to achieve a high downloading rate for PCD. For SLNC, choosing a higher data rate is beneficial because it has better error-tolerance. Since rate that is too high is also undesirable due to very small communication range, the data rate of PCD packets is set to be 12 mbps throughout the chapter.

7.5.7.2 Simulation settings

We consider both highway and urban scenarios (Figure 7.12). We use a VANET mobility generator (http://nile.cise.ufl.edu/important/software.htm.) to generate the movement patterns. Vehicles are placed uniformly at random in the road area; when a vehicle hits the boundary it randomly selects another entry point of the map. This removes the boundary effect; equivalently, the AoI is infinitely large. Table 7.2 is a list of parameters.

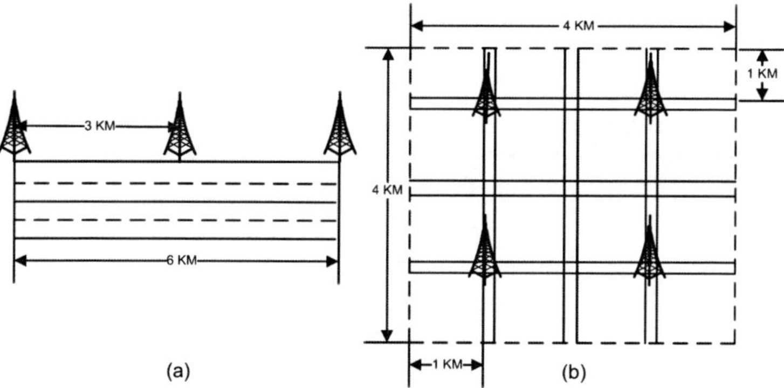

Figure 7.12   (a) Highway scenario. (b) Urban scenario. Reproduced by permission of © 2011 IEEE.

The highway scenario consists of a bi-direction, four lane highway with length 6 km. Vehicles' speeds are randomly drawn from [20,30]m/s with a maximum acceleration of 0.5 m/s^2. The urban scenario is 4 km $\times$ 4 km as shown in Figure 7.12. In order to evaluate the impact of topology and traffic density, we simulate sparse and dense traffic for both scenarios. The sparse settings simulate delay-tolerant network (DTN), where the total number of vehicles is 100 for highway and 160 for urban. The dense highway setting has 300 vehicles while the dense urban has 400 vehicles.

7.5.8   Simulation results

7.5.8.1 Downloading performance

We evaluate the downloading performance from three aspects: 1. downloading progress, which is the change of average downloaded percentage of the file with

Table 7.2   Simulation parameter settings

CodeOn/CodeOnBasic		CodeTorrent	
Δt_{max}	2 ms	Maximum random backoff delay	5 ms
T_J	100 µs	Gossip interval	0.5 s
		Periodic request interval	0.5 s
		Unicast retry limit	7

Common parameters	
$\mid F \mid$	16 MB
L	50
K	32
M	16
q	8
J/N	10 ($J = 10$ KB)
CR, ER	250 m, 700 m
Data rate/base rate	12 Mbps (16QAM)/3 Mbps (BPSK)
SNR thresholds	15 dB, 4 dB
Data capture threshold	20 dB
Data/safety message sizes	1 KB, 256B (without header)
Propagation model	Nakagami $m = 3$

the elapsed time (averaged upon each vehicle); 2. average downloading delay: the average elapsed time from downloading start to 100% completion; 3. average downloading rate, where the downloading rate for each vehicle is the file size divided by its downloading delay.

We present the downloading progress in Figure 7.13–Figure 7.16 for all four scenarios. It can be seen that CodeOn significantly outperforms both CodeOnBasic and CodeTorrent. The downloading progress of CodeOn is the fastest, especially when the average downloaded file percentage is below 90%. The comparison between CodeOnBasic over CodeTorrent demonstrates the effectiveness of our new push-based protocol design and the comparison between CodeOn and CodeOnBasic shows the advantage of the use of SLNC, which we will discuss later.

Next, we evaluate the average downloading delays and rates in Figure 7.17 and Figure 7.18. Some of the average delays are not shown since their downloading progresses cannot reach 100% within the given simulation period. There are two key observations. First, the average downloading rates of CodeOn are much higher than both CodeOnBasic and CodeTorrent, for both highway and urban scenarios and both sparse and dense traffic. Second, CodeOn maintains high downloading rate in all cases shown, especially for the two extremes, i.e., sparse urban scenario and dense highway cases which represent the lowest and highest traffic density, respectively. This means CodeOn is the most robust to variations in topology and vehicle density.

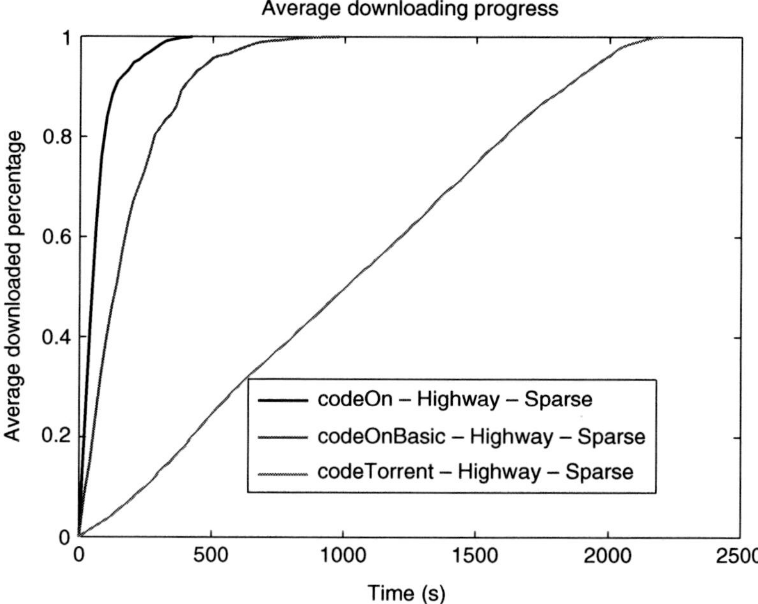

Figure 7.13   Downloading progresses: sparse highway scenario. Reproduced by permission of © 2011 IEEE.

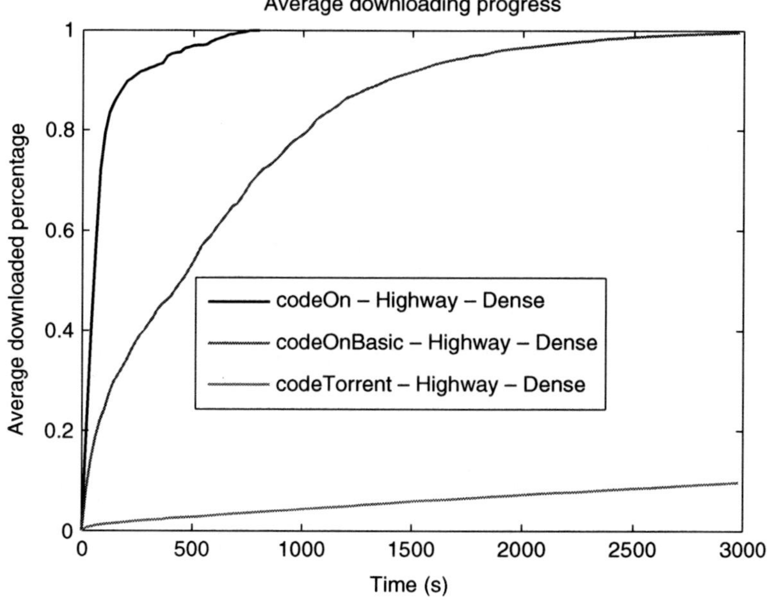

Figure 7.14   Downloading progresses: dense highway scenario. Reproduced by permission of © 2011 IEEE.

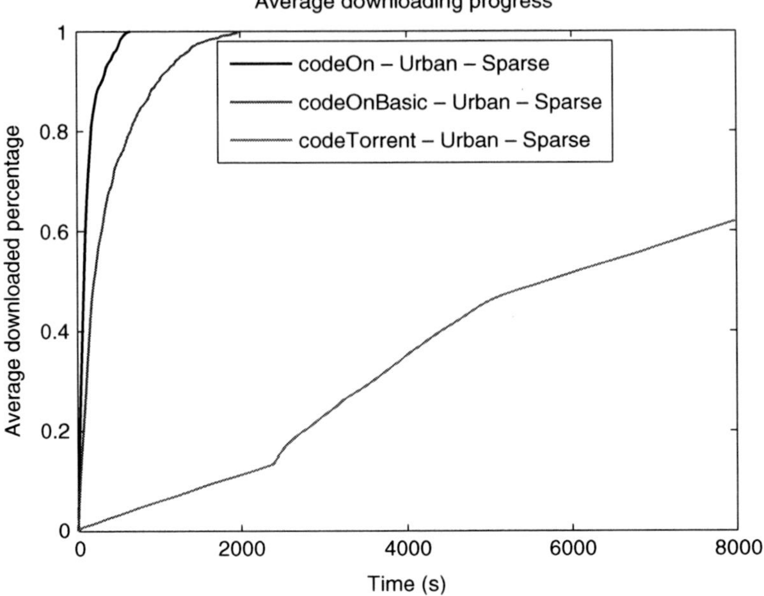

Figure 7.15   Downloading progresses: sparse urban scenario. Reproduced by permission of © 2011 IEEE.

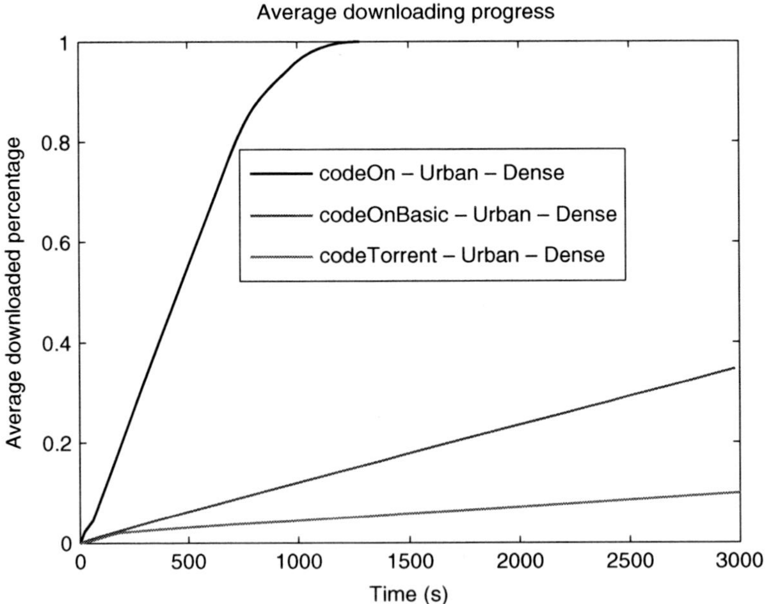

Figure 7.16   Downloading progresses: dense urban scenario. Reproduced by permission of © 2011 IEEE.

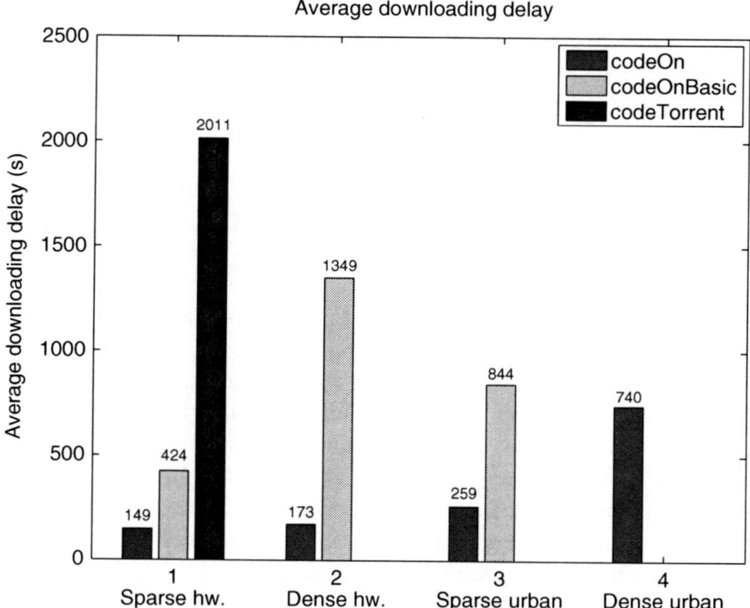

Figure 7.17 Average downloading delay. Reproduced by permission of © 2011 IEEE.

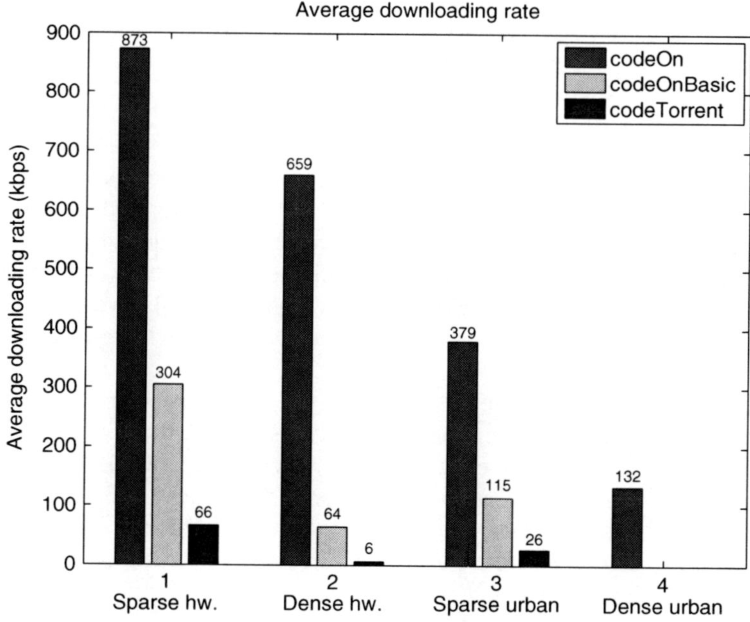

Figure 7.18 Average downloading rate. Reproduced by permission of © 2011 IEEE.

The first phenomenon above is attributed to the push-based protocol design combined with SLNC. In CodeOn, using a prioritized relay selection mechanism with the transmission coordination that avoids heavy packet collisions, the contents can be distributed proactively to the vehicles in the AoI so that the VANET bandwidth is fully utilized. Moreover, each transmitted piece of content has a high probability of being useful to a relay's neighboring vehicles, in terms of increasing the ranks of their decoding matrices. In addition, with SLNC, the symbols in content pieces are received with a higher rate from APs and relays, which results in a higher downloading rate.

The robustness of CodeOn under low traffic density is mainly attributed to the enhanced reception reliability brought by SLNC. Compared with PLNC, SLNC actually enables vehicles in a larger range to receive some useful information in a piece. In the sparse urban setting, although the vehicular contact opportunities are much less, CodeOn is able to mitigate the impact of low traffic density.

On the other hand, CodeOn is less affected under dense VANET. For the dense scenarios, the differences between CodeOn's downloading rates and those of CodeOnBasic and CodeTorrent are both larger than the sparse scenarios (Figure 7.18). For CodeTorrent, the performance degradation is due to lack of coordination and the use of PLNC for a large file. 1. Under dense VANET, the number of requesting vehicles in a node's neighborhood increases. Since there may be more than one responder for each requester, the chance of packet collisions also increases. The unicast-with-overhearing mechanism retransmits packets after they are collided, which aggravates the problem. 2. For both CodeTorrent and CodeOnBasic, the use of PLNC prevents a requester from receiving a whole piece under frequent packet collisions. However, CodeOn alleviates the above problems dramatically through prioritized relay selection and the use of SLNC.

7.5.8.2 Fairness

Fairness is embodied in the distribution of downloading delays of all vehicles, shown in Figure 7.19 – Figure 7.21. We show the distributions for all three cases. The fairest situation has zero variance, i.e, all the delays are equal. From the figures, one can see that the distributions of CodeOn are more concentrated (more fair) than those of CodeOnBasic and CodeTorrent. Few vehicles need a very long time to receive the whole file. Again, the same robustness of CodeOn to variations in traffic density can be observed.

The superiority of CodeOn in fairness is still attributed to the use of SLNC. This enables more reliable reception of the coded symbols, because an overhearing node will buffer any innovative clean symbol it received. In CodeOn, because the granularity of information reception is smaller and vehicles have similar opportunities to contact APs and other vehicles within a time period in the order of 1000 s, their reception progress has low variance. However in CodeOnBasic and CodeTorrent, a vehicle either receives a whole piece or receive nothing, so the variance among reception progress is larger. Again, the results on fairness demonstrate the benefit of using SLNC and the effectiveness of CodeOn's protocol design.

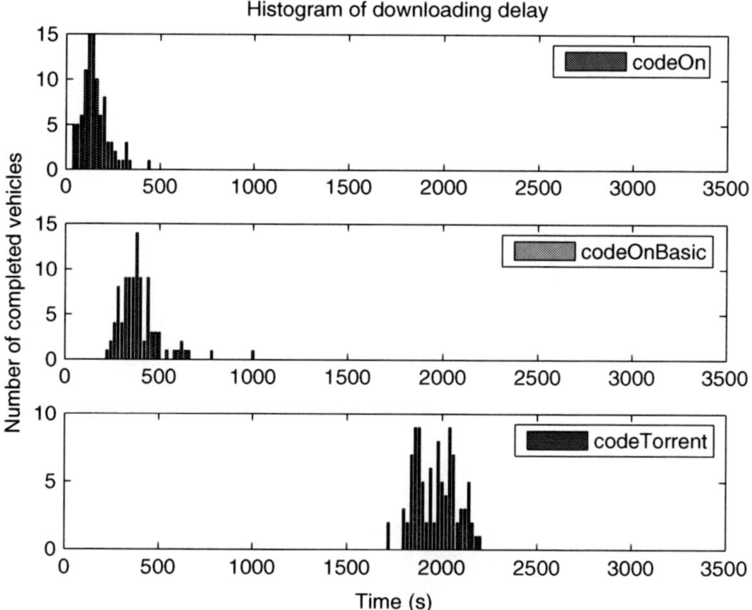

Figure 7.19   *Distribution of downloading delay: sparse highway scenario. Reproduced by permission of © 2011 IEEE.*

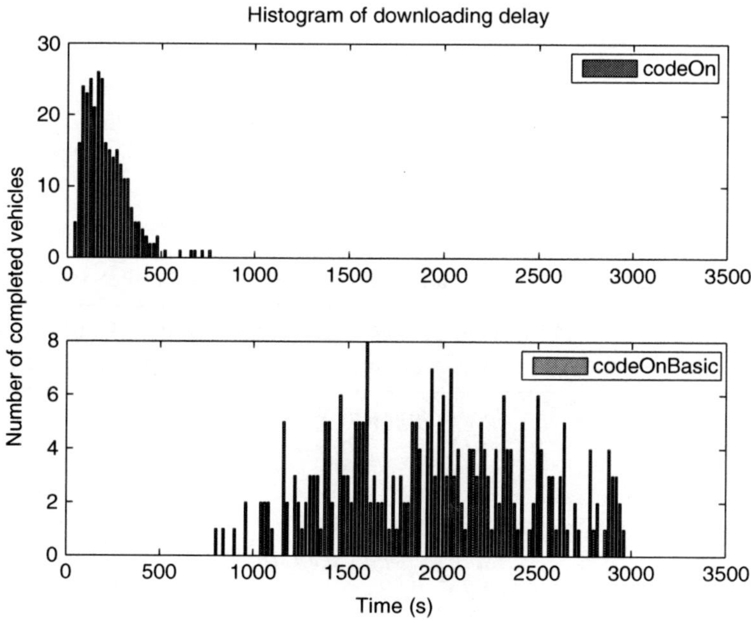

Figure 7.20   *Distribution of downloading delay: dense highway scenario. Reproduced by permission of © 2011 IEEE.*

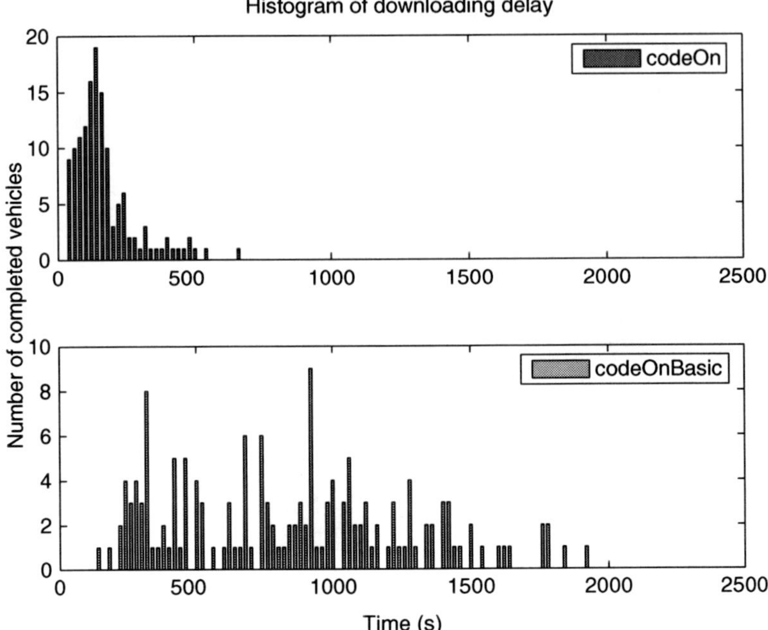

Figure 7.21   Distribution of downloading delay: sparse urban scenario. Reproduced by permission of © 2011 IEEE.

7.5.8.3   Protocol efficiency

One may wonder if CodeOn achieves fast push-based downloading by sacrificing protocol efficiency. To further investigate this issue, we present the results on protocol efficiency in Table 7.3 and Table 7.4.

Table 7.3   Protocol efficiency (total number of pieces in the file: 1600)

Protocols	Percentage of non innovative received pieces	Average no. of failed overheard pieces per received piece
	Sparse highway scenario	
CodeOn	N/A	N/A
CodeOnBasic	0.476	4.26
CodeTorrent	0.325	27.27
	Sparse urban scenario	
CodeOn	N/A	N/A
CodeOnBasic	0.228	3.47
CodeTorrent	0.167	80.74

Table 7.4 Protocol efficiency (continued)

Protocols	Average no. of pieces sent by a vehicle	Average no. of pieces sent by an AP
	Sparse highway scenario	
CodeOn	2202.12	26023.00
CodeOnBasic	4054.87	51578.00
CodeTorrent	32889.87	53665.00
	Sparse urban scenario	
CodeOn	1031.14	43445.25
CodeOnBasic	3525.31	143905.00
CodeTorrent	52465.69	222287.50

As we have shown in the last section, the protocol overhead of CodeOn is small. To evaluate the amount of data traffic experienced, we show the average number of pieces sent by a vehicle and an AP during the whole simulation time (a node will not transmit when all of its neighbors receive 100% of the file). CodeOn has the smallest number among the three protocols. Its high protocol efficiency comes from both the high symbol reception probability due to SLNC and the high usefulness of the transmitted symbols due to relay selection. As CodeOnBasic adopts the same relay selection mechanism, it enjoys similar high protocol efficiency to CodeOn. However, CodeTorrent sends many pieces due to a large number of failed overhears explained below. Note that the APs are always the most advantageous nodes so they transmit a lot in all three protocols.

To further study the role of relay selection we compute the percentage of total number of noninnovative pieces out of the total number of received pieces, which reflects the usefulness of the received content. We also calculate the average number of failed overheard pieces (in which the coding vectors are received but not all the subsequent packets) per received piece. For the former, CodeOnBasic is slightly higher than CodeTorrent but for the latter CodeOnBasic is much lower than CodeTorrent. This is because in CodeTorrent, a responder uses the requester's null-space vector to decide whether to transmit a coded piece, which is definitely innovative to the requester. However, in CodeTorrent, a responder's transmission mainly benefits the requester itself but few others due to uncoordinated transmissions. On the other hand, in CodeOnBasic the selected relays can benefit their whole neighborhood, while the broadcasted contents are still highly useful. As a result, both the downloading rate and efficiency are high.

7.5.8.4 Discussion

Finally, we give some insights that can be obtained from our results.

Push versus pull. First we compare the *push-versus-pull*-based content distribution in VANETs. CodeOnBasic and CodeTorrent are both based on

PLNC, but the former performs much better than the latter for all scenarios in Figure 7.13 – Figure 7.16 and Figure 7.19 – Figure 7.21. An obvious reason is the difference in bandwidth utilization. CodeOnBasic lets the APs and relays broadcast proactively (push), so that the service time slots are almost fully utilized. However, in CodeTorrent each node make requests (pull) periodically and responders transmit passively. Whenever it receives a piece in error, a requester will wait until the next period to make subsequent requests. Due to the lossy property of the wireless channel in VANETs, this happens frequently so the service channel is under utilized.

However, a more fundamental reason that the push method in CodeOn and CodeOnBasic is better is related to to the relay selection mechanism. If there was no transmission coordination between vehicles, the push-based content distribution could easily lead to frequent packet collisions. For CodeTorrent, which is pull-based, its high chance of packet collisions is already evident from the large number of failed overheard pieces of CodeTorrent in Table 7.3. One can imagine that this situation will be aggravated if CodeTorrent is changed to push-based distribution where nodes transmit more aggressively.

Apart from transmission coordination, in designing a push-based protocol, it is always critical to maximize the usefulness of the broadcasted content from each relay nodes. As nodes do not make explicit downloading requests, and as "push" uses broadcast transmission, it is basically impossible to ensure the usefulness of broadcast content of a relay for all its neighbors. In CodeOn and CodeOnBasic, our approach is to select a relay to be the one that can bring the maximum amount of useful content to all its neighbors by implicitly calculating node utilities based on fuzzy average rank differences. In contrast, in CodeTorrent each responder will only ensure the content to be 100% innovative for one requestor, using accurate null-space indicators. Interestingly, as one can see from the number of noninnovative pieces in Table 7.3, the number of CodeOnBasic is quite close to that of CodeTorrent, which can be regarded as a lower bound. This proves the effectiveness of our relay selection approach.

SLNC versus PLNC. The advantage of using SLNC is evident by comparing CodeOn with CodeOnBasic in Figure 7.13 – Figure 7.16, which are only different in the network coding method. With PLNC, in CodeOnBasic a coded packet is discarded whenever it is received in error, which leads to unsuccessful reception of the whole piece. However, with SLNC, CodeOn records every innovative received symbol in a piece and then combines innovative symbols to decode the piece.

As previously mentioned, SLNC is superior in tolerating transmission error. This is a direct reason of why CodeOn has the best robustness under dense traffic scenarios. By both coding and receiving according to a small granularity of symbols (yielding higher content diversity), vehicles have a better chance of receiving some useful information, even when packet collisions are frequent due to dense traffic, or when there are few vehicles or APs around. However, with PLNC, the content diversity is lower. Although our push-based protocol design is able to choose the best relay nodes and alleviate collision, without SLNC, small downloading delays and a high level of fairness are very hard to achieve for all topologies and traffic densities.

7.6   CodePlay: a live multimedia streaming scheme for VANETs based on SLNC

In this section, we study the cooperative broadcast of live multimedia streams (LMS) in VANETs. We present CodePlay, a distributed live multimedia streaming scheme in VANETs based on SLNC. We should note that, unlike general popular content distribution services in VANETs in the last section where the main goal is high average downloading rate, LMS services require not only a high average streaming rate but also demand that the streaming rate keep stable for the purpose of smooth playback. LMS services are also different from nonlive streaming services such as video-on-demand, where various vehicles may be interested in different contents and those contents are not closely related to the real world's time. For LMS services, the streaming contents are usually generated as time progresses and are only useful to vehicles within a short period of time, for example several seconds to tens of seconds. However, these time constraints are usually not as tight as those of real-time services, like intelligent collision avoidance, which usually require a delay smaller than hundreds of milliseconds.

In addition to a stable and high streaming rate for smooth playback and short service delivery delay, LMS also needs high bandwidth efficiency for better coexistence with other competing services. In a dynamic and lossy VANET, the biggest challenge to providing satisfiable LMS services is how to achieve the above multiple objectives simultaneously.

As we already mentioned in Section 7.2, in this section we approach the LMS problem in VANETs using SLNC again, due to its potential to alleviate packet losses and enhance space reusability. However, providing satisfiable LMS services with minimal bandwidth is not a trivial problem even with the help of SLNC. Essentially, this corresponds to the following core design issue: **which vehicles should transmit what content to whom at which service time slots?** In particular, as broadcast is adopted as the basic transmission paradigm and multiple receivers may have different stream reception and playback status, how do we select proper relay nodes to ensure smooth playback of multiple vehicles? How to coordinate the transmission of multiple relays so that spacial reusability is maximized? How can the above be achieved efficiently with low overheads? All these key issues imply that new design considerations are needed.

To this end, the main contribution of CodePlay is to introduce a *coordinated local push* (CLP) mechanism based on SLNC, which features a mixed lightweight transmission coordination strategy. From the global level, the whole AoI is divided into segments of fixed length and within each segment a relay node will be selected. The transmission schedule among the adjacent segments follows a deterministic scheduling method in order to disseminate the streaming content from sources to all the segments smoothly and in a timely manner. From the local level, within each segment a lightweight coordination mechanism is used to ensure the (dynamic) selection of one relay node whose transmissions can bring the most amount of useful information to all the vehicles in its neighborhood and will ensure the LMS delivery before the desired deadline to each neighboring vehicle as much as possible. Again, the CLP mechanism benefits from the use of network coding in simplifying the transmission coordination and scheduling.

Furthermore, to enhance the LMS performance for sparse VANETs, an opportunistic transmission scheduling algorithm is proposed, where the wasted transmission opportunities in the network can be adaptively utilized by the relays, merely based on carrier sensing. The organization of rest of this section is similar to the previous section.

7.6.1   Design objectives

Live multimedia streaming is one typical delay-bounded application with QoS requirements. Thus, the design of CodePlay pursues the following primary objectives.

- Smooth playback at all the interested vehicles, which refers to all vehicles inside AoI. This requirement can be translated into providing a stable and high streaming rate.

- Prompt service delivery, which can be translated into short end-to-end delay for all the receivers. For a receiver, this delay is defined as the elapsed time from the generation of specific LMS content at the source to the start of playback of this content at the receiver. Meanwhile, it is desirable to achieve a high degree of fairness, i.e., the service delays among neighboring receivers should be similar.

- Minimized bandwidth cost, which can be translated into incurring small protocol overhead and data traffic. This is for better coexistence with other possible services.

7.6.2   Overview of codeplay

We will first give an overview of CLP from the macroscopic level and then show a simple example of CodePlay's protocol run.

7.6.2.1   Global view

A random access control protocol such as CSMA/CA is widely used to coordinate local transmissions within a neighborhood. However, from a global point of view, such a protocol cannot ensure the fast and smooth propagation of information flow across the multihop VANET within the AoI. For applications such as live multimedia streaming, where each packet has a playback deadline, such a protocol is inadequate. CodePlay divides the whole AoI into segments of fixed length, which can be referred to as the *logical segmentation of communication path* approach. This is possible because such configuration can be broadcast to vehicles by APs and it is also not difficult for a vehicle to determine which segment it is in with GPS. Within each segment, a relay node will be selected based on the ideas presented in the previous section. The transmission schedule among the adjacent segments must follow some deterministic scheduling method. Figure 7.22 shows the effect of such segmented prorogation idea, where the scheduling is simply round robin. In time slot 1, the AP, selected relay nodes in segment 3 and segment 6 are transmitting. In time slot 2, relay nodes in segments 1, 4, 7 are transmitting. In time slot 3, relay nodes in segments 2, 5, 8 are transmitting. This will ensure the smooth prorogation of the contents along the AoI, with roughly one segment length per time slot.

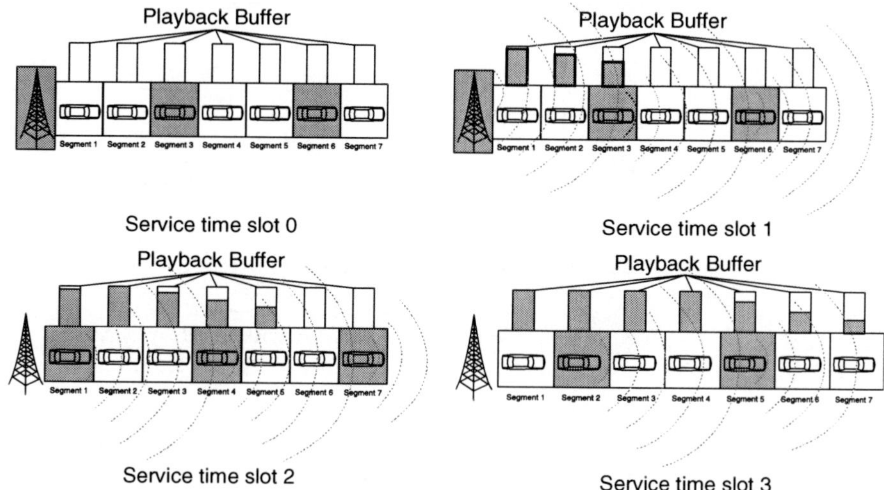

Figure 7.22   Coordination for LMS in VANET: segmentation of communication path. © 2011 IEEE. Reprinted, with permission, from Z. Yang, M. Li and W. Lou, "CodePlay: Live Multimedia Streaming in VANETs using Symbol-Level Network Coding", IEEE Transactions on Mobile Computing (TMC), 2011.

7.6.2.2 Example of codeplay's protocol run

In the following, we illustrate the workflow of CodePlay using a simple example. For proof-of-concept, we describe CodePlay under a one-dimensional highway scenario (Figure 7.23). However, CodePlay can be easily extended to the urban scenario, i.e., a two-dimensional case.

1. *System initialization*. To ensure smooth playback in LMS, node coordination shall be facilitated in an efficient way, yet, in a dynamic VANET. The idea of CLP is that, *we introduce road segmentation during initialization so that coordination decisions can be made locally*[9]. Each road is divided into fixed segments of equal length (SL) and is uniquely numbered. The road

---

[9] We note that a similar segmentation approach has been used for solving different problems in previous works (Johnson *et al.* 2006; Li *et al.* 2009).

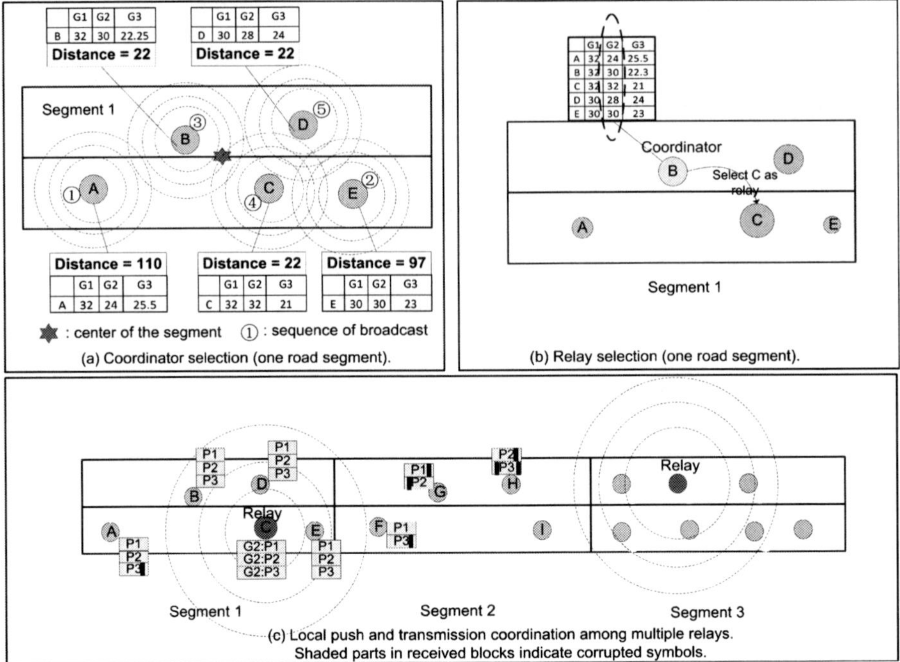

Figure 7.23   The concept of coordinated local push. Reproduced by permission of © 2011 IEEE.

segmentation can be preconfigured and provided by the access points with the help of GPS. Every vehicle is assumed to possess this information before entering the AoI.

2. *Local coordinator selection.* In order to make the relay selection process reliable and efficient, the key method is to let vehicles agree on a local *coordinator*, which selects the relay on behalf of other nodes (Figure 7.23(a)). This is achieved by taking advantage of the obligated safety message service in the control channel required by the IEEE 802.11p standard, where every vehicle has to broadcast a safety message to inform its current location in each control time slot. CodePlay lets each vehicle piggyback a short piece of additional information on the safety message. This information contains the minimum Euclidean distance to the geographical center of the road segment that this vehicle knows (either its own distance to the center or the broadcasted distance overheard from other vehicles in the same segment) and also the vehicle's current LMS content reception and playback statuses. We will introduce an efficient representation of this information later. The vehicle closest to the center of the segment is selected as local coordinator, like vehicle B in Figure 7.23(a).

3. *Distributed relay selection*. The coordinator selects real relay based on the reception and playback status of all nearby vehicles, i.e., what LMS contents each of them have received or are needed for playback in the immediate future. In particular, the coordinator computes the *utility* of each node in its segment as how much useful information can that node provide to its neighbors, and designates that node as relay via unicast. This is shown in Figure 7.23(b), where coordinator B designates vehicle C as relay and the generation G_2 as the broadcasted content. One generation represents a short period of LMS content and the precise definition will be given in the following section.

4. *Local push and transmission coordination of relays*. In order to create a stable and continuous LMS flow, only relays in certain segments are allowed to transmit concurrently in each service time slot. Those relays actively "push" coded LMS blocks to their vicinity, which will be received by neighboring vehicles. To maximize spacial reusability, we exploit SLNC's symbol-level diversity by purposely reducing the distance between two concurrent transmitting relays (thus introducing a proper amount of signal interference). In the snapshot given in Figure 7.23(c), the two relays are separated by two road segments, which maybe too close to be allowed if packet level collision avoidance mechanism is adopted. Specifically, we address the following issues: 1. what is the optimal number of segments between two adjacent transmitting relays? 2. how can we opportunistically schedule the relays' transmission if the density of the VANET is so sparse that some road segments are empty and no relay could be selected for them?

7.6.3 LMS using symbol-level network coding

The integration of SLNC is similar to that of CodeOn in Section 7.5. The definition of generations and pieces, the piece-division, run-length coding algorithm are the same.

The main difference is that, each receiver v maintains a *playback buffer* for generations to be played in the immediate future, which buffers all the received useful coded symbols. Note that v also maintains a decoding matrix for each symbol position j of each generation, which consists of the coding vectors of all the jth symbols. Again, the rank of each matrix is called *symbol rank*. A coded symbol is called *useful* in CodePlay if: 1. it is received correctly (Katti *et al.* 2008); 2. it can increase the corresponding symbol rank (*innovative*); 3. it belongs to a generation that is after v's current playing point. When receiving enough useful symbols for a position, the receiver can decode the original symbols by performing Gaussian elimination on the corresponding matrix.

Like CodeOn, CodePlay also uses the *average symbol rank* reception status representation method. It exploits multichannel capability by piggybacking the node's average symbol rank information of each generation in safety messages in the control channel. Here is a back-of-the-envelope calculation: for ten generations with packet length of 30 symbols and $K = 32$, the piggybacked information is only about ten bytes, which can be easily embedded in a safety message without affecting its reliability (Xu *et al.* 2004).

Each node plays the buffered generations sequentially and keeps eliminating older generations to make room for newer content. Those generations within α seconds after the current playback time are called *priority generations*. The piggybacked reception status, which contains a priority generation with average rank less than K is considered as an implicit *urgent request*. The above definitions are depicted in Figure 7.24. Note that vehicles on the opposite road of the AoI behave exactly the same as described above except that they do not need to playback the received LMS contents.

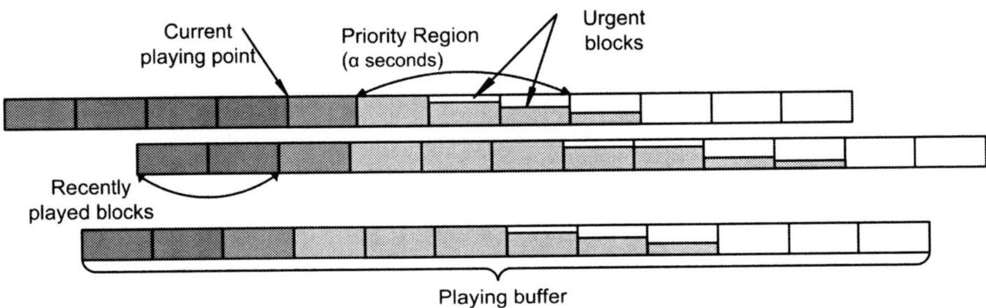

Figure 7.24   Playback buffer and priority generations. Reproduced by permission of © 2011 IEEE.

7.6.4   Coordinated and distributed relay selection

The main purpose of the relay selection is to maximize the utility of each transmission to save the precious bandwidth resource in the VANET. The selected relays should best satisfy all neighbors' smooth playback needs, which can be inferred through vehicles' reception statuses. Here three components are needed: 1. a local coordinator that serves as an arbitrator, with which a consensus on relay selection can be reliably and efficiently achieved; 2. the computation of nodes' "utilities" that represent their capability to satisfy others; 3. The selection of appropriate parameters (such as segment length) for fast LMS propagation and continuous coverage.

7.6.4.1   Distributed coordinator selection

All vehicles in the same road segment agree on an unique local coordinator at the end of each control time slot, based on geographic information. For both reliability and efficiency considerations, we propose an accumulated consensus mechanism based on information piggybacked in the safety messages. We firstly define a *temporary coordinator* as the vehicle closest to the segment center that a vehicle currently knows. Each vehicle considers itself as the default temporary coordinator at the beginning of each control time slot. For each overheard safety message originated

Sequence number of current road segment	Temporary coordinator's ID	Temporary coordinator's distance	Flow ID	Sequence number of generation under playing	Sequence number of the first generation in buffer	Rank1	Rank2	......	RankN
1	4	4	1	2	2			N	

Figure 7.25 *The format of piggybacked information (in bytes), where N is the size of the playback buffer (in generation). Reproduced by permission of © 2011 IEEE.*

from a vehicle in the same segment, the receiver checks if the temporary coordinator piggybacked (Figure 7.25) is closer to the segment center than the one known to itself presently. If yes, the receiver replaces its temporary coordinator with the overheard one. Since the vehicle closest to the segment center will be repeatedly claimed as temporary coordinator by multiple safety messages (like vehicle B in Figure 7.23(a)), this accumulated consensus mechanism makes the probability of selecting multiple coordinators within one segment negligible, no matter there are lossy wireless links or sparse connections.

7.6.4.2 Relay selection

At the beginning of the following service time slot, each coordinator C firstly checks if its segment is scheduled to transmit in this slot or not, where the scheduling algorithm will be introduced in the next section. If yes, C will then calculate the *node utility* for each vehicle in $\mathcal{V}(C)$, the set of all the vehicles in the same segment as C, and designate the one with the highest utility as relay. If a tie appears, the vehicle located in the LMS propagation direction wins. The calculation of node utility consists of two steps:

1. Find the range of *interested generations* for all vehicles in $\mathcal{N}(C)$, which is the neighbor set of C and we require $\mathcal{N}(C) \supseteq \mathcal{V}(C)$. Only the generations representing streaming contents after the earliest playback time among vehicles in $\mathcal{N}(C)$ are regarded as interested ones. If there exists some urgent generations *UrgentGen*, C will give strict priority to the transmission of *UrgentGen* during this time slot to ensure smooth playback at those vehicles. Otherwise, all the interested generations will be considered by C.

2. Calculate node utility for each vehicle in $\mathcal{V}(C)$. If *UrgentGen* $\neq \emptyset$, only the generations in it will be considered in this calculation. With SLNC, the usefulness of a potential relay v's generation G_i is determined by the difference in the symbols' ranks of G_i between v and its neighbors. Due to wireless medium's broadcast nature, G_i's utility to others increases with both the average usefulness of G_i and the number of vehicles it can benefit. Thus, for $v \in \mathcal{V}(C)$, the *generation utility* of G_i is defined as:

$$U(G_i, v) = \sum_{v' \in \mathcal{N}(v)} Step(\lfloor \bar{r}_{v,i} \rfloor - \lfloor \bar{r}_{v',i} \rfloor) \times Urgent(G_i, v') \qquad (7.14)$$

where $\lfloor \bar{r}_{v,i} \rfloor$ is the fussy average rank of node v's generation i. $Step(x) = x$, if $x > 0$; otherwise, $Step(x) = 0$. And $Urgent(G_i, v') = priValue$, if G_i is urgently requested by vehicle v', otherwise, $Urgent(G_i, v') = \frac{priValue}{2^{i-i_0}}$, where i_0 is the index of the urgent generation closest to the physical world's time. The *priValue* is an adjustable system parameter that controls the relative importance of priority generations. Note that, as the coordinator does not know $\mathcal{N}(v)$ under the single-hop piggyback mechanism, we substitute $\mathcal{N}(v)$ by $\mathcal{N}(v) \cap \mathcal{N}(C)$. In fact, if we assume the safety messages are sent at the basic rate that can reach a larger range (e.g. 2×) than normal data packets, then $\mathcal{N}(v)$ can be further reduced to nodes within v's data communication range ($\mathcal{N}'(v)$) (which will be explained later), which can be estimated by C.

This utility measures how much innovative information node v can give to other vehicles in $\mathcal{V}(c)$ in total if it broadcasts coded packets generated from G_i. Currently we do not consider the link qualities between v and the receivers. The *node utility* $U(v)$ of vehicular node v is defined as $max_{G_i \in \text{interested generations}}\{U(G_i, v)\}$, which estimates the maximum amount of innovative information v can provide to other vehicles in $\mathcal{N}(v)$ for one generation. We do not look at the aggregate utility of multiple generations, because transmitting many generations takes a long time, which may cross multiple time slots and the VANET topology will have already changed.

The coordinator C designates R, the vehicle having the maximum $U(R)$, as the relay using a unicast message, which enables R to use the current service time slot. R then actively pushes coded packets generated from G_R with the maximum $U(G_R, R)$. Note that the required number of coded pieces to send during one service time slot can be estimated based on $\frac{1}{|\mathcal{N}(R)|} \sum_{v' \in \mathcal{N}(R)} Step(\lfloor \bar{r}_{R,i} \rfloor - \lfloor \bar{r}_{v',i} \rfloor)$, which will not be elaborated here.

7.6.4.3 Determining the segment length

The length of the segment, SL, is an important parameter that affects the utility of relay selection and propagation speed of the LMS flow. On the one hand, if SL is too large, a relay at one end of a segment may not convey enough information to the neighboring segment in its scheduled time slot and in the next slot the relay in the neighboring segment would have little innovative information to transmit, which affects smooth playback of LMS. On the other hand, if SL is too small, vehicles in adjacent segments tend to have similar reception statuses and their relays probably will transmit duplicate information. Both extremes could lead to low bandwidth efficiency and large service delivery delay.

In general, we should ensure that for a sender and receiver pair of distance SL, the symbol reception probability is sufficiently high. However, with a realistic fading channel, it is hard to define such a range because symbol reception is probabilistic. For a simpler alternative approach, we define an equivalent *data communication range CR* under free space propagation model(Friis) as before: $CR = \sqrt{\frac{T_p G}{Th_{CR}}}$, where T_p is the transmission power, G is the antenna gain and Th_{CR} is the data reception threshold. Thus we set $SL \approx CR$ in this chapter.

7.6.5   Transmission coordination of relays

We have determined which vehicles should transmit what content to whom. In this section, we answer the last question: in which time slots should each relay actively push the coded LMS? This is addressed from both spacial and temporal aspects.

7.6.5.1   Spacial coordination

Due to the use of SLNC, concurrent transmissions of more relays are encouraged to take advantage of spacial reusability (Katti *et al.* 2008). But two transmitting relays that are too close will cause heavy collisions, which in turn degrades the bandwidth efficiency. There exists an optimal average distance between two concurrent transmitting relays, d_{opt}, under which the relays can convey the highest amount of useful information to their neighbors within unit time. In other words, the bandwidth can be used most efficiently.

Next we discuss how to determine the d_{opt}. Consider a straight highway of length L, where vehicles are uniformly distributed. n relays, $v_1, v_2, \ldots\ldots, v_n$, lie on the highway with equal inter distance. From the same analysis in Section 7.4, the "optimal inter relay distance" is derived as the inter relay distance that maximizes the *average symbol reception probability* for all the vehicles in the VANET. The results are referred to in Figures 7.8(a) and 7.8(b).

In CodePlay, the channel access decisions are made largely based on carrier sense. We also exploit this characteristic in the opportunistic scheduling algorithm in the next section.

7.6.5.2   Temporal coordination

To provide continuous streaming coverage and to satisfy the strict time constraint of LMS service, the traditional random medium access mechanisms are not appropriate since their channel access delays are not bounded. We propose to use local round-robin (LRR) scheduling to coordinate the transmissions of neighboring relays. At first, we define the number of separating segments between two adjacent transmitting relays as W_{opt}, which can be calculated as $\lfloor \frac{d_{opt}}{SL} \rfloor$. The round length R in LRR is exactly $W_{opt} + 1$. For a relay in segment i, its scheduled slots T_i are determined as: $T_i \equiv i \bmod (W_{opt} + 1)$. For example, assume $W_{opt} = 2$, then segment 1 is scheduled to use time slots 1, 4, 7, 10, etc. Using this local round-robin schedule, LMS can flow from the source to receivers within the AoI smoothly. From a receiver's point of view, if the VANET is well connected, it is always able to obtain new LMS content for playback within determined waiting time.

7.6.6   OLRR: opportunistic LRR scheduling for sparse VANETs

Due to the highly dynamic nature of VANET, it tends to experience partitions frequently (Park *et al.* 2010), especially when the traffic density is low. In sparse VANET, some road segments will be devoid of relays and the scheduled transmission opportunities would be wasted, which results in low bandwidth efficiency. This is illustrated in Figure 7.26(a), where the segments 4, 7, 10 contain no vehicles, and their scheduled time slots are wasted. To solve this problem, we propose an opportunistic LRR (OLRR) scheduling algorithm by taking advantage of those available slots.

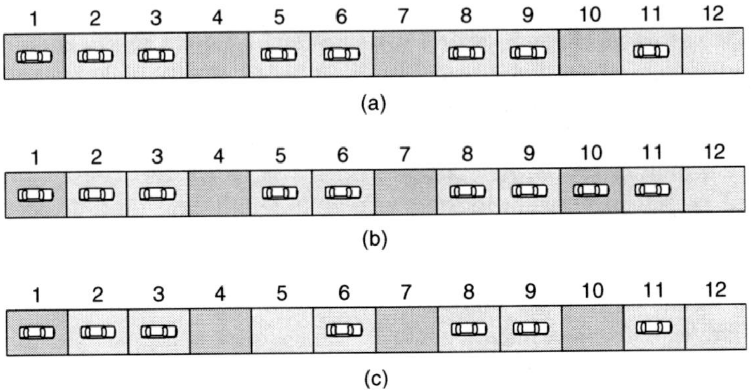

Figure 7.26   *Sparse VANETs (T = 1, R = 3). (a): LRR wastes transmission opportunities. (b)-(c): Using OLRR, secondary segments can take the unused transmission opportunities of primary segments (4,7,10). Reproduced by permission of © 2011 IEEE.*

The OLRR operates in a way resembling cognitive radio, which leverages nodes' carrier sensing capability. Essentially, during each service time slot, the coordinators in each segment will detect if there are relays in the nearby *primary segments*, which are scheduled segments by LRR in that time slot. If not, certain *secondary segments* will gain channel access according to some priority assignment. In order to sense the channel, a few additional rounds ($3 \times (W_{opt} + 1)$ subslots) are allocated before data transmission. Thanks to SLNC, each coordinator/relay does not need to consider the transmitters out of its energy detection capability, which greatly simplifies protocol design.

The algorithm is described in Algorithm 7.1 In line 3, there are two cases where a relay cannot be selected: C_i is the only node in i, or no node can provide innovative information to others. *ConflictSet*(i) is the set of coordinators (also segments) that has higher transmission priority than i. The nearer a segment is to a primary segment (with lower ID), the higher its priority. If two secondary segments happen to have the same distance to their primary segments, they will both access the channel as is the case in LRR.

We use the examples in Figure 7.26 to illustrate the basic idea of OLRR. Suppose $W_{opt} = 2$ and C_1, C_4, C_7, C_{10} are scheduled to use the channel simultaneously in the current service-time slot. In Figure 7.26(a), C_5 will decide to take this time slot since it senses that C_4 and C_7 do not exist. The same for C_8 and C_{11}. For Figure 7.26(b), C_8 will give up this opportunity since otherwise it will incur unnecessary interference to the transmission of C_{10}. The situation in Figure 7.26(c)

Algorithm 7.1 Opportunistic LRR scheduling at each coordinator (at the beginning of a service channel slot)

1: **Input**: Coordinator C_i, segment ID i, round length $R = W_{opt} + 1$
2: **Output**: Whether to allow the relay access channel
3: If C_i is able to select a relay from i
4:      Broadcast a short signal in the subslot $i' \leftarrow i \bmod 3R$
        $ConflictSet(i) \leftarrow \emptyset$
5: For subslot j' from 0 to $3R - 1$//determine which segments have relays
6:      If sensed signal during j'
7:          $ConflictSet(i) \leftarrow ConflictSet(i) \cup C_{j'}, C_{j'} \in Segment j,$
          where $Segment j$ is the nearest one to i between the two:
          $j' + i - i'$ and $j' + i - i' \pm 3R$ //the most probable segment
8: Prune from $ConflictSet(i)$ the segments that are more than R
   segments away from i //regarded as not conflicting
9: Prune from $ConflictSet(i)$ segments j with $j \bmod R > i \bmod R$
   //the one nearer to a primary segment has higher priority
10: If $ConflictSet(i) \neq \emptyset$
11:      C_i tells relay in i to abort transmission
12: Else, C_i tells relay in i to access the channel in current service time slot

---

is a little different. Now C_6 and C_8 will try to take extra transmission opportunities left by empty segments 4 and 7 respectively. To avoid heavy collision between them, OLRR assigns each secondary segment a priority based on its distance to the primary segment with lower ID. In this case, C_8 has higher priority and will take this transmitting opportunity.

Finally, the reason we have $3 \times (W_{opt} + 1)$ subslots is to ensure that each coordinator will be able to determine a unique segment (w.h.p) that is transmitting in each subslot. Since the sensing process is purely based on detecting the energy, the time overhead can be negligible. In CodePlay, we set the sensing signal length to be 50 bytes and the length of each sub slot to be 100 μs, which takes preamble, SIFS, etc. into consideration. For $W_{opt} = 2$, the total extra time is $3 \times (2 + 1) \times 100 = 900$ μs, which is less than 2% of a service time slot with length of 50 ms.

7.6.7 Performance evaluation

We implemented and evaluated CodePlay by simulations using NS-2.34. The SLNC is implemented based on Katti *et al.* (2008), with an enhanced run-length coding technique, which is more suitable for consecutively broadcasting a generation of coded pieces in CodePlay. To ensure unique coordinator selection within the same segment, at the beginning of service time slots use an additional broadcast round (shorter than 1 ms) to resolve collisions between potential coordinators. The simulation scenario consists of a straight four-lane highway and one or two LMS source(s) (e.g., access points) can be located at one or both ends of the highway. The upper part of the highway (west bound) is regarded as the AoI. We simulate both dense and sparse VANETs by using two traffic densities: 66.7 cars/km and

Table 7.5 Parameter settings

Data rates for LMS and safety msg.	12 mbps, 3 mbps
Data communication range	$CR = 250$ m
Time per generation, piece size	2 s, 1 KB
Safety message length (with piggyback)	130B
Buffer capacity	15 generations
PriValue	32
No. of generations in priority region	$\alpha = 1$

Reproduced by permission of © 2011 IEEE.

35.5 cars/km. The vehicular speeds are randomly selected from 20–30 m/s. The simulation parameters are shown in Table 7.5.

The protocol for comparison is the PLNC version of CodePlay (CodePlay+PLNC) and the W_{opt} for PLNC is used. The closest state-of-the-art LMS scheme to ours is emergency video dissemination in VANETs using PLNC (NCDD), Park *et al.* 2010). However NCDD was not designed to meet the practical application layer requirements defined in this chapter, and it is hard to evaluate those metrics based on NCDD protocol. Thus we chose to not implement NCDD, but we have compared our results with the reported ones in Park *et al.* (2010).

The performance of CodePlay is evaluated by multiple metrics. 1. Initial buffering delay, which is the user experienced service delay. In the simulation, we impose the same initial buffering delay for all receiving vehicles. 2. Source rate, which reflects the supported LMS generation rate from the application layer. 3. Skip ratio, the fraction of generations skipped due to incomplete reception before playback time over all the generations that are played. Buffering level, the percentage of the buffered LMS contents between current playback time and physical world time. They both reflect the playback quality, i.e., smoothness (Wang and Li 2007).

7.6.7.1 Effect of number of LMS sources

We first consider how the LMS performance is affected by the number of sources (AP), i.e., only one AP, which is placed on one end of the highway, or two at both ends of it. Our main finding is that, the two-source case significantly outperforms the single-source case. Figure 7.27 shows the difference between using one and two APs under the dense highway with length $L = 2250$ m. Both protocols, CodePlay+SLNC and CodePlay+PLNC perform much better under the two-AP case than the single-AP case. Another observation is that CodePlay+PLNC cannot work well even in the two-AP case, the skip ratio of which is as high as 24%. However, the adoption of SLNC can reduce the skip ratio to less than 8%, which enables a much better playback experience.

This can be explained as follows. Because of lossy wireless links, a single flow is not able to sustain smooth playback of the LMS content after traversing a

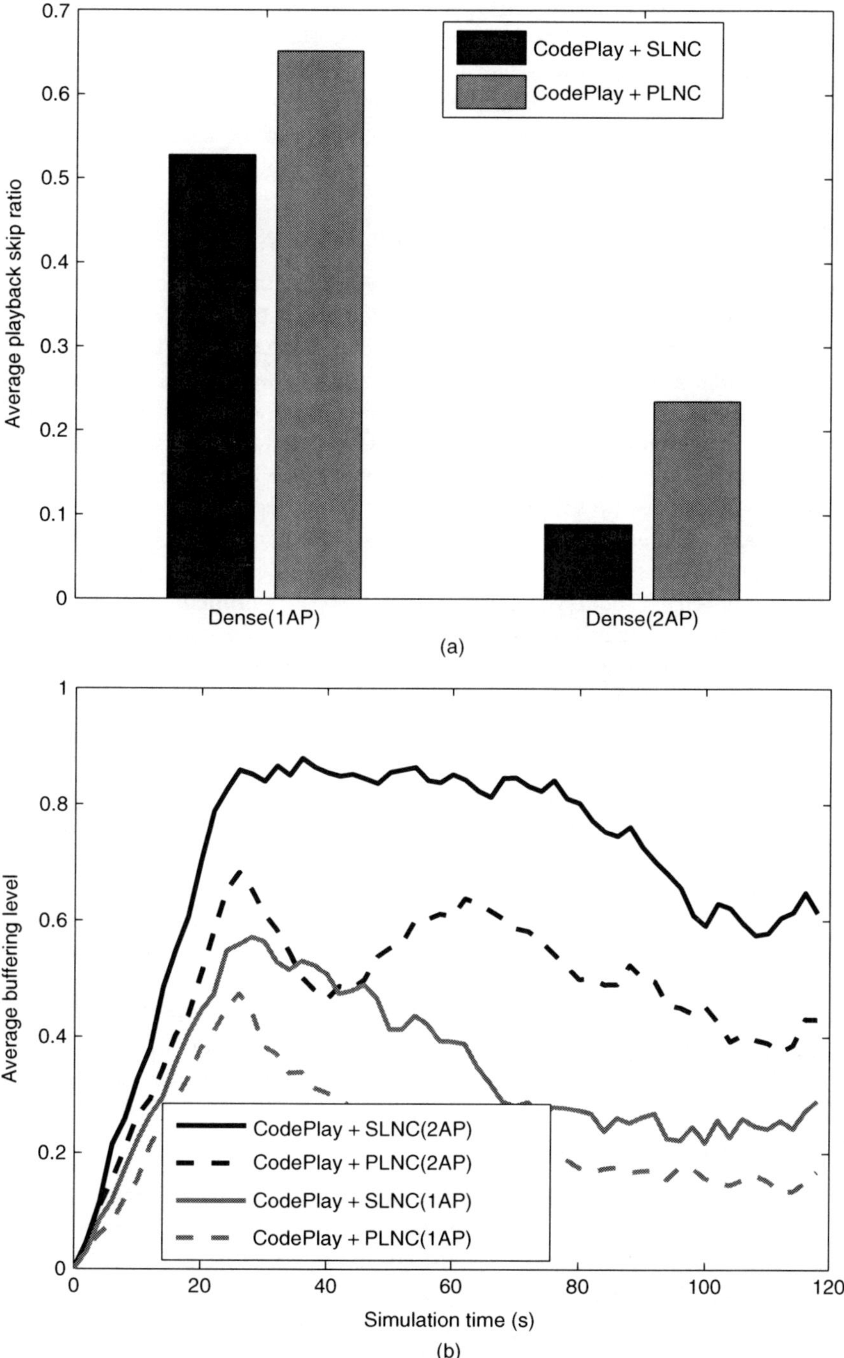

Figure 7.27 Comparison between using one and two APs, dense highway, source rate = 12 KB/s, initial buffering delay = 16 s. Reproduced by permission of © 2011 IEEE.

large number of hops in the VANET, which is also in line with the conclusions of routing throughput in multihop wireless networks. For two crossing flows with the same content, the packet losses are compensated by innovative symbols/packets from both directions. This can also be proved by the higher buffering levels in the two-AP case shown in Figure 7.27(b). Therefore, in the following we evaluate CodePlay based on the two-AP case.

7.6.7.2 Initial buffering delay and smooth playback

To further illustrate the advantage of CodePlay in providing better LMS services in VANET, we investigate the relationship between initial buffering delay, source rate and the metrics for smooth playback under a relatively sparse highway scenario. In the first simulation set, we fix initial the buffering delay as 16 s and increase the source rate from 24 KB/s to 30 KB/s. The results are presented in Figure 7.28. We can see that the skip ratio for CodePlay+SLNC is much lower that its PLNC-based component, where the former's skip ratio is 0 under 24 KB/s and 6% under 30 KB/s. This suggest that rates up to 30 KB/s could be supported without affecting smooth playback. Also, for each rate CodePlay+PLNC's buffering level decreases faster over time and is less stable compared with that of CodePlay+SLNC. This reflects that CodePlay+SLNC achieves a more stable flow of multimedia streaming, which shows the effectiveness of the integration of SLNC with the coordinated local push mechanism. We note that, the NCDD protocol only provided 10 KB/s source rate for video dissemination (Park *et al.* 2010).

In the second simulation set, we fix the source rate as 30 KB/s and increase the initial buffering delay from 16 to 24 s. From Figure 7.29, we can see an obvious reduction in the skip ratio for the CodePlay+SLNC, from 6% to 0.8%, and an increase in the buffering level for both protocols. This result is consistent with intuitions and implies that initial buffering delay plays an important role in VANET LMS services.

The CodePlay+SLNC works well through all source rates no greater than 30 KB/s, and for buffering delays of 16 s and 24 s. We argue that those delays are acceptable in VANETs. For example, for a delay equal to 16 s and vehicular velocity of 30 m/s, a car will travel about 500 m after it enters the AoI to begin playing an emergency multimedia content. For $L = 2250$ m, the car will be at 1750 m from the accident spot and may still have enough time to take actions.

7.6.7.3 Effect of traffic density

Next we study the performance of CodePlay under the dense traffic condition. Figure 7.30 shows the whole set of simulation results with various source rates and buffering delays. Though CodePlay+SLNC still outperforms CodePlay+PLNC, compared with the sparse case, the skip ratios of both protocols are higher and buffering levels lower. Especially, the skip ratio reaches up to more than 10%, which could be unacceptable from application layer. We have observed (not shown) that the relay selection is almost always unique and is highly reliable, therefore

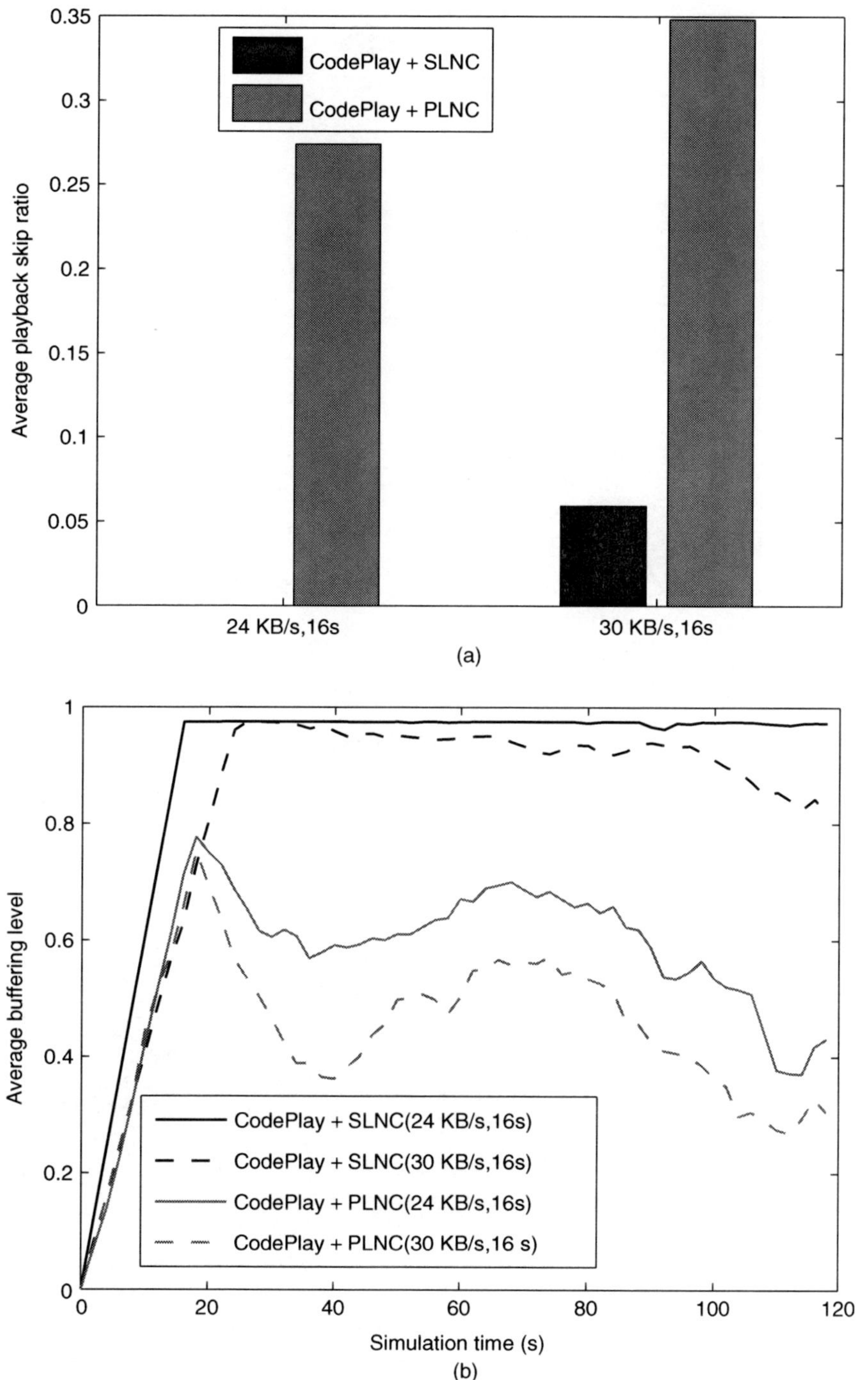

*Figure 7.28   Fixed initial buffering delay, varying source rates. Sparse highway.
Reproduced by permission of © 2011 IEEE.*

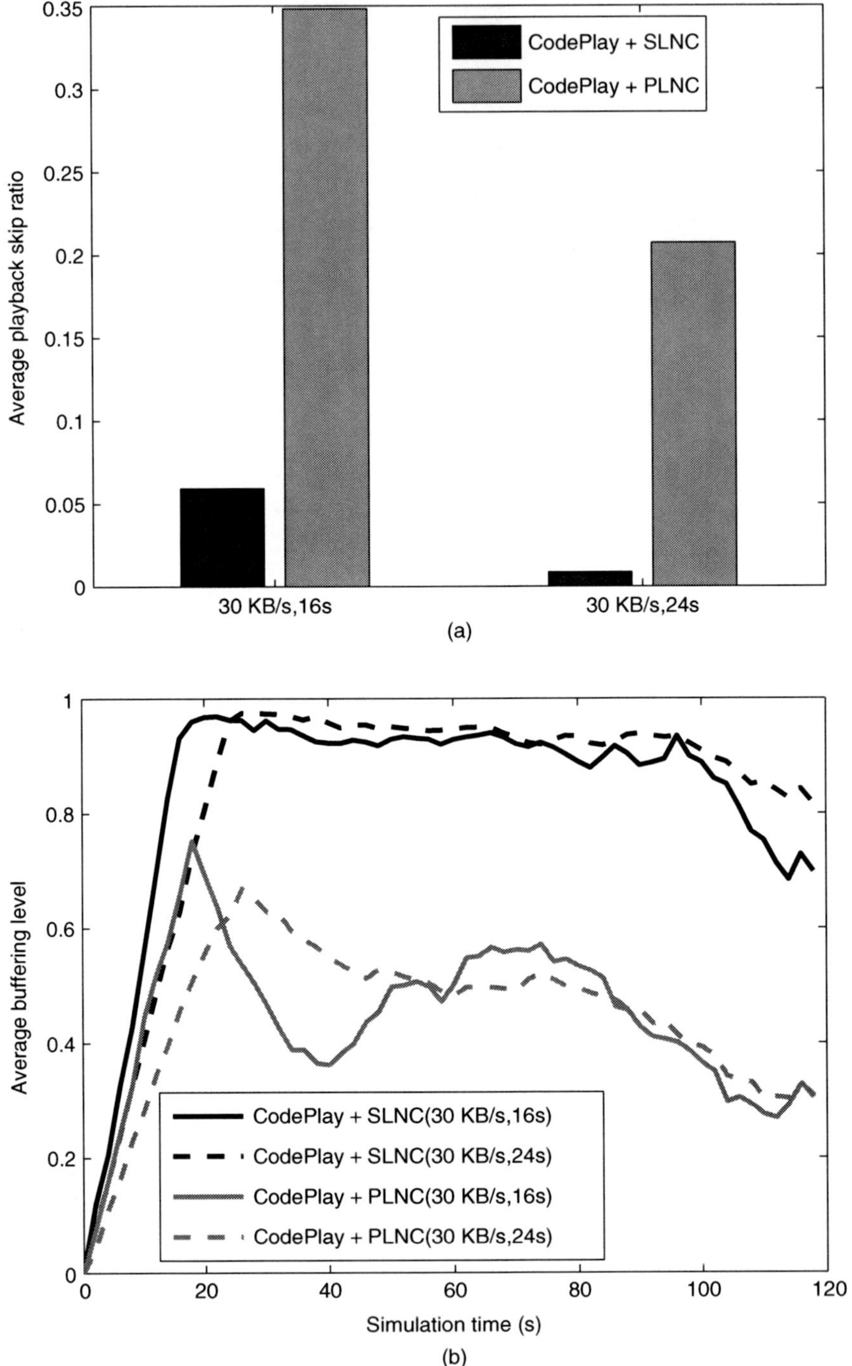

Figure 7.29   Fixed rate, varying initial buffering delay. Sparse highway. Reproduced by permission of © 2011 IEEE.

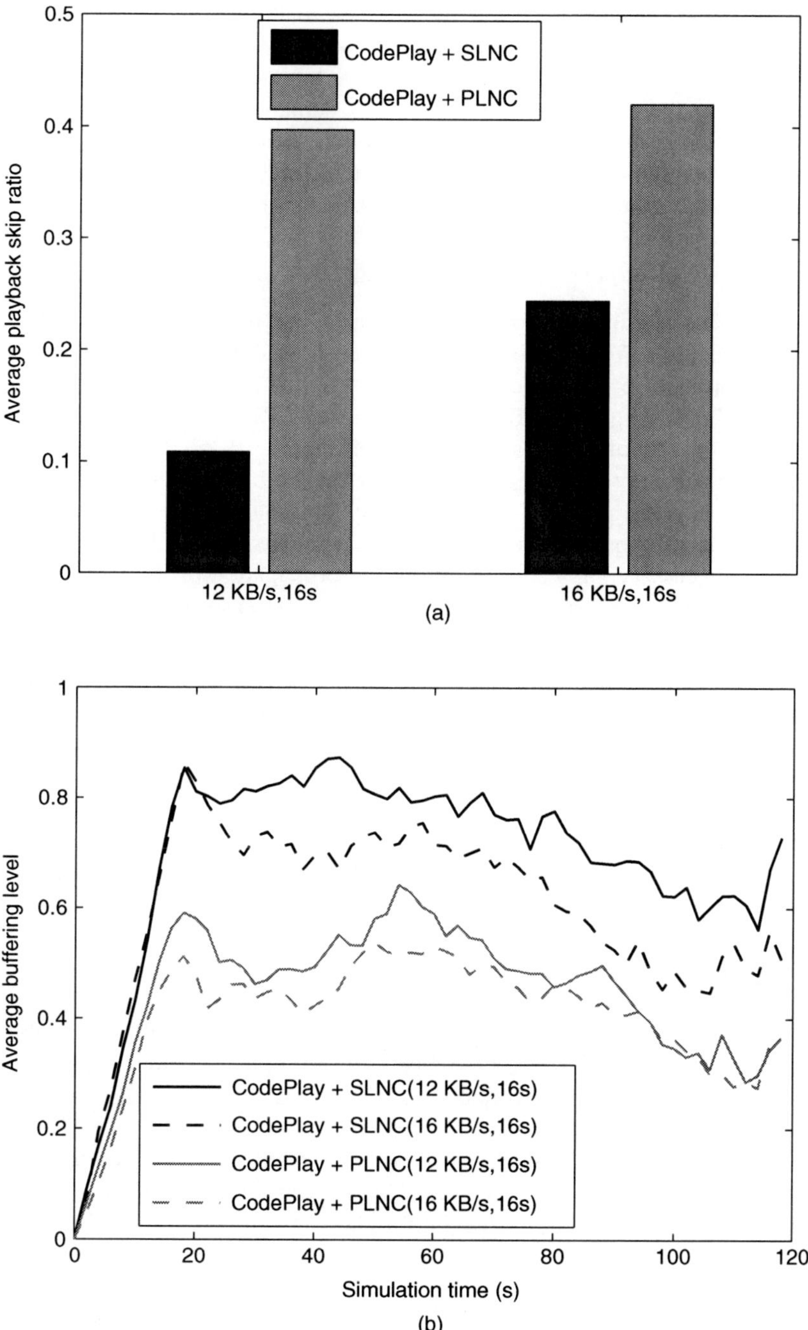

Figure 7.30   Impact of traffic density. Dense highway. Reproduced by permission of © 2011 IEEE.

the worse performance can be mainly ascribed to limitations in the node utility functions, which are directly associated with how much innovative information a relay can deliver to all neighboring nodes. For broadcasting in a dense VANET, since there could be too many vehicles urgently demanding different portions of the LMS content, it is intrinsically hard to satisfy all their needs in a short time. Due to the time constraints of LMS applications this leads to more frequent playback skips than in the sparse VANETs.

7.6.7.4   Effect of opportunistic scheduling

In the previous simulations for sparse scenario, we have the OLRR scheduling enabled by default. Yet it is interesting to see how the opportunistic scheduling affects the protocol performance. Thus, we presented in Figure 7.31 the results of enabling and disabling the OLRR algorithm (using LRR instead). All the protocols run with source rate of 30 KB/s and initial buffering delay of 16 s. We can see that the OLRR much improves the performance over the basic LRR algorithm, which reduces the skip ratio from 20% to 6%. By opportunistically utilizing the idle scheduled transmission slots left by primary segments, the OLRR can adaptively "fill" the unnecessary gaps created during the propagation of the LMS flow. And this mechanism works especially well for SLNC, because the transmission tends to be more reliable over larger distances.

Remark. To summarize, from the above results of CodePlay, we can obtain the following main implications: 1. LMS services in VANET with high source rates are hard, yet feasible to provide with satisfiable user experience. Even using SLNC, we may need the help of few additional infrastructure (APs) along the road to facilitate the dissemination of LMS. 2. Using CodePlay with SLNC, the playback smoothness can be greatly enhanced over traditional protocols for source rates up to 30 KB/s, and with acceptable buffering delay, especially in sparse VANETs.

7.7   Conclusion

In this chapter we further investigated the transmission coordination problem in opportunistic routing and studied how the strict coordination requirement in ExOR can be greatly eased using network coding. First we reviewed MORE, a well-known state-of-the-art MAC-independent opportunistic routing protocol. After identifying the inadequacy of MORE in dealing with multicast and broadcast, we demonstrated how network coding can be integrated with opportunistic listening in wireless broadcast. In particular, we looked into a class of challenging problem – mobile content distribution (MCD) in VANET, where large files are broadcast proactively from a few APs to vehicles inside an interested area. To combat the lossy wireless transmissions in VANETs we leverage symbol level network coding (SLNC), which exploits symbol-level diversity to achieve better error-tolerance compared with traditional packet level network coding (PLNC). We then qualitatively characterize the advantages of SLNC compared with PLNC from two aspects, namely higher throughput and spacial-reusability. Using two typical MCD applications as

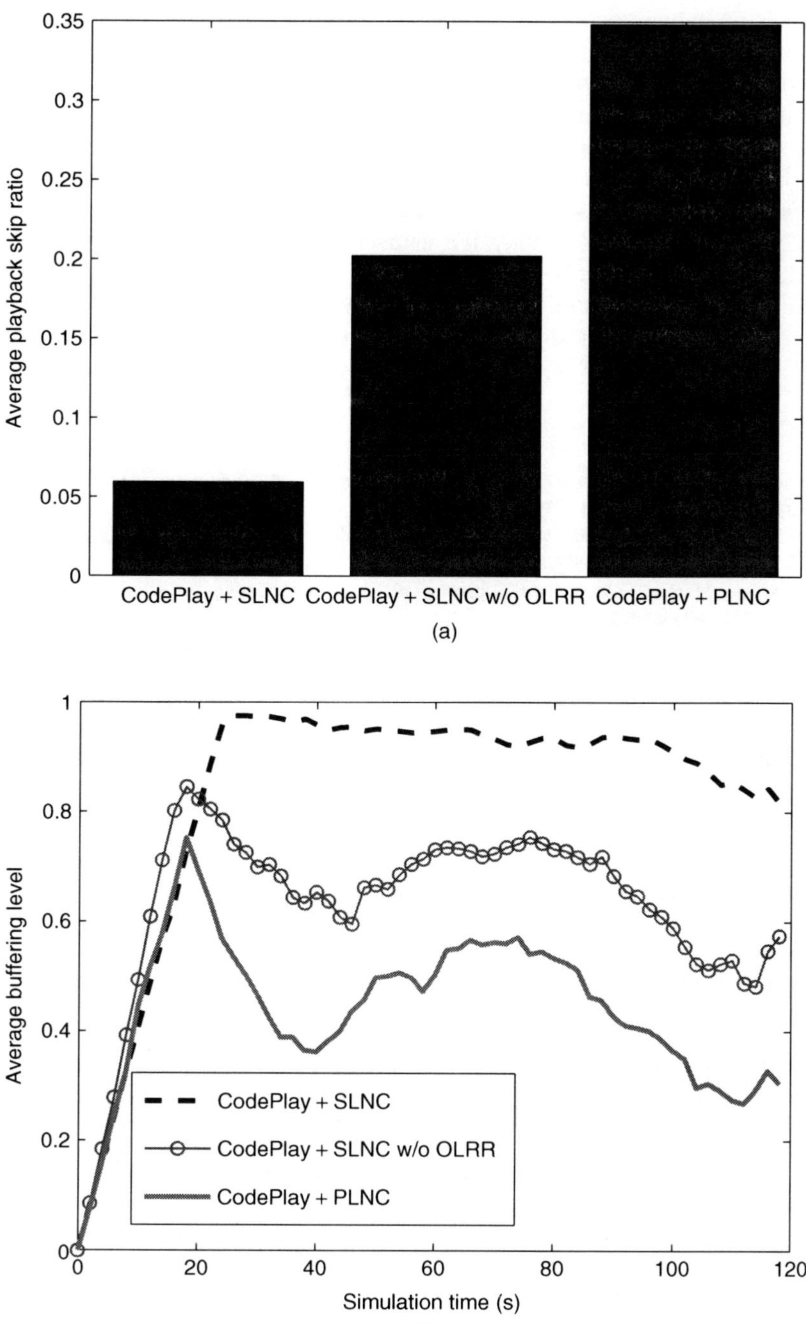

Figure 7.31   Effect of opportunistic transmission scheduling. Reproduced by permission of © 2011 IEEE.

examples – popular content distribution and live multimedia streaming, we present two novel push-based broadcast schemes – CodeOn and CodePlay, respectively. The common ideas underlying the two schemes include a prioritized relay selection algorithm that opportunistically maximizes the usefulness of transmitted content, and a simple transmission coordination mechanism that exploits the higher spacial reusability brought by SLNC. Compared with state-of-the-art pull-based contend distribution protocols, CodeOn and CodePlay both achieve significant performance gains. Through simulation study the gain can be partly attributed to the use of SLNC, and partly attributed to the new push-based protocol design. Finally, thanks to the use of opportunistic listening and network coding (especially SLNC), the challenging problem of designing a MCD protocol in VANETs is solved elegantly.

References

Ahlswede R, Cai N, Li SY and Yeung R 2000 Network information flow. *IEEE Transactions on Information Theory* **46**(4), 1204–1216.

Ahmed S and Kanhere SS 2006 Vanetcode: network coding to enhance cooperative downloading in vehicular ad-hoc networks *IWCMC '06*, pp. 527–532.

Biswas S and Morris R 2005 Exor: Opportunistic multi-hop routing for wireless networks *SIGCOMM'05*, Philadelphia, PA.

Bonuccelli M, Giunta G, Lonetti F and Martelli F 2007 Real-time video transmission in vehicular networks. *MoVE*.

Brodsky MZ and Morris RT 2009 In defense of wireless carrier sense *SIGCOMM '09*, pp. 147–158.

Bucciol P, Masala E, Kawaguchi N, Takeda K and De Martin J 2005 Performance evaluation of h. 264 video streaming over inter-vehicular 802.11 ad hoc networks *PIMRC 2005*, pp. 1936–1940.

Chachulski S, Jennings M, Katti S and Katabi D 2007 Trading structure for randomness in wireless opportunistic routing *Proceedings of the 2007 Conference on Applications, Technologies, Architectures, and Protocols for Computer Communications*, pp. 169–180 SIGCOMM '07.

De Couto DSJ, Aguayo D, Bicket J and Morris R 2003 A high-throughput path metric for multi-hop wireless routing *Proceedings of the 9th Annual International Conference on Mobile Computing and Networking*, pp. 134–146 MobiCom '03.

Dutta A, Saha D, Grunwald D and Sicker D 2009 Smack: a smart acknowledgment scheme for broadcast messages in wireless networks. *SIGCOMM Computing Communication Review* **39**, 15–26.

Fiore M and Barcelo-Ordinas JM 2009 Cooperative download in urban vehicular networks *IEEE MASS '09*.

Guo M, Ammar M and Zegura E 2005 V3: a vehicle-to-vehicle live video streaming architecture *PerCom '05*, pp. 171–180.

Ho T, Medard M, Koetter R, Karger D, Effros M, Shi J and Leong B 2006 A random linear network coding approach to multicast. *IEEE Transactions on Information Theory* **52**(10), 4413–4430.

Jain K, Padhye J, Padmanabhan VN and Qiu L 2005 Impact of interference on multi-hop wireless network performance. *Wireless Networks* **11**, 471–487.

Jiang D, Taliwal V, Meier A, Holfelder W and Herrtwich R 2006 Design of 5.9ghz dsrc-based vehicular safety communication. *IEEE Wireless Communications* **13**(5), 36–43.

Johnson M, Nardis LD and Ramch K 2006 Collaborative content distribution for vehicular ad hoc networks. Allerton Conference Communication, Control, and Computing, Monticello, IL, September 2006.

Katti S, Katabi D, Balakrishnan H and Medard M 2008 Symbol-level network coding for wireless mesh networks *SIGCOMM '08*, pp. 401–412.

Koetter R and Médard M 2003 An algebraic approach to network coding. *IEEE/ACM Trans. Netw.* **11**, 782–795.

Koutsonikolas D, Hu Y and Wang CC 2009 Pacifier: High-throughput, reliable multicast without "crying babies" in wireless mesh networks *INFOCOM 2009, IEEE*, pp. 2473–2481.

Lee SH, Lee U, Lee KW and Gerla M 2008 Content distribution in vanets using network coding: The effect of disk i/o and processing o/h *Sensor, Mesh and Ad Hoc Communications and Networks, 2008. SECON '08. 5th Annual IEEE Communications Society Conference on*, pp. 117–125.

Lee U, Park JS, Yeh J, Pau G and Gerla M 2006 Code torrent: content distribution using network coding in vanet *MobiShare '06*, pp. 1–5.

Li M, Lou W and Zeng K 2009 Oppcast: Opportunistic broadcast ofwarning messages in vanets with unreliable links *IEEE MASS'09*.

Li M, Yang Z and Lou W 2010 Codeon: Cooperative popular content distribution for vehicular networks using symbol level network coding. *Technical Report, ECE, WPI*.

Li M, Yang Z and Lou W 2011 Codeon: Cooperative popular content distribution for vehicular networks using symbol level network coding. *IEEE Journal on Selected Areas in Communications(JSAC)*. **20**(1), 223–235.

Mak T, Laberteaux K, Sengupta R and Ergen M 2009 Multichannel medium access control for dedicated short-range communications. *IEEE Transactions on Vehicular Technology* **58**(1), 349–366.

Nandan A, Das S, Pau G, Gerla M and Sanadidi MY 2005 Co-operative downloading in vehicular ad-hoc wireless networks *WONS '05*, pp. 32–41.

Nandan A, Tewari S, Das S, Gerla M and Kleinrock L 2006 Adtorrent: Delivering location cognizant advertisements to car networks. The Third International Conference on Wireless On Demand Network Systems and Services (WONS 2006), Les Menuires, France, January.

Ni SY, Tseng YC, Chen YS and Sheu JP 1999 The broadcast storm problem in a mobile ad hoc network *Proceedings of the 5th Annual ACM/IEEE International Conference on Mobile Computing and Networking*, pp. 151–162 MobiCom '99. ACM, New York, NY, USA.

Park JS, Lee U and Gerla M 2010 Vehicular communications: emergency video streams and network coding. *Journal of Internet Services and Applications* **1**, 57–68.

Park JS, Lee U, Oh SY, Gerla M and Lun DS 2006 Emergency related video streaming in vanet using network coding *ACM VANET '06*, pp. 102–103.

Qadri N, Altaf M, Fleury M, Ghanbari M and Sammak H 2009 Robust video streaming over an urban vanet. *IEEE Wireless and Mobile Computing, Networking and Communication*, pp. 429–434.

Ramachandran K, Gruteser M, Onishi R and Hikita T 2007 Experimental analysis of broadcast reliability in dense vehicular networks. *IEEE Vehicular Technology Magazine* **2**(4), 26–32.

Seferoglu H and Markopoulou A 2009 Video-aware opportunistic network coding over wireless networks. *IEEE JSAC* **27**(5), 713–728.

Soldo F, Casetti C, Chiasserini CF and Chaparro P 2008 Streaming media distribution in vanets. *GLOBECOM '08* pp. 1–6.

Taliwal V, Jiang D, Mangold H, Chen C and Sengupta R 2004 Empirical determination of channel characteristics for dsrc vehicle-to-vehicle communication *ACM VANET '04*. ACM.

Talucci F, Gerla M and Fratta L 1997 Maca-bi (maca by invitation)-a receiver oriented access protocol for wireless multihop networks *The 8th IEEE International Symposium on Personal, Indoor and Mobile Radio Communications, 1997. 'Waves of the Year 2000'. PIMRC '97.*, vol. 2, pp. 435–439, vol.2.

Torrent-Moreno M, Jiang D and Hartenstein H 2004 Broadcast reception rates and effects of priority access in 802.11-based vehicular ad-hoc networks *ACM VANET '04*. ACM.

Torrent-Moreno M, Killat M and Hartenstein H 28-25 Sept., 2005 The challenges of robust inter-vehicle communications. *IEEE VTC 2005* **1**, 319–323.

Torrent-Moreno M, Schmidt-Eisenlohr F, Fussler H and Hartenstein H 2006 Effects of a realistic channel model on packet forwarding in vehicular ad hoc networks *IEEE WCNC*, pp. 385–391.

Wang M and Li B 2007 R2: Random push with random network coding in live peer-to-peer streaming. *IEEE JSAC* **25**(9), 1655–1666.

WAVE 2006 Ieee trial-use standard for wireless access in vehicular environments (wave)–multi-channel operation. *IEEE Std 1609.4-2006* pp. c1–74.

Xu Q, Mak T, Ko J and Sengupta R 2004 Vehicle-to-vehicle safety messaging in dsrc *ACM VANET '04*, pp. 19–28.

Yang Z, Li M and Lou W 2009 R-code: network coding based reliable broadcast in wireless mesh networks with unreliable links *IEEE GLOBECOM'09*, pp. 2168–2173.

Yang Z, Li M and Lou W 2010 Codeplay: Live multimedia streaming in vanets using symbol-level network coding *Eighteenth International Conference on Network Protocols, 2010. (ICNP '2010) Proceedings*.

Yang Z, Li M and Lou W 2011 R-code: Network coding-based reliable broadcast in wireless mesh networks. *Elsevier Ad Hoc Networks*.

Ye W, Heidemann J and Estrin D 2002 An energy-efficient mac protocol for wireless sensor networks *INFOCOM 2002. Twenty-First Annual Joint Conference of the IEEE Computer and Communications Societies. Proceedings. IEEE*.

Zhang J, Zhang Q and Jia W 2009a Vc-mac: A cooperative mac protocol in vehicular networks. *IEEE Transactions on Vehicular Technology* **58**(3), 1561–1571.

Zhang Y, Zhao J and Cao G 2009b Roadcast: A popularity aware content sharing scheme in vanets. *IEEE ICDCS '09* pp. 223–230.

Zhao J, Arnold T, Zhang Y and Cao G 2008 Extending drive-thru data access by vehicle-to-vehicle relay *ACM VANET '08*, pp. 66–75.

Zhao J, Zhang Y and Cao G 2007 Data pouring and buffering on the road: A new data dissemination paradigm for vehicular ad hoc networks. *IEEE Transactions on Vehicular Technology* **56**(6), 3266–3277.

Zimu Liu, Chuan Wu BLSZ 2010 Uusee: Large-scale operational on-demand streaming with random network coding *INFOCOM '10*.

8

Multirate geographic opportunistic routing protocol design

In this chapter, we carry out a study on the impacts of multiple rates, as well as candidate selection, prioritization and coordination, on the performance of GOR. We propose a new local metric, **opportunistic effective one-hop throughput (OEOT)**, to characterize the tradeoff between the packet advancement and one-hop packet forwarding time. We further propose a local rate adaptation and candidate selection algorithm to approach the optimum of this metric. Simulation results show that the multirate GOR (MGOR) incorporating the rate adaptation and candidate selection algorithm achieved higher throughput and lower delay than the corresponding single-rate and multirate traditional geographic routing and opportunistic routing protocols.

The rest of this chapter is organized as follows. We discuss the impacts of multirate capability, forwarding strategy and candidate coordination delay on the throughput of opportunistic routing in Section 8.2. The local metric is introduced in Section 8.3. We propose the heuristic algorithm in Section 8.4. A multirate link quality measurement mechanism is proposed in Section 8.5. Simulation results are presented and analyzed in Section 8.6. Section 8.7 concludes the chapter.

8.1 System model

In this chapter, we consider the local MGOR scenario as in the example in Figure 8.1. Assume node S, i.e., the sender, is forwarding a packet to a remote

Multihop Wireless Networks: Opportunistic Routing, First Edition.
Kai Zeng, Wenjing Lou and Ming Li.
© 2011 John Wiley & Sons, Ltd. Published 2011 by John Wiley & Sons, Ltd.

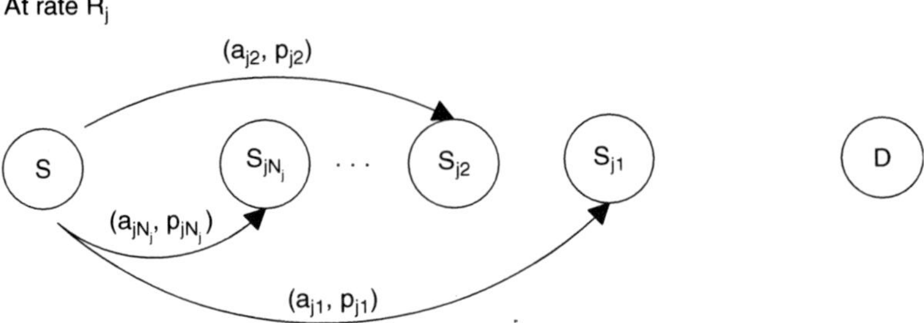

Figure 8.1   Node S is forwarding a packet to a remote destination D with transmission rate R_j. Reproduced by permission of © 2009 IEEE.

destination D. S can transmit the packet at k different rates $R_1, R_2, \ldots, R_k$. Each rate corresponds to a **communication range**, within which the nodes can receive the packet sent by S with some non-negligible probability, which is larger than a threshold, e.g., 0.1. The **available next-hop node set** C_j $(1 \le j \le k)$ of node S under a particular transmission rate R_j is defined as all the nodes in the communication range of S that are closer to D than S. We denote the nodes in C_j as $s_{j_1}, s_{j_2}, \ldots, s_{j_{N_j}}$, where $N_j = |C_j|$. In a similar way to geographic routing (Karp and Kung 2000; Lee *et al.* 2005; Seada *et al.* 2004), we assume S is aware of the location information of itself, its one-hop neighbors and the destination D. Define the **packet advancement** as $a_{j_m} 1 \le m \le N_j$ in Equation (8.1), which is the Euclidian distance between the sender and destination ($d(S, D)$) minus the Euclidian distance between the neighbor s_{j_m} and destination ($d(s_{j_m}, D)$).

$$a_{j_m} = d(S, D) - d(s_{j_m}, D) \tag{8.1}$$

Then at each rate R_j, each node in C_j is associated with one pair, (a_{j_m}, p_{j_m}), where p_{j_m} is the data packet reception ratio (PRR) from node S to s_{j_m}. Note that for different data rates, the PRR from node S to the same neighbor may be different. Let $\mathcal{F}_j$ denote the **forwarding candidate set** of node S at rate R_j, which contains the nodes that participate in the local opportunistic forwarding. Note that here $\mathcal{F}_j$ is a subset of C_j, while in the existing pure opportunistic routing schemes (Biswas and Morris 2005; Zorzi and Rao 2003), $\mathcal{F}_j = C_j$.

The multirate GOR (MGOR) procedure is as follows: node S decides a transmission rate R_j, and selects $\mathcal{F}_j$ based on its knowledge of C_j (a_{j_m}'s and p_{j_m}'s); then it broadcasts the data packet to the forwarding candidates in $\mathcal{F}_j$ at rate R_j after detecting the channel is idle for a while. Candidates in $\mathcal{F}_j$ follow a specific priority to relay the packet – that is, a forwarding candidate will only relay the packet if it has received the packet correctly and all the nodes with higher priorities failed to do so. The actual forwarder will become a new sender and suppress all the other potential forwarders in $\mathcal{F}_j$. When no forwarding candidate has successfully received the packet, the sender will retransmit the packet if retransmission is enabled. The sender will drop the packet when the retransmissions reach the limit. This procedure iterates until the packet arrives at the destination.

In this chapter we use a contention-based MAC protocol like 802.11 and apply a compressed slotted acknowledgement mechanism similar to that in Zubow *et al.* (2007) to coordinate the relay priority among the candidates, which is described as follows. After sensing the channel has been idle for a DIFS (distributed interframe space), the sender broadcasts the data packet at the selected rate. In the header of the packet, the intended MAC addresses of the forwarding candidates and the corresponding relay priorities are identified. If the first-priority candidate receives the packet correctly, it broadcasts an ACK with a delay of SIFS (short interframe space) after the successful data reception. The ACK is used for informing the sender of the data packet reception as well as suppressing lower priority candidates from forwarding duplicated copies. If the first-priority candidate does not receive the packet, it just remains silent. For the second-priority candidate, it sets a waiting period of $2T_{SIFS} - T_{rx/tx}$ after it received the data packet correctly, where T_{SIFS} and $T_{rx/tx}$ is the time duration of SIFS and radio receive/transmit status turnaround delay, respectively. If within the waiting period, it detects a transmission emerged (e.g. a significant signal strength increase) in the channel, the ACK packet is considered as sent. Then it just drops the received packet. On the other hand, if no transmission emergence is detected, the second-priority candidate concludes that the highest prioritized candidate did miss the data packet. So the second-priority candidate will turn around its radio from receiving status to transmitting status and send out the ACK with $2T_{SIFS}$ delay after it received the packet. Generally, the ith-priority ($i > 1$) candidate that receives the data packet will set a waiting period as $i \times T_{SIFS} - T_{rx/tx}$ after the data packet reception. If it detects a transmission emerging in this period, it will suppress itself from forwarding the packet; otherwise, it will send out an ACK at $i \times T_{SIFS}$ to claim its reception. In Section 8.2.4, we will further elaborate on the impact of reliability of this ACK technique on the performance of OR.

8.2   Impact of transmission rate and forwarding strategy on OR performance

In this section, we discuss the factors that affect the one-hop performance in terms of throughput and delay of OR. These factors include rate and forwarding strategy, which further includes candidate selection, prioritization and coordination.

The impacts of transmission rate on the performance of opportunistic routing are twofold. On the one hand, different rates achieve different transmission ranges, which lead to different neighborhood diversity. Explicitly, high-rate causes short transmission range, then in one hop, there are few neighbors around the sender, which presents low neighborhood diversity. Low-rate is likely to have long transmission range, therefore achieves high neighborhood diversity. So from the diversity point of view, low rate may be better. On the other hand, although low rate brings the benefit of larger one-hop distance, which results in higher neighborhood diversity and fewer hop counts to reach the destination, it is still possible to achieve a low effective end-to-end throughput or high delay because it needs more time to transmit a packet at lower rate. So it is nontrivial to decide which rate is indeed better.

Besides the inherent rate-distance, rate-diversity and rate-hop tradeoffs, which affect the performance of opportunistic routing, the forwarding strategy will also have an impact on the performance. That is, for a given transmission rate, different candidate forwarding sets, relay priority assignments, and candidate coordinations will all affect the OR performance.

In the following subsections, we will examine the impact of transmission rate and forwarding strategy on the one-hop performance of opportunistic routing, which leads us to the design of an efficient local rate adaptation and candidate selection scheme. First we will analyze the one-hop packet forwarding time introduced by opportunistic routing.

8.2.1 One-hop packet forwarding time of opportunistic routing

We define the one-hop packet forwarding time cost by the ith candidate as the period from the time when the sender is going to transmit the packet to the time when the ith candidate becomes the actual forwarder. Although the one-hop packet forwarding time varies for different MAC protocols, for any protocol, it can be divided into two parts. One part is introduced from the sender and the other part is introduced from the candidate coordination, which are defined as follows:

- T_s: the sender delay, which can be further divided into three parts: channel contention delay (T_c), data transmission time (T_d) and propagation delay (T_p):

$$T_s = T_c + T_d + T_p \tag{8.2}$$

  For a contention-based MAC protocol (like 802.11), T_c is the time needed for the sender to acquire the channel before it transmits the data packet, which includes the backoff time and Distributed Interframe Space (DIFS). T_d is equal to protocol header transmission time (T_h) plus data payload transmission time (T_{pl}), which is

$$T_d = T_h + T_{pl} \tag{8.3}$$

  where T_h is determined by physical layer preamble and MAC header transmitting time, and T_{pl} is decided by the data payload length L_{pl} and the data transmission rate. The payload may be transmitted at different rates.
  T_p is the time for the signal propagating from the sender to the candidates, which can be ignored when electromagnetic wave is transmitted in the air.

- $T_f(i)$: the ith forwarding candidate coordination delay which is the time needed for the ith candidate to acknowledge the sender and suppress other potential forwarders. Note that $T_f(i)$ is an increasing function of i, since the lower priority forwarding candidates always need to wait and confirm that no higher priority candidates have relayed the packet before it takes its turn to relay the packet. For the protocol we introduced in Section 8.1, $T_f(i) = i \times T_{SIFS} + T_{ACK}$, where T_{ACK} is the ACK transmission time.

Thus, the total medium time needed for a packet forwarding from the sender to the ith forwarding candidate is

$$t_i = T_s + T_f(i) \tag{8.4}$$

8.2.2   Impact of transmission rate

We examine the impact of transmission rate on the one-hop throughput of OR by using two examples. In one example transmission at higher rate is better, while in the other example lower rate achieves higher throughput. The one-hop throughput is defined as bit-meters successfully delivered per second with unit bmps. The one-hop delay per bit-meter is the inverse of the throughput. So higher throughput implies lower delay in this context.

Assume the data payload $L_{pl} = 1000$ bytes, $T_{SIFS} = 10\,\mu s$, $T_{ACK} = 192\,\mu s$, $T_h = 200\,\mu s$, and the sender delay only includes the data transmission time (T_d). According to Equations (8.2), (8.3), (8.4) and the MAC protocol we discussed in Section 8.1, $t_i = \frac{8000}{R_j} + 10i + 392\,\mu s$. In Figure 8.2, assume at each rate, the neighbor closer to the destination is assigned higher relay priority. Suppose S sends out N packets. Then when $R_j = 11\,mbps$, there are $L_{pl}(300 \cdot 0.7N + 200 \cdot 0.95 \cdot 0.3N) = 2.136N$ megabit-meters are delivered, and the corresponding total packet forwarding time is $(t_1 \cdot 0.7N + t_2 \cdot 0.3N) = 1132.27N\,\mu s$. So the one-hop throughput is $1.886G$ bmps. Similarly, the one-hop throughput at $5.5\,mbps$ is $1.651G$ bmps, which is smaller than the throughput at $11\,mbps$. That is, in this example, although lower rate introduces more spacial diversity (more neighbors), this benefit does not make up the cost on the longer medium time. Now let's assume the neighbor s_3 is removed from Figure 8.2 for each rate. Then the one-hop throughput is $1.60G$ bmps and $1.49G$ bmps at $5.5\,mbps$ and $11\,mbps$, respectively. So transmitting at lower rate is better than higher rate in this case, because the extra spacial diversity brought by lower rate does help to improve the packet advancement but only introduce moderate extra packet forwarding time.

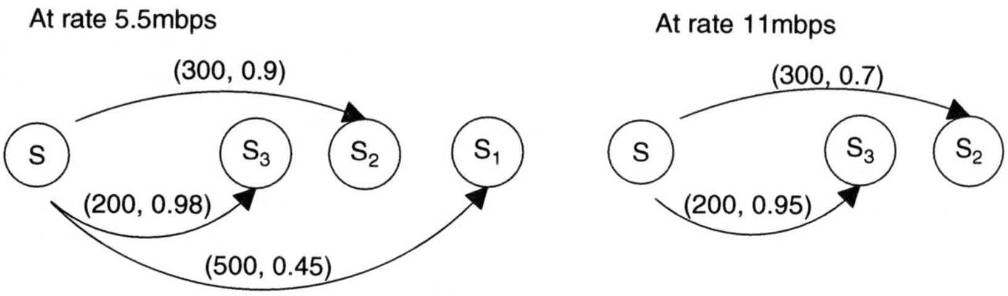

Figure 8.2   Different transmission rates result in different next-hop neighbor sets. Reproduced by permission of © 2009 IEEE.

8.2.3 Impact of forwarding strategy

We have seen that multirate capability has an impact on throughput and delay. Other than this factor, for any given rate, different candidate prioritization also results in different throughput and delay in opportunistic routing. Still using the example in Figure 8.2 at a rate of 5.5 mbps, if we assign s_2 the highest priority, then s_1, then s_3, the one-hop throughput is 1.306G bmps, which is lower than that achieved by assigning higher priority to the candidate closer to the destination. Actually, it has been proved in (Zeng *et al.* 2007) that giving candidates closer to the destination higher priorities achieves maximum expected packet advancement (EPA).

8.2.4 Impact of candidate coordination

The coordination delay is another key factor affecting the packet forwarding time and one-hop throughput. When this delay is much larger than the sender delay, then it would be better to retransmit the packet instead of waiting for other forwarding candidates to relay the packet in order to save the packet forwarding time. When this delay is negligible, we should involve all the available next-hop neighbors into opportunistic forwarding because any extra candidates would help to improve the relay reliability but without introducing any extra delay. We should also give candidates closer to the destination higher relay priorities, since larger advancement candidates should always try first in order to maximize the EPA. If they fail to relay the packet, the lower-priority candidates could instantaneously relay the correctly received packet without having to wait. The coordination delay therefore has a great impact on throughput. Since we use the compressed slotted acknowledgement, which introduces a small coordination delay among candidates, it would be better to give candidates closer to the destination higher relay priorities.

In the compressed slotted acknowledgement mechanism, ACK plays two roles: one is to acknowledge the sender of data reception, the other is to suppress other candidates from forwarding duplicated packets. We discuss the reliability of this mechanism according to these two ACK roles. Firstly, following the collision-avoidance rule, each node should sense the channel to be clear for at least DIFS before transmission. Since the ith-priority candidate broadcasts the ACK with a short delay ($i \times T_{SIFS}$, which is usually shorter than DIFS in our scheme) after successful packet reception, the ACK is unlikely to collide with other transmissions at the sender side. The empirical results in (Sang *et al.* 2007) also confirm that ACK can be received by the sender with high probability. Furthermore, since the ACK is transmitted at the basic rate (1 mbps), the ACK link from the candidate to the sender should be more reliable than the data link from the sender to the candidate. So when the candidate correctly receives the data packet from the sender, the ACK can usually be correctly received by the sender with high probability. Secondly, since all the forwarding candidates are in the data transmission range of the sender, the longest possible distance between any two candidates is twice that of the data transmission range. Typically, carrier sensing range is around double the data transmission range. So any two forwarding candidates will be in the carrier sensing range of each other. Then lower prioritized candidates should be able to detect a transmission appearing in the channel if a higher prioritized candidate does send out an ACK. A false positive could occur when a lower priority candidate senses a transmission emergence but it is from other transmission source. In this case, the lower priority candidate would drop its received packet. If all the lower priority candidates that have received the packet correctly believe there is a higher priority candidate that has received the packet but actually there is not, no ACK would be sent back to the sender, then the sender would retransmit the packet. However, the probability of other transmissions emerging in the short coordination period (multiple SIFS) and suppressing all the potential forwarding candidates should be relatively low.

8.3 Opportunistic effective one-hop throughput (OEOT)

According to the analysis above, for a given next-hop neighbor set C_j, we now introduce the local metric, *Opportunistic Effective One-hop Throughput* (OEOT) (in Equation (8.5)), to characterize the local behavior of GOR in terms of bit-meter advancement per second.

$$OEOT(\mathcal{F}_j) = L_{pl} \cdot \frac{\sum_{i=1}^{r} a_{j_i} p_{j_i} \cdot \prod_{w=0}^{i-1} \overline{p}_{jw}}{t_r \overline{P}_{\mathcal{F}_j} + \sum_{i=1}^{r} t_i p_{j_i} \cdot \prod_{w=0}^{i-1} \overline{p}_{jw}} \tag{8.5}$$

where $\mathcal{F}_j = \langle s_{j_1}, \ldots, s_{j_r} \rangle$, which is an ordered subset of C_j with priority $s_{j_1} > \ldots > s_{j_r}$; $r = |\mathcal{F}_j|$; $p_{j_0} := 0$; $\overline{p}_{jw} = 1 - p_{jw}$; and

$$\overline{P}_{\mathcal{F}_j} = \prod_{i=1}^{r}(1 - p_{j_i}) \tag{8.6}$$

which is the probability of none of the forwarding candidates in $\mathcal{F}_j$ successfully receiving the packet in one physical transmission from the sender.

The physical meaning of the OEOT defined in Equation (8.5) is the expected bit advancement per second for a local GOR procedure when the sender S transmits the packet at rate R_j. Opportunistic Effective One-hop Throughput integrates the factors of packet advancement, relay reliability, and one-hop packet forwarding time. Now for multirate GOR, our goal is to select an R_j and the corresponding $\mathcal{F}_j$ to locally maximize this metric. The intuitions to locally maximize the OEOT are as follows. 1. As the end-to-end achievable throughput is smaller than per-hop throughput on each link, to maximize the local OEOT is likely to increase the path throughput. 2. The path delay is the summation of per-hop delay, which is actually relative to the delay introduced by transmitting the packet and coordinating the candidates. As the per-hop delay factors (T_s and $T_f(i)$) are integrated in the denominators of OEOT, to maximize OEOT is also implicitly to decrease per-hop delay, which may further decrease the path delay. 3. As the transmission reliability of $\mathcal{F}_j$ is also implicitly embedded in OEOT, maximizing OEOT also tends to improve the reliability. Reliability is a key factor affecting throughout and delay for the following reason. If a packet is transmitted on a low reliable link, several retransmissions are needed to make a successful packet forwarding at one hop.

These retransmissions not only harm the throughput and delay performance of the flow that the packet belongs to but also introduce huge medium contentions to other flows, thus further decreasing the whole system performance. However, maximizing the one-hop reliability does not necessarily lead to better end-to-end throughput. Because reliable links likely have short hop distance, this short hop distance may result in taking many hops to deliver a packet from the source to the destination, which may also introduce a large delay or more medium contention to other flows. Our OEOT metric jointly takes into account the hop advancement, reliability and packet-forwarding time.

8.4   Heuristic candidate selection algorithm

A straightforward way to get the optimal R_j and $\mathcal{F}_j$ to maximize the OEOT is to try all the ordered subsets of C_j for each R_j, which runs in $O(keN!)$ time, where k is the number of different rates, e is the base of natural logarithm, and N is the largest number of neighbors at all rates. It is, however, not feasible when N is large. In this section, we propose a heuristic algorithm to obtain a solution approaching the optimum.

As there is a finite number of transmission rates, a natural approach is to decompose the optimization problem into two parts. First, we find the optimal solution for each R_j; then, we pick the maximum one among them. So we only need to discuss how to find the solution approaching the optimum for a given rate, R_j, and the corresponding available next-hop neighbor set, C_j. Lemma 8.1 guides us to design the heuristic algorithm.

Lemma 8.1 *For given R_j and C_j, define $\mathcal{F}_j^r$ as one feasible candidate set that achieves the maximum OEOT by selecting r nodes, then $\forall\, r\ (1 \leq r \leq |C_j|)$, $\exists\, \mathcal{F}_j^r$, s.t. $\mathcal{F}_j^1 \subseteq \mathcal{F}_j^r$.*

Proof. We prove this Lemma by contradiction. Assume $\forall\, r\ (1 \leq r \leq |C_j|)$, we could find a feasible $\mathcal{F}_j^r$, s.t. $\mathcal{F}_j^1 \nsubseteq \mathcal{F}_j^r$. Then from that $\mathcal{F}_j^r$, we can obtain a new ordered set by substituting the lowest-priority candidate in $\mathcal{F}_j^r$ as the node in $\mathcal{F}_j^1$. According to Equation (8.5) and from the fact that $\mathcal{F}_j^1$ achieves the maximum OEOT by selecting 1 node, we can derive that the OEOT of the new set is larger than that of the $\mathcal{F}_j^r$. It is a contradiction, so the assumption is false, then the Lemma is true.

Lemma 8.1 basically indicates that, for a given R_j and C_j, the candidate achieving the maximum OEOT by selecting 1 node from C_j is contained in the candidate set achieving the maximum OEOT by selecting more nodes from C_j.

Actually, the numerator of OEOT is the EPA defined in Zeng *et al.* (2007). The EPA has three nice properties: priority rule, containing property and concavity, as we demonstrated in Chapter 2.

The concavity property indicates that involving more forwarding candidates will increase EPA but the gained EPA becomes marginal when we keep doing so. It was shown in Zeng *et al.* (2007) that the maximum EPA hardly increases at all when the number of forwarding candidates is larger than four. Furthermore, involving more forwarding candidates may increase the probability of false positives, that is, lower priority candidates are more likely to be falsely suppressed by other transmissions in the network. So in our algorithm design, we set a maximum allowable forwarding candidate number, r_{max}.

Now we examine the denominator of the OEOT in Equation (8.5). For the compressed slotted ACK mechanism, the denominator can be further simplified as $T_s(j) + T_{ACK} + T_{SIFS}(\sum_{i=1}^{r} i \cdot p_{j_i} \prod_{w=0}^{i-1} \overline{p}_{j_w} + r \cdot \overline{P}_{\mathcal{F}_j})$, where $T_s(j)$ is the delay at the sender side when the data packet is transmitted at rate R_j. The third part of this summation is the expected time introduced by candidate coordination, which is upper bounded by $r \cdot T_{SIFS}$. As $T_{SIFS} \ll T_s(j) + T_{ACK}$ and r is a small number, the denominator can be seen as a constant at a fixed rate R_j. So maximizing the OEOT is equivalent to maximizing its numerator, EPA.

Therefore, according to the three properties and the analysis above, we propose a heuristic greedy algorithm, which finds the transmission rate and the corresponding forwarding candidates approaching the maximum OEOT. This heuristic algorithm FindMOEOT is described in Algorithm 8.1, where the input is the multirates, R_j's, the corresponding C_j's and the maximum allowable forwarding candidate number r_{max}, and the output is the selected rate R^* and forwarding candidate set $\mathcal{F}^*$. For each rate R_j, this algorithm first finds the set $\mathcal{F}_m$ with one candidate that maximizes the OEOT, then it augments the current $\mathcal{F}_m$ by one more candidate in each iteration (line 6). Whenever adding a new candidate, it calculates the *OEOT* (line 7), then updates the $\mathcal{F}_m$ when finding a new set achieving higher OEOT than the existing one. Note that, according to Lemma 8.1, when the final returned set contains no more than 2 nodes, it is indeed the global optimum. Otherwise, it is an

---

Algorithm 8.1 FindMOEOT(C_j's, R_j's, r_{max})

1: $R^* \leftarrow 0$; $\mathcal{F}^* \leftarrow \emptyset$; $OEOT^* \leftarrow 0$;
2: **for** each C_j **do**
3:      $\mathcal{F}_m \leftarrow \emptyset$; $OEOT_m \leftarrow 0$; $\mathcal{A} \leftarrow C_j - \mathcal{F}_m$;
4:      **while** ($\mathcal{A} \neq \emptyset$ && $|\mathcal{F}_m| < r_{max}$) **do**
5:          **for** each node $s_n \in \mathcal{A}$ **do**
6:              $\mathcal{F}_t \leftarrow$ Insert s_n into $\mathcal{F}_m$ according to **Relay Priority Rule**;
7:              Get *OEOT* on $\mathcal{F}_t$ according to Equation (8.5);
8:              **if** ($OEOT > OEOT_m$) **then**
9:                  $OEOT_m \leftarrow OEOT$; $\mathcal{F}_m \leftarrow \mathcal{F}_t$;
10:             **end if**
11:         **end for**
12:         $\mathcal{A} \leftarrow C_j - \mathcal{F}_m$;
13:     **end while**
14:     **if** ($OEOT_m > OEOT^*$) **then**
15:         $R^* \leftarrow R_j$; $\mathcal{F}^* \leftarrow \mathcal{F}_m$; $EOT^* \leftarrow EOT_m$;
16:     **end if**
17: **end for**
18: **return** (R^*, $\mathcal{F}^*$);

---

approximate optimal solution. An interesting finding is that this algorithm almost certainly returns the global optimal solution even when the returned set contains more than two candidates.

8.5   Multirate link-quality measurement

To make our MGOR protocol work, we need to estimate the link quality (PRR) at different data rates. We propose a broadcast-based multirate link-quality measurement scheme in this section. This link quality measurement scheme also serves for multirate neighborhood management.

Recall that there are k different data rates. Each node maintains k neighbor tables corresponding to the k data rates. The jth table stores the bidirectional PRR information about its neighbors at rate R_j. For every τ second, each node broadcasts k "Hello" messages with each transmitted at a different data rate, e.g. 11 mbps, 5.5 mbps, and 2 mbps. Whenever a node n receives a "Hello" message sent from a node m at rate R_j, it will include node m into the corresponding neighbor table. Two events drive the updating of PRR_{mn} at R_j on node n: one is the periodical updating event set by node n. For example, every t_u seconds node n will update PRR_{mn}. We denote this event as T. The other is the event that node n receives a "Hello" packet sent from m at rate R_j. We denote this event as H.

The exponentially weighted moving average (EWMA) method (Woo and Culler 2003) is used to update PRR information. Since at each rate, the PRR is updated according to the same EWMA mechanism, we only describe the EWMA at a particular rate as follows. Let PRR_{mn} be the current estimation made by node n, *lastHello* be the time stamp of the last event H, N_m be the number of known missed "Hello" packets between the current event H and last event H based on "Hello" message sequence number difference, and N_g be a guess on the number of missed packets based on "Hello" message broadcast frequency $\frac{1}{\tau}$ over a time window between the current T event and last H or T event. N_l and N_g are initialized to be 0, and FDR_{mn} is initialized to be 1.

This technique allows node n to measure PRR_{mn} and m to measure PRR_{nm}. Each "Hello" message sent at rate R_j by a node n contains PRR measured by n from each of its neighbors N_n at that rate during the last period of time. Then each neighbor of n, N_n, gets the PRR to n whenever it receives a "Hello" message from n.

The pseudocode of EWMA algorithm for node n to estimate PRR_{mn} at rate R_j is described in Table 8.1, where $current_{Seq}$ and $last_{Seq}$ denote the sequence numbers of the current received "Hello" message and the last received "Hello" message, respectively, and $0 < \gamma < 1$ be the tunable parameter.

8.6   Performance evaluation

In this section, we evaluate the performance of MGOR by simulation, and compare the performance of MGOR with multirate geographic routing (MGR), single-rate geographic routing (GR), and single-rate opportunistic routing. Our MGOR degenerates into MGR, when we choose only one forwarding candidate and further degenerates into GR, when we also fix the transmission rate. For all the OR

Table 8.1   Pseudocode of EWMA for a particular data rate

For node n:
When H event happens
$\quad N_l = current_{Seq} - last_{Seq} - 1$
$\quad last_{Seq} = current_{Seq}$
$\quad lastHello = $ current time
$\quad l = Max(N_l - N_g, 0)$
$\quad N_g = 0$
$\quad PRR_{mn} = PRR_{mn} \cdot \gamma^{l+1} + (1 - \gamma)$
When T event happens
$\quad N_g = (currenttime - lastHello) \times \frac{1}{\tau}$
$\quad l = N_g$
$\quad PRR_{mn} = PRR_{mn} \cdot \gamma^{l}$

With kind permission from Springer Science+Business Media: Energy aware efficient geographic routing in lossy wireless sensor networks with environmental energy supply, ACM Wireless Networks (WINET), vol. 15 , no. 1, Jan 2009, pp. 39–51, K. Zeng, K. Ren, W. Lou, and P. J. Moran.

protocols, candidates closer to the destination are assigned higher relay priorities. The performance metrics we evaluate include: throughput, delay and hop count. In order to get insight into our rate and candidate selection algorithm, for MGOR, we show the number of packets transmitted at each rate in the whole network and the average number of forwarding candidates used at each node on each data rate.

8.6.1   Simulation setup

We implement the multirate link quality measurement mechanism and MGOR protocol with compressed slotted ACK in GlomoSim. The FindMOEOT algorithm proposed in Section 8.4 is used to select transmission rate and forwarding candidates for MGOR. This algorithm is also used to select forwarding candidates for single-rate GOR by fixing the transmission rate. According to the analysis in Section 8.4 and considering the candidate coordination overhead, the maximum allowable forwarding candidate number (r_{max}) is set as 3. Other than the candidate coordination scheme, our OR protocol follows the same CSMA/CA medium access mechanism as that in 802.11b. The simulated network has 50 stationary nodes randomly uniformly distributed in a $d \times d m^2$ square region. When the SNR is larger than a defined threshold and the signal receiving power is above the corresponding threshold, the packet is received without error. Otherwise the packet is dropped. Table 8.2 lists the related simulation parameters. According to the findings in (Sang *et al.* 2007) and the discussion in Section 8.2.4, we assume the candidate coordination can be ensured by the compressed slotted ACK mechanism.

We examine the impact of node density on the performance by setting $d = 1500, 1800, 2100, 2400$. The corresponding network density in terms of average number of neighbors per node at each rate is summarized in Table 8.3. We randomly choose 25 communication pairs in the network. The sources are CBR (constant bit

Table 8.2 Simulation parameters

Simulation parameter	Value
Nodes number	50
Transmission power	15 dbm
Data transmission rates	11 mbps, 5.5 mbps, 2 mbps
ACK transmission rate	1 mbps
Retry limit	5
Carrier sensing threshold	−100 dbm
11 mbps receiving threshold	−83 dbm
5.5 mbps receiving threshold	−87 dbm
2 mbps receiving threshold	−91 dbm
1 mbps receiving threshold	−94 dbm
Pathloss model	Two-ray
Fading model	Ricean with $K = 4$
Hello packet interval	1s

Table 8.3 Average number of neighbors per node at each rate under different network densities

Data rate (mbps)	Terrain side length			
	1500	1800	2100	2400
2	19.7	14.4	11.3	8.8
5.5	16.3	11.9	8.8	6.8
11	11.1	7.9	5.8	4.3

rate). We examine two different packet sizes. All the results shown in Section 8.6.2 are under 512-byte packet size, and Section 8.6.2 discusses the performance with packet size of 1024 bytes. We examine two traffic demands with CBR interval at 60 ms and 75 ms. User Datagram Protocol is used as the transport layer protocol. Each communication session continues for 40 s. All the simulation results are averaged over 25 flows under five simulation runs with different seeds.

8.6.2 Simulation results and analysis

8.6.2.1 Throughput and delay

The throughput is measured as the average throughput per flow in the communication period. We first set the CBR packet interval as 60 ms in order to push the traffic demand approaching to the capacity of MGOR. Figure 8.3 shows the throughput of MGOR, single-rate GOR, MGR, and single-rate GR. We can see that MGOR achieves the highest throughput among all the protocols and yields up to 20% higher throughput than MGR (when the terrain side length is 2400 m). Generally,

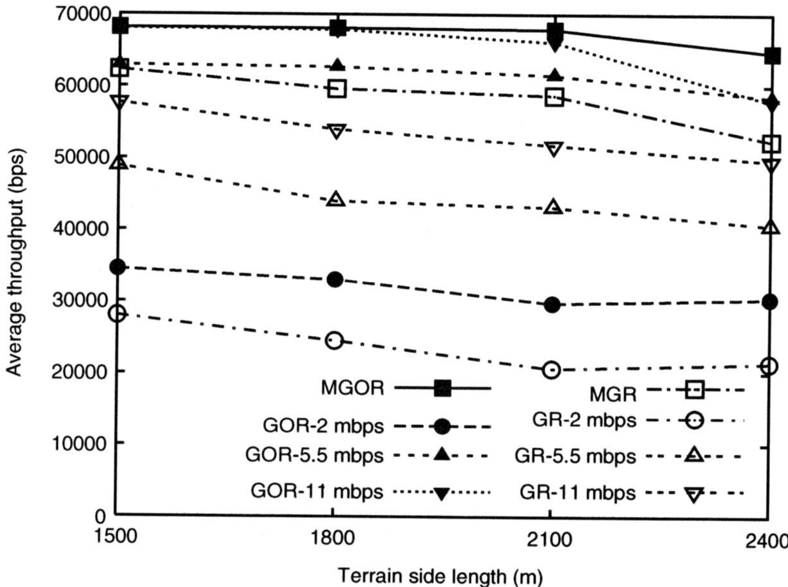

Figure 8.3   Throughput of MGOR, single-rate GOR, MGR, and single-rate GR under different network densities with CBR interval at 60 ms. Reproduced by permission of © 2009 IEEE.

opportunistic routing protocols achieve higher throughput than the corresponding traditional routing protocols at each rate. The spacial diversity gain introduced by involving multiple forwarding candidates in opportunistic routing does increase the probability of a successful transmission at each hop, which reduces the retransmission overhead. The reduction of retransmission can alleviate the medium contention and allow more packets to get through in the network and result in higher throughput. We would like to point out that due to the randomness of the network topology and limited transmission range, the packet lost in 11 mbps GOR and GR is partially due to the communication void where a forwarding node cannot find any neighbor that is geographically closer to the destination. Solving the communication void problem in geographic routing is out of the scope of this chapter. However, we note that lowering the transmission rate (from 11 mbps to 5.5 mbps) increases the transmission range and improves the network connectivity, which in turn alleviates the void problem. This can be seem as a side effect or advantage of multirate geographic routing protocols over single rate ones. That is, by using our local candidate selection and rate adaptation schemes, the multirate protocols take advantage of higher transmission rate (11 mbps) whenever there is sufficient spacial diversity or node density, but switch to lower rate to improve spacial diversity and connectivity in sparser area. Similar observation is obtained in an 18-node 802.11b indoor mesh network tested that the throughput of multirate opportunistic routing is higher than any single-rate one (Laufer and Kleinrock 2008).

The delay performance of these protocols with CBR interval at 60 ms is shown in Figure 8.4. We can see that all the opportunistic routing protocols achieve much

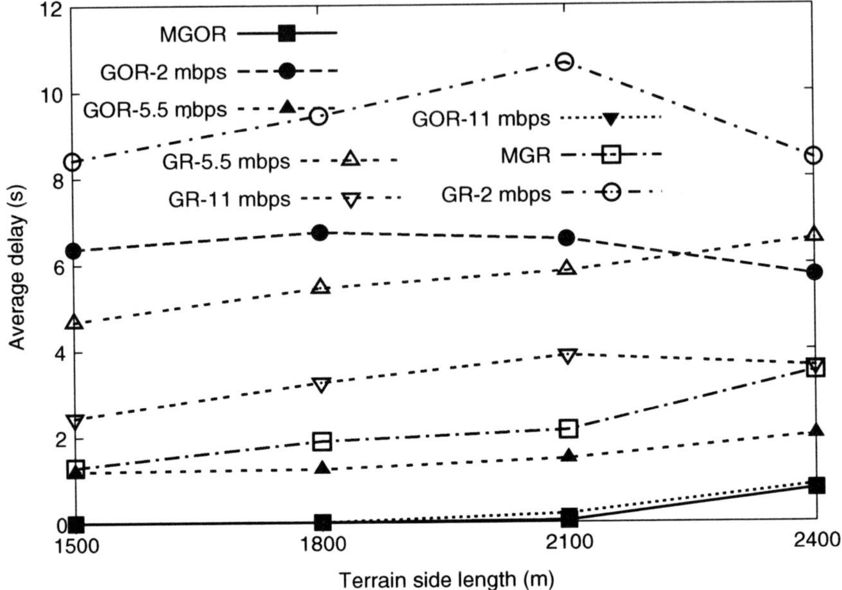

Figure 8.4   Delay of MGOR, single-rate GOR, MGR, and single-rate GR under different network densities with CBR interval at 60 ms. Reproduced by permission of © 2009 IEEE.

lower delay than the corresponding traditional ones. Generally, MGOR achieves the lowest delay among all the protocols. When the network density is high, 11 mbps GOR achieves almost the same delay (0.01 s and 0.015 s with terrain side length being 1500 m and 1800 m, respectively) as MGOR. When the network becomes sparser, MGOR outperforms 11 mbps GOR. In the saturated network, the end-to-end delay consists of per-hop queuing delay, data transmission and retransmission delay, and medium access delay. Opportunistic routing makes use of multiple forwarding candidates to relay packets, thus improves per transmission reliability. This enhancement of reliability reduces retransmission delay, which in turn reduces the queuing and medium access delay, thus reduces end-to-end delay.

In order to conduct a "fairer" comparison between MGOR and GOR at 11 mbps and separate the impact of the transmission reliability on the end-to-end delay from other factors (such as excessive medium contention and long queuing delay due to high traffic demand, and communication voids), we run another simulation with lower traffic demand where the CBR interval is set as 75 ms and only count the cases without communication voids. This traffic demand is below the capacity of MGOR and GOR at 11 mbps and 5.5 mbps, so they achieve nearly the same throughput as shown in Figure 8.5. Figure 8.6 shows the delay performance of these three protocols. We can see that MGOR achieves lower delay than the other two protocols, especially when the network becomes sparser. MGOR can tune its transmission rate at each hop according to different network conditions to maximize OEOT. When the number of neighbors at 11 mbps is small, MGOR transmits

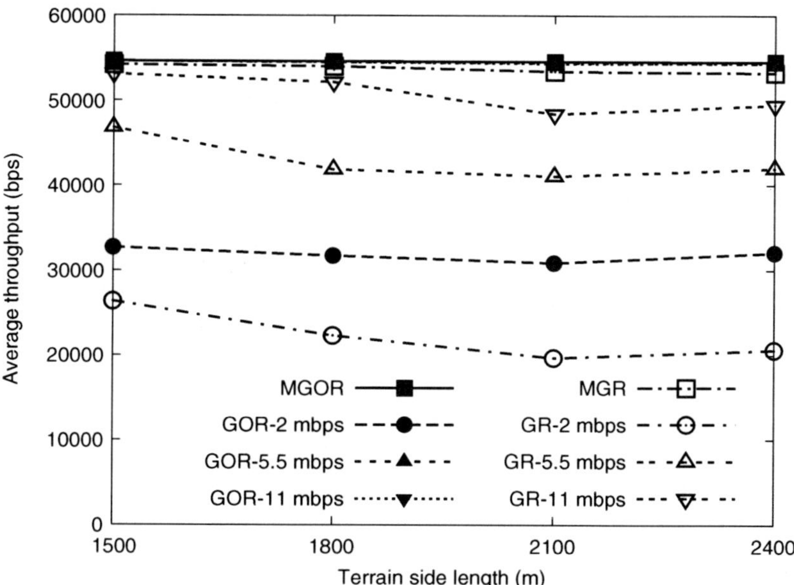

Figure 8.5   Throughput of MGOR, single-rate GOR, MGR, and single-rate GR under different network densities with CBR interval at 75 ms. Reproduced by permission of © 2009 IEEE.

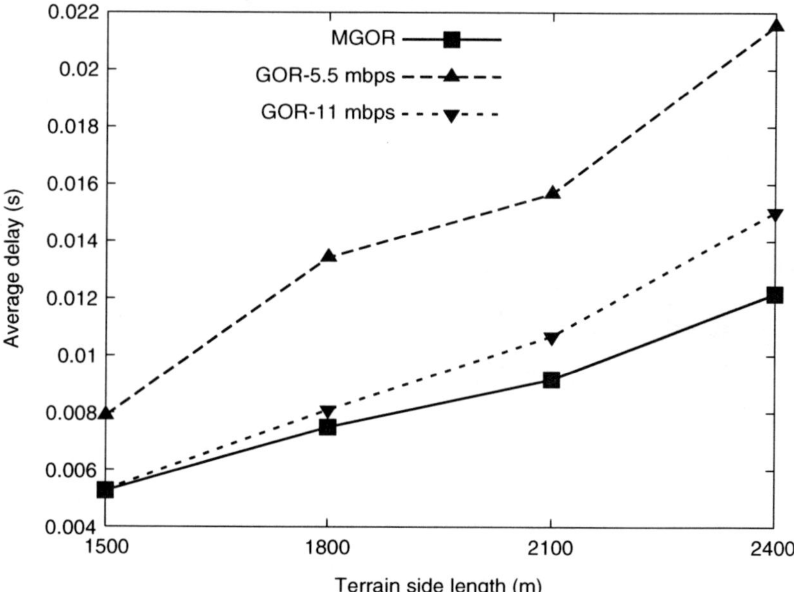

Figure 8.6   Delay of MGOR, single-rate GOR, MGR, and single-rate GR under different network densities with CBR interval at 75 ms. Reproduced by permission of © 2009 IEEE.

packets at 5.5 mbps in order to involve more forwarding candidates to harvest the opportunistic gain (e.g. achieve higher transmission advancement and reliability). When transmitting at 11 mbps already introduces sufficient spacial diversity, MGOR chooses to transmit at higher rate (11 mbps). We will show the proportion of packets transmitted at each rate in MGOR later.

We also find that although MGR can support at least 96% of this lower traffic demand, it still presents one or two orders of longer delay than MGOR. The difference of transmission reliability is the essential reason of this observation. That is, MGR has only one predefined forwarding candidate, so it usually needs more than one transmission to deliver a packet at each hop, whereas MGOR usually needs only one transmission since it introduces multiple forwarding candidates and improves transmission reliability.

Since the relative performance of hop count, average number of forwarding candidates and proportion of packets transmitted at each rate of each protocol is similar under these two traffic demands, we only show the simulation results with CBR interval at 75 ms in the following discussions.

8.6.2.2 Hop count

From Figure 8.7, we can see that GOR has larger hop count than GR at each single rate. Although GOR allows packets to be forwarded on long-distance links,

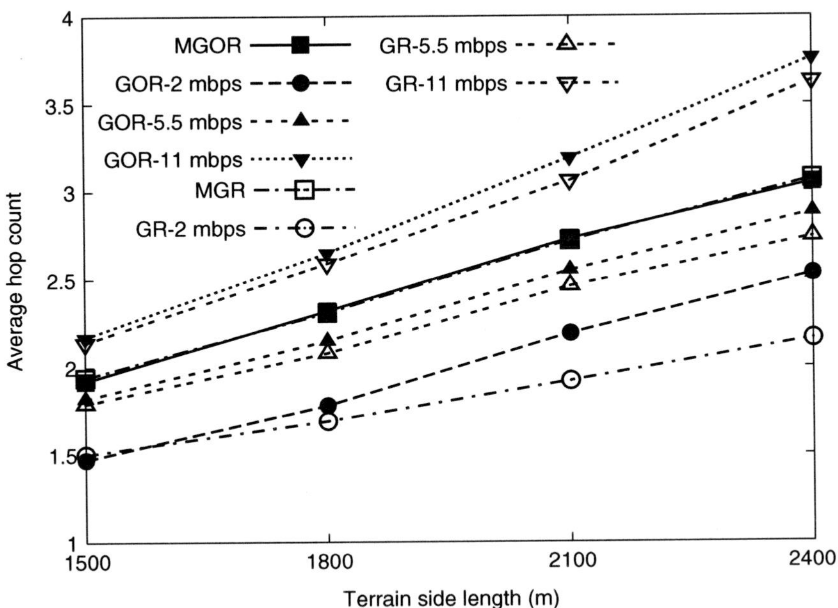

Figure 8.7   Hop count of MGOR, single-rate GOR, MGR, and single-rate GR under different network densities with CBR interval at 75 ms. Reproduced by permission of © 2009 IEEE.

some forwarding candidates with smaller advancement may also be chosen as the actual forwarders, which results in larger hop count. The hop count of MGOR is nearly the same as MGR, and is between those of GOR at 11 mbps and 5.5 mbps, but closer to that at 5.5 mbps. The rate-distance tradeoff is explicitly shown in the figure for both GR and GOR, that is, the hop count of lower rate is smaller than that of higher rate, since lower rates results in longer transmission ranges.

8.6.2.3   Average number of forwarding candidates

Figure 8.8 shows the average number of forwarding candidates at each rate for MGOR. We can see that the number of forwarding candidates at each rate decreases when the network density is decreased. Furthermore, transmission at lower rate (5.5 mbps) results in more forwarding candidates than at higher rate (11 mbps). In our MGOR, we do not choose 2 mbps transmission rate, as the traffic demand is already larger than the capacity that 2 mbps can provide.

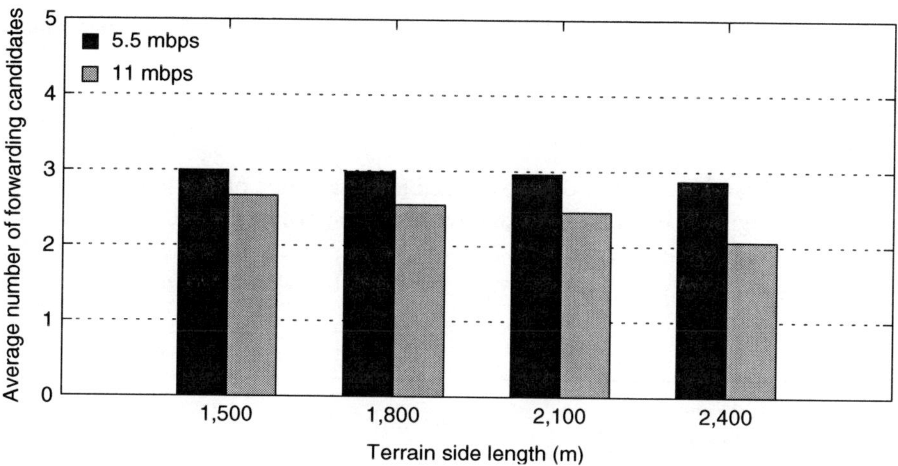

Figure 8.8   Average number of forwarding candidates of MGOR under different network densities with CBR interval at 75 ms. Reproduced by permission of © 2009 IEEE.

8.6.2.4   Proportion of packets transmitted at each rate per node

Figure 8.9 shows the proportion of packets transmitted at each rate per node. We can observe that when the network becomes sparser, more packets are selected to be transmitted at 5.5 mbps in our MGOR protocol than when the network is dense. Lower transmission rate results in longer transmission range, which leads to a greater number of neighbors (shown in Figure 8.8) and increases spacial diversity. The increased diversity gain does improve the probability of a successful transmission, which reduces the retransmission overhead, then improves the throughput (shown in Figure 8.3) and decreases the delay (shown in Figure 8.6).

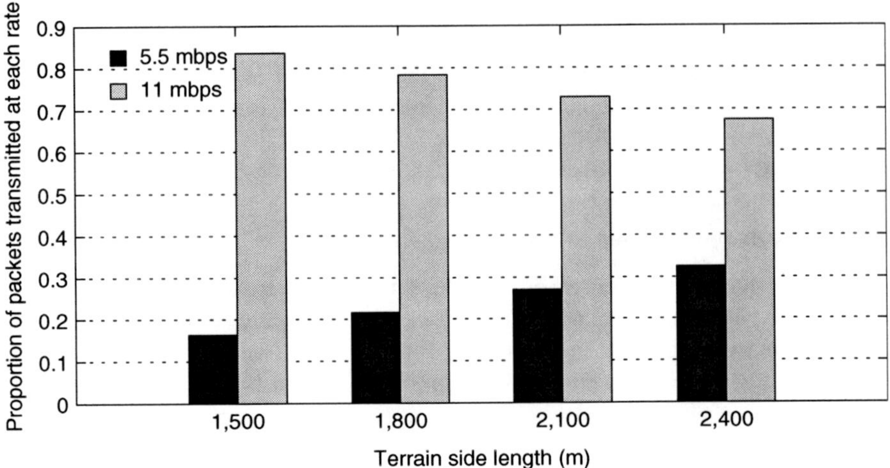

Figure 8.9   Proportion of packets transmitted at each rate of MGOR under different network densities with CBR interval at 75 ms, packet size = 512 bytes. Reproduced by permission of © 2009 IEEE.

8.6.2.5   Impact of packet size

We also evaluated the impact of packet size on the selection of transmission rate. By comparing Figure 8.10 with Figure 8.9, we notice that when the packet size is larger (such as 1500 bytes in contrast to 512 bytes), more packets are transmitted at higher data rate (i.e. 11 mbps). Because when the packet payload size is increased, the

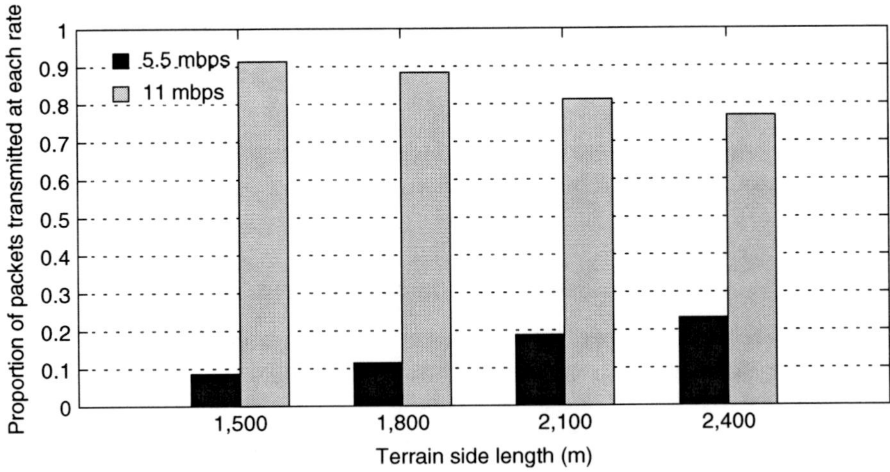

Figure 8.10   Proportion of packets transmitted at each rate of MGOR under different network densities with CBR interval at 75 ms, packet size = 1500 bytes. Reproduced by permission of © 2009 IEEE.

time of protocol overhead (such as packet header, preamble and ACK transmission time) becomes relatively smaller compared to the payload transmission time. So higher transmission rate will be more favorable when packet size becomes larger.

8.7 Conclusion

In this chapter, we studied MGOR, and examined the factors that affect its performance, which include multirate capability, candidate selection, prioritization, and coordination. Based on our analysis, we proposed the local metric, the OEOT, to characterize the tradeoff between packet advancement and medium time cost under different data rates. We further proposed a rate and candidate selection algorithm to approach the local optimum of this metric. We presented a multirate link-quality measurement mechanism to provide the link packet reception ratio information for the network layer to assist routing decision. We compared the performance of MGOR with single-rate GOR, single-rate GR and multirate GR. Simulation results show that the MGOR incorporating the rate adaptation and candidate selection algorithm achieves the highest throughput and lowest delay among all the protocols.

References

Biswas S and Morris R 2005 Exor: Opportunistic multi-hop routing for wireless networks *SIG-COMM'05*, Philadelphia, Pennsylvania.

Karp B and Kung H 2000 Gpsr: Greedy perimeter stateless routing for wireless networks *ACM MOBI-COM*, Boston.

Laufer RP and Kleinrock L 2008 Multirate anypath routing in wireless mesh networks. Technical Report UCLA-CSD-TR080025, UCLA Computer Science Department.

Lee S, Bhattacharjee B and Banerjee S 2005 Efficient geographic routing in multihop wireless networks *MobiHoc*.

Sang L, Arora A and Zhang H 2007 On exploiting asymmetric wireless links via one-way estimation *MobiHoc '07: Proceedings of the 8th ACM International Symposium on Mobile ad hoc Networking and Computing*, pp. 11–21. ACM, New York, USA.

Seada K, Zuniga M, Helmy A and Krishnamachari B 2004 Energy efficient forwarding strategies for geographic routing in wireless sensor networks *ACM Sensys'04*, Baltimore, MD.

Woo A and Culler D 2003 Evaluation of efficient link reliability estimators for low-power wireless networks. Technical report, University of California, Berkeley.

Zeng K, Lou W, Yang J and Brown DR 2007 On geographic collaborative forwarding in wireless ad hoc and sensor networks *WASA'07*, Chicago, IL.

Zorzi M and Rao RR 2003 Geographic random forwarding (geraf) for ad hoc and sensor networks: energy and latency performance. *IEEE Transactions on Mobile Computing* 2(4), 349–365.

Zubow A, Kurth M and Redlich JP 2007 Multi-channel opportunistic routing *IEEE European Wireless Conference*, Paris, France.

9

Opportunistic routing security

In this chapter, we identify several security attacks on opportunistic routing and propose countermeasures. The first security vulnerability we identify is the attack on link-quality measurement. In this attack, malicious attackers lie on the measured link quality (i.e. packet reception ratio), in order to disturb the routing decision of link-state-based routing protocols (including traditional and opportunistic routing). We demonstrate the performance degradation of the traditional and opportunistic routing under this type of attack. We find that opportunistic routing is more resilient to link quality bluffing attack than traditional single path routing. We then propose an efficient broadcast-based secure link quality measurement (SLQM) mechanism, which prevents the malicious attacker from reporting a higher packet reception ratio than the actual one. We also identify several attacks on the current opportunistic routing MAC protocols and discuss possible countermeasures to these attacks. Finally, we evaluate the resilience of opportunistic routing to packet dropping attacks. We find that in some scenarios, opportunistic routing is more vulnerable to packet dropping attacks than traditional routing when the attackers are randomly distributed in the network. However, in the scenario when an attacker is on the path of the traditional routing, no packets can be delivered whereas in this scenario, opportunistic routing can still deliver a portion of packets owing to its multipath nature.

9.1 Attack on link quality measurement

The packet reception ratio (PRR) has been widely used as an indicator of the link reliability in multihop wireless networks. It has been shown that routing performance is significantly improved by considering the link PRR information. For example, *expected transmission count* (ETX)-based routing achieves much higher

Multihop Wireless Networks: Opportunistic Routing, First Edition.
Kai Zeng, Wenjing Lou and Ming Li.
© 2011 John Wiley & Sons, Ltd. Published 2011 by John Wiley & Sons, Ltd.

throughput than traditional minimum-hop routing protocols in wireless mesh networks (Couto *et al.* 2003). The ETX is defined as $\frac{1}{p_f \cdot p_r}$, where p_f and p_r is the forward and reverse link PRR, respectively. Recent work in sensor networks (Sang *et al.* 2007) suggests a link metric (ETF), *expected number of transmissions over forward links*, which only considers forward link PRR. State-of-the-art geographic routing protocols (Seada *et al.* 2004; Zeng *et al.* 2007b) and most opportunistic routing protocols (Biswas and Morris 2005; Zeng *et al.* 2007a) rely on link quality information to make routing decisions.

Providing accurate link quality measurement (LQM)[1] is essential to ensure the correct operation of the above protocols/schemes. Furthermore, LQM is also important in supporting QoS guarantees in multihop wireless networks. Lastly, accurate long-term statistics about link-quality information are necessary to diagnose a network to identify the source of network failures and reduce the management overhead.

The existing LQM mechanisms proposed in the literature (Couto *et al.* 2003; Kim and Shin 2006; Sang *et al.* 2007) can be generally classified into three types: active, passive, and cooperative (Kim and Shin 2006). For broadcast-based active probing (Couto *et al.* 2003), each node periodically broadcasts hello/probing packets and its neighbors record the number of received packets to calculate the PRRs from the node to themselves. In passive probing (Kim and Shin 2006), the real traffic generated in the network is used as probing packets without introducing extra overhead. For cooperative probing (Kim and Shin 2006), a node overhears the transmissions of its neighbor to estimate the link quality from the neighbor to itself.

However, for any of the existing LQM mechanisms, the inherent common fact is that a node's knowledge about the forward PRR from itself to its neighbor is informed by the neighbor. Since multihop wireless networks are generally deployed in an ad hoc style or in untrusty environments, nodes may be compromised and act maliciously. This receiver-dependent measurement opens up a door for malicious attackers to report a false measurement result, thus disturbing the routing decision for all the PRR-based protocols. For example, in Figure 9.1, suppose A is the source and D is the destination, and the actual PRR is indicated above each link in Figure 9.1(a). The ETF-based shortest path routing would select the path $A \rightarrow B \rightarrow D$, because it has the lowest ETF path cost. However, if C is a malicious node, and reports to A that the PRR from A to itself is 0.9 (indicated below the link in Figure 9.1(b)), then A would select path $A \rightarrow C \rightarrow D$. In such a way, a suboptimal path is selected between A and D, thus degrading routing performance. More seriously, C attracts all the traffic from A, then with the control of the traffic, it can further maliciously drop the packets.

To the best of our knowledge, none of the existing work addresses security vulnerabilities in the existing LQM mechanisms. As LQM is becoming an indispensable component in multihop wireless networks, it is necessary to make this component work securely and provide actual and accurate PRR information.

---

[1] In this chapter, we mainly focus on PRR measurement. Without specification the link quality indicates PRR.

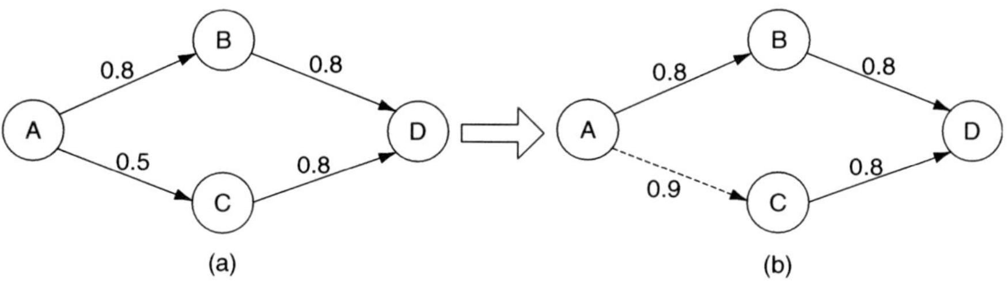

Figure 9.1   A four-node example. (a) The actual PRR on each link is indicated, and the ETF-based routing selects the optimal path $A \to B \to D$. (b) The malicious node C bluffs A into believing that the PRR from A to C is 0.9, then the ETF-based routing would select the suboptimal path $A \to C \to D$. Reproduced by permission of © 2008.

In this chapter, we analyze the security vulnerabilities in the existing LQM mechanisms. We then propose a broadcast-based secure LQM mechanism, which prevents the malicious attacker from reporting a higher PRR than the actual one. This framework can be easily applied to unicast-based and cooperative LQM mechanisms.

The rest of this section is organized as follows. Section 9.1.1 introduces the existing link quality measurement mechanisms and point out their security pitfalls. The performance degradation of opportunistic routing as well as traditional routing is demonstrated in Section 9.1.2. We then propose a broadcast-based secure LQM (SLQM) mechanism and analyze its security strength and overhead in Section 9.1.3.

9.1.1   Existing link quality measurement mechanisms and vulnerabilities

This section gives an overview of the existing LQM mechanisms and analyzes their security vulnerabilities. According to the type of probing packets, LQM can be classified into broadcast-based and unicast-based probing. While based on the generation source of probing packets, LQM can also be categorized into active, passive, and cooperative probing (Kim and Shin 2006).

9.1.1.1   Broadcast-based active probing

For broadcast-based active probing (Couto *et al.* 2003), each node broadcasts link probes of a fixed size, at an average period τ (e.g. 1 second). Every node remembers the probes it receives during the last w seconds (e.g. 10 seconds), allowing it to calculate the PRR from the measuring node at any time t as: $r(t) = \frac{count(t-w,t)}{w/\tau}$, where $count(t - w, t)$ is the number of probes received during the window w, and w/τ is the number of probes that should have been received. In the case of two neighboring nodes A and B, this technique allows A to measure the PRR from B to A, and B to measure the PRR from A to B. Each probe sent by a node A contains the number of probing packets received by A from each of its neighbors during the last w seconds. This allows each neighbor of A to calculate the forward link PRR to A whenever it receives a probe from A.

The security vulnerability in the broadcast-based active probing is that a malicious node can easily report a false measurement result. For example, if node B is an attacker, it can bluff A into believing that the PRR from A to itself is 1 by claiming that it received w/τ packets in the last probing window w.

9.1.1.2 Unicast-based passive probing

Unicast-based passive probing (Kim and Shin 2006) makes use of the real unicast traffic as the "natural" probing packets without incurring extra overhead. It is applicable when there is enough unicast traffic on a measured unidirectional link. It runs as follows: for instance, suppose node A has enough traffic to node B. Then, A gets the information about the number of successful transmissions (N_s) and the total number of transmissions (N_t) from its MAC's MIB (Management Information Base) for the traffic. At the end of an update period, the PRR is derived as $\frac{N_s}{N_t}$, and is further smoothed by moving average (Kim and Shin 2006).

For unicast-based passive probing, it is hard but not impossible for an attacker to cheat on the link quality. In 802.11 (802.11b 1999 n.d.), the Distributed Coordination Function (DCF) defines two access mechanisms for packet transmissions: basic access mechanism, and RTS/CTS access mechanism. We analyze the security vulnerability of the unicast-based passive probing under these two access mechanisms as follows.

In the basic access mechanism, a sender starts the transmission of a DATA frame after it senses the channel is idle for a while. Upon successful decoding the whole DATA frame, the receiver sends an ACK frame back to the sender, indicating successful reception of the DATA frame. In this case, even when it can not decode the whole data frame, a receiver may decode some parts of it (Jamieson and Balakrishnan 2007). So it is possible for a malicious receiver to figure out the sender's address and send back an ACK to claim a correct reception even when it receives a corrupted data frame.

The RTS/CTS access mechanism uses a four-way handshake in order to reduce bandwidth loss due to the hidden terminal problem. Unlike the case with the basic access mechanism, a sender will send a RTS frame to the receiver before it sends out the DATA frame. Upon successful reception of the RTS frame, the receiver then sends a CTS frame back to the sender. The sender can start sending the DATA frame after the reception of the CTS frame. As in the basic access mechanism, upon successful reception of the DATA frame, the receiver sends an ACK frame back to the sender. In this case, by receiving the RTS, a malicious receiver can figure out the sender's address, so even it receives a corrupted data frame, it can still claim a successful reception by sending back an ACK.

In summary, although a sender estimates the link quality based on its own MIB information in the unicast-based passive probing, this information is still dependent on the feedback (ACK) from the receiver. A malicious receiver may still be able to make use of the ACK to bluff the sender into believing that there exists a high-quality link from the sender to the receiver.

9.1.1.3   Cooperative probing

Cooperative probing (Kim and Shin 2006) is used when there is not enough unicast traffic from a measuring node to its neighbor, but there is enough to others. For example, a measuring node A has two one-hop neighbors, B and C. A has no egress traffic to C, but it has egress traffic to B. The neighbor node (C) with no traffic to it from the measuring node (A) is called a "cooperative" node. Due to the broadcast nature of wireless media, the node C can overhear the traffic from the measuring node A to B. This traffic is called *cross traffic*. The overhearing result is then used for the measuring node to derive the quality of link $A \rightarrow C$. Kim and Shin (2006) assume that node C cannot receive duplicate frames from its MAC layer even in the promiscuous mode, the retransmitted packets are not used for measurements. So node A counts first-time successful transmissions (C_c) within the cross traffic. In the update period, a report of overheard results (C_a) from C is sent to A, and then the PRR in this period is calculated as $\frac{C_a}{C_c}$.

To attack cooperative probing, similar to the unicast-based passive probing, a malicious "cooperative" node does not need to decode the whole data frame correctly. As long as it can figure out the sender's address and the status (0/1) of the "retry" bit in the data frame, it can increase its count of C_a.

9.1.1.4   Unicast-based active probing

When there is no egress/cross traffic, unicast-based active probing can be applied (Kim and Shin 2006). For example, if node A has no traffic to B or C, A initiates a unicast-based active probing on link $A \rightarrow B$ by generating unicast probing packets. Then, the link quality from A to B is measured by the same way as passive probing. At the same time, the quality of link $A \rightarrow C$ can be measured by cooperative probing. In this way, unicast-based active probing acts similarly as the broadcast-based active probing, with difference in that in unicast-based probing the receiver need send back an ACK to the sender when it receives the data frame correctly and the sender will retransmit data frames when no ACK receives, while in broadcast-based active probing, any node does not need to send ACK.

For unicast-based active probing, the security vulnerabilities in measuring the link quality from the measuring node (e.g. A) to the intended receiver (e.g. B) and to the "cooperative" node (e.g. C) are the same as the that in unicast-based passive probing and "cooperative" probing, respectively.

To sum up, all the existing LQM mechanisms cannot prevent a receiver cheating on the PRR. The inherent fact is that the receiver can claim a correct data frame reception without showing any evidence. To fix this vulnerability, we propose a broadcast-based secure LQM (SLQM) mechanism based on the challenge-response mechanism in Section 9.1.3. We will show that this broadcast-based mechanism can be easily applied to unicast-based and cooperative SLQM mechanisms. Before we show the solution, we would like to demonstrate the performance degradation of opportunistic routing and traditional routing under the LQM attacks.

9.1.2 Performance demonstration

In this section, we simulate shortest anypath routing (Laufer and Kleinrock 2008), and ETF-based traditional shortest path routing (Sang *et al.* 2007) in a static multihop wireless network scenario to demonstrate the performance degradation of PRR-based routing protocols when malicious nodes intend to report a false PRR. The simulations are implemented within the GloMoSim simulator (Zeng *et al.* 1998). The candidate coordination protocol, FSA, introduced in Chapter 6 is used for opportunistic routing. IEEE 802.11*b* (*http://standards.ieee.org/*). is used as the MAC layer protocol for traditional routing. The simulated network has 50 stationary nodes randomly uniformly distributed in a $d \times d$ m^2 square region. We vary d as 1600, 1800, 2000, and 2200 to examine the routing performance under different node densities. We set the carrier sensing threshold as -100 dbm. The data transmission rate is 11 mbps, and the signal receiving threshold is -83 dbm. The ACK transmission rate is 1 mbps with a signal-receiving threshold of -94 dbm. The frame reception decision is based on the SNR threshold and receiving threshold. When the SNR is larger than a threshold and the signal receiving power is above the corresponding threshold, the frame is received without error. Otherwise the frame is corrupted and dropped. To simulate a randomly lossy channel, we assume a Ground Reflection (Two-Ray) path loss model and Ricean fading model with $K = 4$ (Rappaport 1996) for signal propagation.

The broadcast-based LQM is used to measure the PRR on each link, such that each node broadcasts one probing packet per second, and for every 10 s (measurement period), each node reports to its neighbors the PRR on the link from the neighbors to itself. Each node updates the PRR value to its neighbors according to Equation (9.2) when it receives the report from its neighbors. The α in Equation (9.2) is chosen to be 0.9.

For normal operation (without attacks) we assume that nodes faithfully report the measured PRR. For nonsecured LQM, we assume malicious nodes will always report a 0.95 PRR no matter what the actual one is. The malicious nodes are randomly distributed in the network. To study the impact of number of malicious nodes on the performance, we vary the proportion (P_m) of malicious nodes as 0.1 and 0.3. We randomly choose ten communication pairs in the network. The application traffic takes the form of CBR (constant bit rate) flows. The packet interval of each CBR flow is 200 ms with each packet of 512 bytes. The whole simulation duration is 55 s with 30 s warm-up time, and the CBR flows start around the 30th second and end at the 50th second. Each point in the plotted results represents an average of 20 simulation runs with different seeds. Since the traffic demand is below the network capacity, the throughput performance under attack is almost the same as that under normal case. However, because some packets are routed through a suboptimal route, the end-to-end delay under attack is longer than the normal case.

Figure 9.2 shows the average end-to-end delay of the ten flows under different node densities and different portions of malicious nodes. We can see that for traditional routing, the end-to-end delay can be increased by up to 20% compared

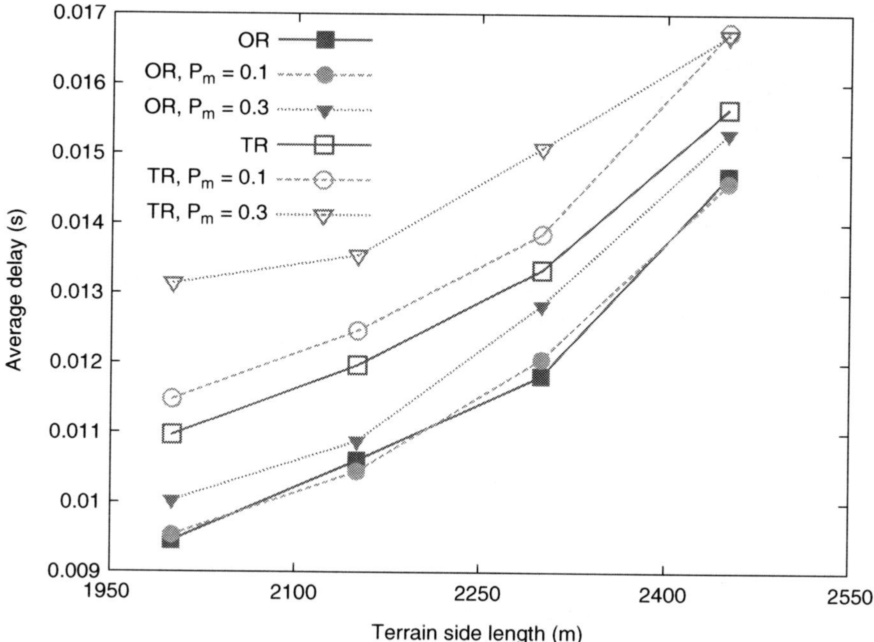

Figure 9.2   Average end-to-end delay of opportunistic routing and traditional routing under link quality bluffing attack.

with the normal case even when there are 30% malicious nodes in the network. The opportunistic routing shows more resilience to the link quality bluffing attack. The delay performance is nearly not changed when 10% of the nodes (five nodes) are malicious in the network. When 30% of the nodes are malicious, the delay is increased by up to 10%. For both routing protocols, the delay increases when the portion of malicious nodes increases. The simulation results indicate the importance of securing LQM mechanism in order to ensure an optimal performance of traditional routing and opportunistic routing. Next, we propose a broadcast-based link-quality measurement mechanism.

9.1.3   Broadcast-based secure link quality measurement

In this section, we propose a broadcast-based secure LQM mechanism, and then analyze its security strength and its computation, storage, and communication overhead. In this volume, we assume that a malicious node always wants to report a higher PRR than the actual measured one, thus disturbing PRR-based routing performance. We also assume that a unique pairwise key has been established between each pair of neighbors. The neighborhood pairwise key establishment mechanisms have been extensively studied in multihop wireless networks (Eschenauer and Gligor 2002).

9.1.3.1 Broadcast-based SLQM framework

Assume a node A has N one-hop neighbors $A_1, A_2, \ldots, A_N$ and needs to measure the link PRR (p_i) to each of its neighbors (A_i). As in Kim and Shin (2006), the measurement is done periodically. Each measurement period consists of three consecutive phases: probing, reporting, and updating phases, which are described as follows.

Probing phase: In this phase, A broadcasts N_s packets to its neighbors. In the jth packet r_j, it embeds a random number. It keeps the broadcasted packets in its buffer within this measurement period. Receiver A_i only stores the XOR-ed result (R_i) of all the correctly received packets, and the corresponding indicator vector V_i defined in Equation (9.1) that indicates the index of the received packet. Note that A_i can compute the XOR-ed result on the fly whenever it receives a new probing packet.

$$V_i(j) = \begin{cases} 1, & A_i \text{ received the } j^{th} \text{ packet correctly} \\ 0, & \text{otherwise} \end{cases} \tag{9.1}$$

where $V_i(j)$ is the jth bit from the higher (left) end of the vector V_i.

Reporting phase: When the probing phase is ended, each neighbor A_i sends A a report $Rep_i := \{H_i, V_i\}$, where $H_i = h_{\mathcal{K}_i}(R_i)$ is a keyed hash of R_i with the pairwise key $\mathcal{K}_i$ shared between A and A_i. The hash function can be any of the existing cryptographic hash functions, such as MD5(http://tools.ietf.org/html/rfc1321).

Updating phase: On receiving A_i's report, A figures out how many and which packets A_i receives by examining the positions of bit '1's in vector V_i. Since A keeps all the packets that it broadcasted, it computes R_i' by doing XOR of the packets that A_i claims it received. A then computes $H_i' = h_{\mathcal{K}_i}(R_i')$. If $H_i' = H_i$, A accepts this report; otherwise, it rejects the report. Suppose A counts there are N_{r_i} bit '1's in V_i, after A accepts the report, A calculates the PRR $p_i = \frac{N_{r_i}}{N_s}$ in this measurement period. A moving average method is further used to smooth the measured result. Denote the measured result in the kth measurement period as $p_i[k]$, the smoothed PRR, $\widetilde{p}_i(k)$, at the end of the kth period is calculated as

$$\widetilde{p}_i(k) = (1 - \alpha)\widetilde{p}_i[k - 1] + \alpha p_i[k] \tag{9.2}$$

where α is a smoothing constant in the range of $(0,1)$.

Figure 9.3 shows an example of the broadcast-based SLQM mechanism in a measurement period. Suppose in the measuring phase, A broadcasts five probing packets $(r_1, \ldots, r_5)$, and A_i receives the packets r_1, r_3, and r_5. In the reporting phase, A_i calculates $H_i = h_{\mathcal{K}_i}(r_1 \oplus r_3 \oplus r_5)$, then sends H_i and a five-bit vector $V_i = 10101$ back to A. When it receives the H_i and V_i, A examines V_i and gets the indices $(u_1, \ldots, u_c)$ of the packets A_i claims it receives, then calculates $H_i' = hash_{\mathcal{K}_i}(r_{u_1} \oplus \ldots \oplus r_{u_c})$. If $H_i = H_i'$, A accepts A_i's report; otherwise, it rejects it.

9.1.3.2 Security strength

We now analyze the security strength of our broadcast-based SLQM mechanism. This mechanism achieves the security goal that prevents a malicious attacker from

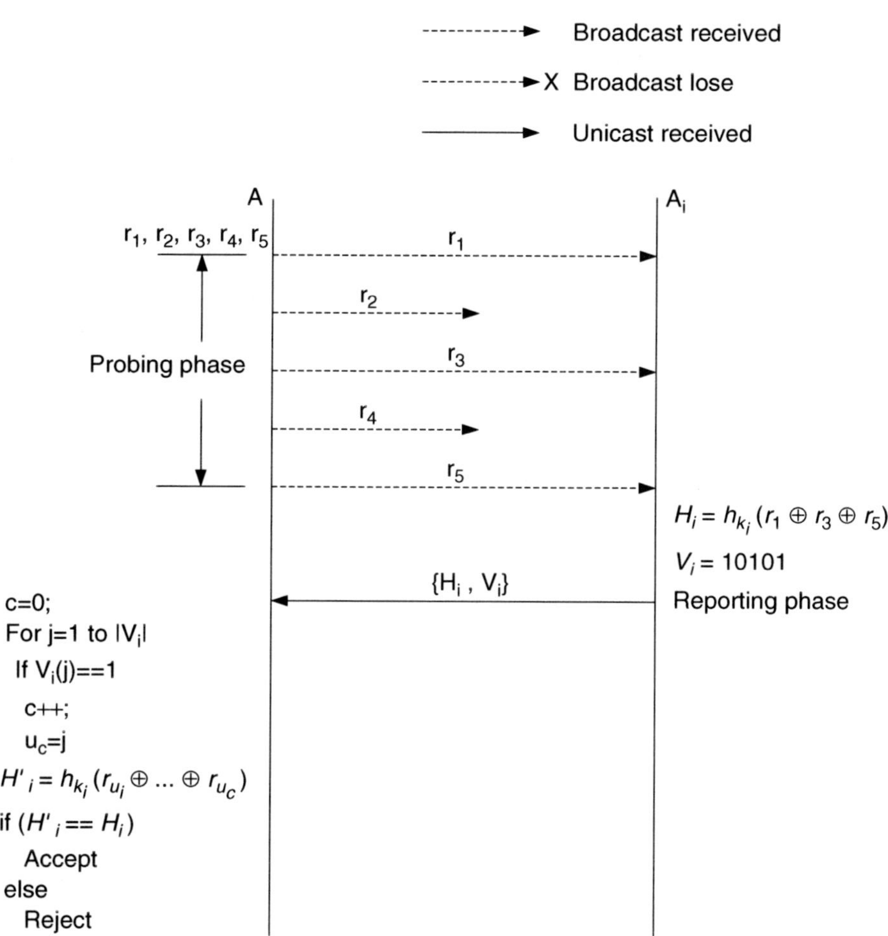

Figure 9.3   Probing and reporting phases of secure link quality measurement between A and A_i in a measurement period. Reproduced by permission of © 2008.

reporting a higher PRR than the actual one. We assume A_i is malicious in the following discussion.

First, it is computationally impossible for A_i to guess the packets that it does not receive, even when A_i overhears the other's report. For example, in Figure 9.3, if A_i wants to claim it receives r_1, r_3, r_4, r_5, it needs to create a hash value $H_i = h_{K_i}(r_1 \oplus r_3 \oplus r_4 \oplus r_5)$. Since it has no idea what r_4 is, the only thing it can do is to guess r_4. However, it is hard to make a correct guess due to the weak collision resistance property of the hash function. Given $x = (r_1 \oplus r_3 \oplus r_4 \oplus r_5)$, it is hard to find a $y = r_1 \oplus r_3 \oplus r'_4 \oplus r_5$, such that $h_{K_i}(x) = h_{K_i}(y)$. Even A_i overhears A_j's report indicating that A_j receives r_4, A_i still cannot get any information about r_4 because of the one-way property of the hash function.

Second, our mechanism prevents A_i from replaying its own or another neighbor's report. According to the randomness embedded in each probing packet, even if A_i receives all the probing packets in some measurement period, it cannot replay this report in the following measurement period. Furthermore, if A_i replays A_j's report, this report cannot pass the verification by A, because A uses $\mathcal{K}_i$ instead of $\mathcal{K}_j$ to verify A_i's report.

9.1.3.3 Computation, storage and communication overhead

Computation overhead: On the sender side, A needs to generate a random number sequence. According to its computation and storage capability, A can generate a large random number sequence to be used for several measurement periods, and refresh this sequence when it is used up. Any of the existing efficient pseudorandom number generators, such as the linear congruential generator (Greenberger 1961), can serve this purpose. To do the verification, A only needs to do XOR and hash operations, which are computationally efficient. On the receiver side, to create the report digest, each neighbor only needs to do a hash computation.

Storage overhead: On the sender side, A only needs to store the generated random numbers. Suppose the length of each random number is L_r bytes, the probing packet broadcast rate is B packet/second, and the probing phase is P seconds. Then, in a measurement period, A needs $S = L_r \cdot B \cdot P$ bytes storage space. For example, if $L_r = 16$, $B = 1$, and $P = 10$, $S = 160$ bytes, which is supportable even on sensor nodes.

Communication overhead: The communication overhead of our SLQM mechanism is comparable to any existing broadcast-based probing mechanism, such as that in Couto *et al.* (2003). As the probing packet broadcast rate is usually low, e.g. $B = 1$, SLQM introduces very light local traffic into the network.

9.1.3.4 Applicability

As discussed above, our SLQM mechanism has very low computation, storage and communication overheads, so it is applicable to resource-constraint networks, such as wireless sensor networks, as well as more powerful networks, such as wireless mesh networks. Basically, broadcast-based SLQM can be implemented at application, networking and MAC layers. Our SLQM framework can also be easily applied to unicast-based and cooperative LQM with a slight modification: we embed a random number in each unicast packet (including retransmitted packets at MAC layer). For unicast-based SLQM, we can ask the receiver to attach a hash value of the received packet in the corresponding ACK. For cooperative probing, the cooperative receiver does the same thing as the broadcast-based SLQM.

9.2 Attacks on opportunistic coordination protocols

In addition to bluffing on the link quality to degrade the performance of opportunistic routing, an attacker can attack the candidate coordination protocols. We can basically classify the opportunistic routing MAC coordination protocols discussed in Chapter 6 into two categories: implicitly prioritized and explicitly prioritized. The implicitly prioritized ones include GeRaF MAC and the contention-based forwarding, where the transmitter does not include a candidate list in the header of the data packets, and the forwarding candidates contend for the right of the packet relay. On the opposite side, the explicitly prioritized ones include ExOR batch-based forwarding, slotted ACK, compressed slotted ACK, and fast slotted ACK (FSA), where the transmitter explicitly includes the candidate list information in the data packet header and the forwarding candidates follow an order to forward the packets. In the following discussion we assume the goal of the attacker is to claim the relay responsibility for as many as packets it can whether it correctly received the packets or not. After the attacker obtains the relay responsibility of the packets, if it indeed received the packets, it can either drop them (black hole attack), selectively relay them, or forward them normally (but this may lead to suboptimal routing). If it does not correctly receive the packets, those packets are implicitly dropped.

9.2.1 Attack on implicit-prioritized coordination protocol

9.2.1.1 Attack on GeRaF MAC

We first discuss possible attacks on GeRaF MAC. Recall that in GeRaf, when a node has a packet to send, it broadcasts an RTS frame. Forwarding candidate nodes that are geographically closer to the destination will contend for the relay responsibility by sending back CTSs. If a clear CTS is received by the transmitter, it will select the node that sent this CTS as the forwarder. Other neighbors will suppress their forwarding. The attacker can therefore send a CTS whenever it overhears a RTS to claim the forwarding responsibility.

This vulnerability comes from the inherent fact that neither the sender nor other neighbors can identify the location of the attacker. If the sender and neighbors can identify the location of the attacker, this attack can be substantially limited. For example, GeRaF MAC only allows neighbors that are closer to the destination than the transmitter to participate in the opportunistic forwarding. If the sender can verify the location information of the node that sends the CTS, it can detect the attack if the attacker is actually further away from the destination. It is possible that the attacker is actually closer to the destination than the transmitter but gives little advancement. In this case, if other legitimate neighbors can overhear the attacker's CTS and can identify the location the attacker, this attack can also be flagged.

9.2.1.2 Attack on CBF

For CBF, similar vulnerability exists in all the three suppression schemes.

In the basic suppression scheme, a forwarding candidate that receives the packet will suppress itself if it overhears any other node that forwarded the packet. An attacker that receives the packet but gives little or even negative advancement can rebroadcast the packets with very short delay, thus suppressing other nodes that may advance the packet further toward the destination. In this case, the CBF is operated in a suboptimal status. Although the location information should be included in the packet header, which may give the transmitter or other neighbor nodes opportunity to detect this attack, the attacker can lie on the location information easily to indicate a large advancement to the destination.

In the area-based suppression scheme, only the forwarding candidates who are in a particular area are supposed to participate in the packet forwarding. However, the attacker can still lie on its location information to fool other nodes.

The active selection scheme is very similar to the GeRaF MAC that the transmitter broadcasts a ready-to-forward (RTF), and forwarding candidates send a clear-to-forward (CTF) frame with nodes closer to the destination shorter delay. Whenever a forwarding candidate overhears a CTF, it suppress itself from forwarding. It is possible that, some nodes may not overhear the CTF sent by other nodes and the transmitter may receive multiple CTF frames. Then it unicasts the data packet to one of the senders of the CTF frames to finally forward the packet. Although the multiple CTF frames can somehow alleviate the attacking possibility, the attacker can raise its transmission power to suppress a large area to increase its chance of being selected as the final forwarder.

In summary, the vulnerability in CBF is also rooted from the fact that the transmitter and other legitimate neighbors cannot verify the location of the attacker. If the attacker's location can be verified, the above attack can be largely prevented or detected out.

9.2.2   Attack on explicit-prioritized coordination protocol

The ExOR MAC, slotted ACK, compressed slotted ACK and fast slotted ACK described in Chapter 6 belong to the explicitly prioritized coordination protocol. We discuss the possible attacks on these protocols as follows.

9.2.2.1   Attack on ExOR MAC

In ExOR MAC, the sender broadcasts a batch of packets. When a forwarding candidate receives the packets, it waits for its turn to rebroadcast the packets according the priority assigned by the sender. This priority information is included in each data packet header. An attacker may not therefore be able to disturb the forwarding order, otherwise it can be easily detected, for example, if an attacker has a lower forwarding priority but it rebroadcasts the packets instantly after the completion of the batch transmission from the sender. Either the sender or other forwarding candidates should be able to overhear the rebroadcast packets, then detect this abnormal behavior.

When it comes to the attacker's turn to forward the packet, the attacker can bluff that it receives all the packets by manipulating its batch map. However, this attack may be easily detected, since the sender knows the packet reception ratio from itself to the attacker. If this packet reception ratio is correct, which can be ensured by our secure link quality measurement mechanism, the sender can estimate how many new packets (those have not been received by the higher priority forwarding candidates) the attacker should be able to receive. If the number of the received packets claimed by the attacker is much larger than the expected one, the sender is able to detect this attack.

The attacker may try to rebroadcast false packets that it does not receive by modifying its batch map but only rebroadcast a certain number of the packets. In this way, these packets are implicitly dropped at this hop, which may introduce large retransmission overheads at the transport layer if TCP is applied. However, if the sender can overhear those packets and check their integrity, this attack can be detected out.

In short, although the attacker can launch the attacks discussed above, it is not too difficult to modify the protocol in order to detect these attacks with reasonable overhead.

9.2.3 Attack on slotted ACK

In the slotted ACK, the sender broadcasts the packet to its forwarding candidates. Each forwarding candidate that receives the packet correctly will send an ACK back to the sender at different slots. That is, the ith candidate sends an ACK at the ith slot if it receives the data packet. Each ACK contains the ID of the highest priority successful recipient known to the ACK's sender. Finally, only the highest priority successful recipient is supposed to forward the data packet; other forwarding candidate should suppress themselves from forwarding. The sender will retransmit the packet if no clear ACK is received.

If an attacker is not in the forwarding candidate set, it may try to send the ACK at the first slot to claim the forwarding responsibility. The attacker can also send the ACK at later slots but in order to maximize the attacking gain it should try from the first slot. There are two possible situations: 1. The ACK collides with the legitimate one, then a legitimate lower priority candidate may also consider itself as a forwarder because it does not receive a clear ACK from the higher priority candidate. So both the forwarding candidates will forward the packet, which introduces duplication and leads to suboptimal operation of the opportunistic forwarding. 2. The first-priority candidate does not send an ACK, then the attacker may be able to obtain the forwarding authority. However, in the second situation, the legitimate first-priority candidate may be able to overhear this ACK, thus can raise an alarm since only itself is supposed to send an ACK in this slot. Furthermore, the sender may also apply the advanced physical layer identification techniques (Brik *et al.* 2008; Demirbas and Song 2006; Faria and Cheriton 2006; Sheng *et al.* 2008; Xiao *et al.* 2008ab, 2007, 2008c; Yang *et al.* 2009; Zeng *et al.* 2010, 2011) to detect this impersonation attack.

If an attacker is in the forwarding candidate set, but does not receive the packet correctly, it can still send an ACK to claim the forwarding responsibility at its slot. It can achieve this only when no higher priority candidate received the packet. To detect if an attacker indeed receives the packet, the sender can ask the attacker to provide evidence of the packet reception, such as a keyed hash of the data packet. It is similar to the philosophy in our secure link quality-measurement mechanism.

9.2.4 Attack on compressed slotted ACK

Unlike the situation with the slotted ACK, in the compressed slotted ACK, when a lower priority candidate does not sense the ACK transmission from the higher priority candidate, it will send out an ACK without waiting for the entire ACK transmission slot of the higher priority candidate. The attack on the slotted ACK can also be applied to the compressed slotted ACK. The possible solutions are also the same.

9.2.5 Attack on fast slotted ACK

In the fast-slotted ACK, a forwarding candidate will sense the channel to see if there is any possible ACK sent from higher priority nodes. If it senses that there is an energy increase in channel during a specific time window, it will assume there is a higher priority candidate sending the ACK, even if it does not clearly receive the ACK. The attack on this coordination scheme is also similar to the attack on slotted ACK. Since the lower priority candidate will not forward the packet if it senses an ACK transmission, the packet duplication will be avoided when the first priority candidate sends an ACK but this collides with the attacker's one. Similarly, if the first-priority node does not send an ACK, then it may be able to detect this attack if it senses an energy change in the time window during which it alone is supposed to send an ACK.

9.3 Resilience to packet-dropping attack

A packet-dropping attack is a common attack in multihop wireless networks, where malicious nodes simply drop all the packets routed through themselves. It is intuitive that, because of its multipath nature, opportunistic routing should be more resilient to the packet dropping attack than the traditional single path routing. We conduct a simulation to validate this intuition. The simulation settings are the same as the simulation in Section 9.1.2 except that the malicious attackers do lie on the link quality but drop packets. For each run, a certain proportion (P_m) of nodes in the network are set as malicious nodes, which drop all the packets routing through them. We assume the source and destination nodes are not malicious. The simulation result is shown in Figure 9.4. Each point on the figure is the average value of five random runs. It shows that generally opportunistic routing is more resilient than traditional single-path routing when faced with a packet-dropping attack. The throughput deceases when more nodes in the network are malicious.

However, in some case, for example, when terrain side length is 2000 m and $P_m = 0.2$, single-path routing can deliver more packets. The reason behind it can be as follows. When the network area is not large, nodes travel fewer hops from the source to the destination. Then when the malicious nodes are few, the probability at which a malicious node is on the path of single-path traditional routing is lower than that of opportunistic routing, because opportunistic routing involves more nodes. Therefore, in this situation, the multipath nature of opportunistic routing may adversely harm opportunistic routing. Consider a scenario in which a

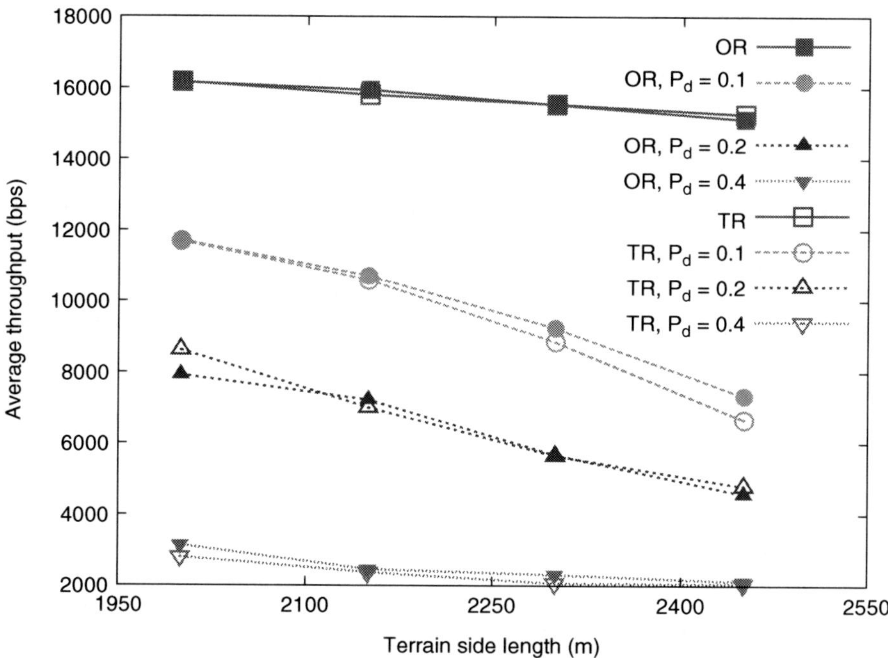

Figure 9.4   Average end-to-end throughput of opportunistic routing and traditional routing under packet dropping attack.

single-path routing only involves one intermediate node, whereas opportunistic routing may involve two intermediate nodes, one of which is shared with the single path routing. When the shared node is malicious, neither of the two routing protocols can deliver packets. When the other node is malicious, however, opportunistic routing may loose some packets due to malicious dropping but traditional routing can deliver all the packets. Then statistically single-path traditional routing can achieve higher throughput than opportunistic routing. It is worth investigating further the mathematical foundations of the performance degradation of opportunistic routing in the presence of a packet-dropping attack, and identifying the situations when opportunistic routing performs better or worse than traditional single-path routing.

9.4   Conclusion

In this chapter, we investigated the existing link-quality measurement mechanisms, and analyzed the security vulnerabilities in them. A common fact inherent in all the existing LQM mechanisms is that they use receiver-dependent measuring; that is, a node's knowledge about the forward PRR from itself to its neighbors is informed by its neighbors. We proposed a broadcast-based secure LQM mechanism that prevents a neighboring node from maliciously claiming a higher measurement result.

Our mechanism has very low computation, storage and communication overheads and thus can be implemented in resource-constraint sensor networks as well as mesh networks. Our SLQM mechanism can be easily applied to unicast-based and cooperative LQM with slight modifications.

We demonstrated that opportunistic routing is more resilient than traditional routing in the presence of a link-quality bluffing attack. We also discussed possible attacks on the existing opportunistic routing coordination protocols and proposed countermeasures. For implicitly prioritized coordination protocols, the inherent vulnerability comes from the fact that the location of the attacker cannot be verified. For explicitly prioritized protocols, it is harder for the attacker to obtain the packet-forwarding authority because there is a predefined packet forwarding priority. We finally compare the resilience of the opportunistic routing and traditional routing under packet dropping attack. Generally, opportunistic routing is more resilient than single-path traditional routing under packet dropping attacks. However, in some cases, single-path routing may be better.

References

Biswas S and Morris R 2005 Exor: Opportunistic multi-hop routing for wireless networks *SIG-COMM'05*, Philadelphia, Pennsylvania.

Brik V, Banerjee S, Gruteser M and Oh S 2008 Wireless device identification with radiometric signatures *Proceedings of the 14th ACM International Conference on Mobile Computing and Networking MobiCom '08*, pp. 116–127.

Couto DD, Aguayo D, Bicket J and Morris R 2003 A high-throughput path metic for multi-hop wireless routing *ACM MobiCom '03*, San Diego, California.

Demirbas M and Song Y 2006 An rssi-based scheme for sybil attack detection in wireless sensor networks *Proceedings of the 2006 International Symposium on World of Wireless, Mobile and Multimedia Networks WOWMOM*, pp. 564–570.

Eschenauer L and Gligor VD 2002 A key-management scheme for distributed sensor networks *CCS '02: Proceedings of the 9th ACM Conference on Computer and Communications Security*, pp. 41–47. ACM, New York, NY, USA.

Faria DB and Cheriton DR 2006 Detecting identity-based attacks in wireless networks using signalprints *Proceedings of the 5th ACM Workshop on Wireless Security, WiSe*, pp. 43–52.

Greenberger M 1961 Notes on a new pseudo-random number generator. *Journal of the Association of Computing Machinery* **8**(2), 163–167.

Jamieson K and Balakrishnan H 2007 Ppr: Partial packet recovery for wireless networks *ACM SIG-COMM*, Kyoto, Japan.

Kim KH and Shin KG 2006 On accurate measurement of link quality in multi-hop wireless mesh networks *MobiCom '06: Proceedings of the 12th Annual International Conference on Mobile Computing and Networking*, pp. 38–49. ACM Press, New York, NY, USA.

Laufer RP and Kleinrock L 2008 Multirate anypath routing in wireless mesh networks. Technical Report UCLA-CSD-TR080025, UCLA Computer Science Department.

Rappaport TS 1996 *Wireless Communications: Principles and Practice*. Prentice Hall, Lower Saddle River, NJ.

Sang L, Arora A and Zhang H 2007 On exploiting asymmetric wireless links via one-way estimation *MobiHoc '07: Proceedings of the 8th ACM International Symposium on Mobile ad hoc Networking and Computing*, pp. 11-21. ACM, New York, NY.

Seada K, Zuniga M, Helmy A and Krishnamachari B 2004 Energy efficient forwarding strategies for geographic routing in wireless sensor networks *ACM Sensys '04*, Baltimore, MD.

Sheng Y, Tan K, Chen G, Kotz D and Campbell A 2008 Detecting 802.11 mac layer spoofing using received signal strength *The 27th Conference on Computer Communications INFOCOM*, pp. 1768–1776.

Xiao L, Greenstein L, Mandayam N and Trappe W 2007 Fingerprints in the ether: Using the physical layer for wireless authentication *IEEE International Conference on Communications, ICC*, pp. 4646–4651.

Xiao L, Greenstein L, Mandayam N and Trappe W 2008a A Physical-Layer Technique to Enhance Authentication for Mobile Terminals *IEEE International Conference on Communications, ICC*, pp. 1520–1524.

Xiao L, Greenstein L, Mandayam N and Trappe W 2008b Using the physical layer for wireless authentication in time-variant channels *IEEE Transactions on Wireless Communication* **7**(7), 2571–2579.

Xiao L, Greenstein LJ, Mandayam NB and Trappe W 2008c MIMO-assisted channel-based authentication in wireless networks *42nd Annual Conference on Information Sciences and Systems, CISS*, pp. 642–646.

Yang J, Chen Y, Trappe W and Cheng J 2009 Determining the number of attackers and localizing multiple adversaries in wireless spoofing attacks *The 28th Conference on Computer Communications INFOCOM*, pp. 666–674.

Zeng K, Govindan K and Mohapatra P 2010 Non-cryptographic authentication and identification in wireless networks. *IEEE Wireless Communications Magazine* **17**(5), 56–62.

Zeng K, Govindan K, Wu D and Mohapatra P 2011 Identity-based attack detection in mobile wireless networks *IEEE Infocom 2011*, Shanghai, China.

Zeng K, Lou W, Yang J and Brown DR 2007a On throughput efficiency of geographic opportunistic routing in multihop wireless networks *QShine '07*, Vancouver, British Columbia, Canada.

Zeng K, Ren K, Lou W and Moran PJ 2007b Energy aware efficient geographic routing in lossy wireless sensor networks with environmental energy supply. *Wireless Networks (WINET)* pp. 477–486.

Zeng X, Bagrodia R and Gerla M 1998 Glomosim: a library for parallel simulation of large-scale wireless networks *Proceedings of PADS '98*, Banff, Canada.

10

Opportunistic broadcasts in vehicular networks

In this chapter, we apply the concept of opportunistic routing in broadcasts in vehicular networks. Vehicular ad hoc network (VANET) is a special type of mobile MWN designed to provide a wide range of road applications such as safety warning (Torrent-Moreno *et al.* 2009; Xu *et al.* 2004), congestion avoidance or mobile infotainment (Li *et al.* 2011). One important function of VANET is the broadcast of event-driven warning messages (WMs) like accident and hazard warning, for example, after two vehicles collided with each other on a highway, or traffic congestion happens because of heavy rain or snow, the upcoming vehicles need to be notified immediately. In both cases, the WMs should be disseminated with only a short delay to vehicles that are up to several kilometers away, not only to prevent more possible accidents but also to enable the vehicles to make a detour as early as possible to avoid congestion. While Dedicated Short Range Communication (DSRC) (http://www.standards.its.dot.gov/Documents/advisories/dsrc advisory.htm) allows the data transmission range of vehicles to be up to a few hundred meters, a single-hop broadcast is not sufficient to provide the desired warning message coverage. So a multihop broadcast is necessary to disseminate time-sensitive warning messages in VANETs.

There are three main performance goals in WM broadcast in VANETs. 1. High reliability, which is usually measured as the percentage of vehicles that received the warning message. 2. Fast dissemination–that is the warning messages should be delivered to the vehicles with short end-to-end delay. 3. High scalability, which means the WM's propagation should only incur a small transmission overhead (especially when the network is dense) because unnecessary transmissions waste

Multihop Wireless Networks: Opportunistic Routing, First Edition.
Kai Zeng, Wenjing Lou and Ming Li.
© 2011 John Wiley & Sons, Ltd. Published 2011 by John Wiley & Sons, Ltd.

precious bandwidth resource in VANET. In contrast to the mobile content distribution applications introduced in Chapter 7, we need to ensure a certain QoS requirement for each individual warning message, for example that message reception probability should be larger than a threshold (e.g., 95%), and maximum reception delay should be smaller than 200 ms for a distance within two propagation hops.

However, in real VANETs these goals are hard to achieve simultaneously. The major challenge comes from the lossy wireless transmissions (Torrent-Moreno *et al.* 28–25 September 2005, 2006), which would undermine the reliability of a one-hop broadcast. According to studies on the DSRC (Torrent-Moreno *et al.* 2004), the one-hop broadcast reception rate is low due to both channel fading and packet collisions caused by hidden terminals. There is also no channel resource reservation mechanism in 802.11 for broadcasts, which could experience severe packet collisions in a dense network with congested channels. Unlike unicast, in VANET it is difficult to let every vehicle acknowledge the reception of each broadcast message, mainly due to the *ACK implosion problem* (Tang and Gerla 2000). Therefore, there is hardly a reception guarantee for one-hop link layer broadcast[1].

Since enhancing the reliability of a broadcast from the link layer is often highly complex most previous works have focused on broadcast strategies that use redundant network layer retransmissions. The blind flooding leads to the well-known *broadcast storm* problem (Ni *et al.* 1999) where packet collisions could arise due to uncoordinated simultaneous rebroadcasts.

Traditionally, connected dominant set-based (CDS-based) algorithms (Dai and Wu 2004; Lim and Kim 2000; Lou and Wu 2002; Qayyum *et al.* 2002) have been adopted in static multihop wireless networks (MWNs) to solve the broadcast storm problem by pre selecting a minimized set of relay nodes from all the nodes in the network to cover the rest of the nodes. They were developed under the unit disk graph (UDG) model in which a packet can be received if a node is within the transmission range of a relay node, otherwise it will be lost. However, this model does not take into account the characteristics of the lossy wireless medium and its broadcast nature. In the presence of lossy links, the pre-defined relay nodes may not receive a broadcast packet from a relay node in the upstream and the packet could die out from further propagation, thus the broadcast performance will be far from optimal in this case. There are also some works investigating optimized broadcasts in MWNs with unreliable links but they still follow the traditional way of pre-selecting multi point relays (MPRs), which is similar to traditional routing. As we have seen in the previous chapters, opportunistic routing takes advantage of the broadcast nature of the wireless medium and the spacial diversity, which can increase the throughput of routing significantly. Thus, by introducing the concept of opportunistic routing/opportunistic forwarding, the broadcast performance is also expected to improve dramatically.

Most broadcast algorithms in static MWNs are stateful, i.e., they require global or local topology information and need extra overheads for neighbor information

---

[1] Although a recent technique (Dutta *et al.* 2009) can solve this problem in wireless networks with fixed topology; it is less suitable for VANETs with dynamic topology.

exchange. However, VANET is a mobile network with a highly dynamic topology, where the neighbor information is often outdated; and stateful algorithms always incur a large overhead in acquiring neighbor information, which in turn causes performance degradation. Thus, in VANETs, various stateless broadcast schemes were also proposed to mitigate the broadcast storm problem, such as probability-based broadcast schemes (Wisitpongphan *et al.* December 2007b) and timer-based broadcast schemes (Alshaer and Horlait 2005; Mangharam *et al.* 2007; Oh *et al.* 2006; T-Moreno 2007). Although these schemes enjoy high reliability when the channel load is moderate, they do not have a well-coordinated broadcast mechanism, which still makes the amount of redundant transmissions prohibitively large in a dense network or where there is heavy data traffic. This drawback heavily degrades the broadcast performance and limits the scalability to be deployed in a real VANET.

In this chapter, we study the WM broadcast in VANETs, and propose an *opportunistic broadcast* (OppCast) protocol, a fully distributed protocol that simultaneously achieves high reliability and fast WM propagation while incurring low transmission overheads. OppCast achieves message reception reliability from two aspects: one is the link-layer, and the other is network layer. For the link layer, we propose a distributed *opportunistic broadcast coordination function* (OBCF), a reliable and efficient MAC-layer broadcast primitive for the recipients of a single broadcast to select the "best" relay nodes in a localized manner. An OBCF exploits the idea of opportunistic forwarding to enhance the reception reliability and reduce the hop delay in each single transmission; by transmitting a long-range, short broadcast acknowledgement (BACK) before each rebroadcast, OBCF effectively minimizes the redundant transmissions and alleviates the hidden terminal problem in a lossy environment.

From the network layer, the broadcast of each WM consists of two types of broadcast phases, where one phase quickly propagates a WM using relatively long hops and the other phase uses additional make-up transmissions between the long hops to ensure a certain PRR. The designs of both phases are optimized to minimize the total number of transmissions.

In addition, we investigate the WM broadcast problem under sparse VANETs with frequent network partitions. To maintain both high WM reception reliability and small end-to-end delay, OppCast switches adaptively between the fast propagation mode and the store-carry-and-forward mode, where the first one is used within continuous vehicle platoons and the second is used between platoons. The optimal switching condition between the two modes is characterized via both theoretical analysis and simulations.

The rest of the chapter is organized as follows. Sections 10.1 and 10.2 review the existing broadcast techniques in multihop wireless networks and VANETs. In Section 10.3 we give the problem statement. This is followed by an overview of the proposed opportunistic broadcast protocol (OppCast) in Section 10.4, and then the main design part is presented in Section 10.5. After that, Section 10.6 presents the theoretical analysis on parameter optimization, and Section 10.7 is the performance evaluation. Section 10.8 wraps up the chapter.

10.1 Related works on broadcasts in general MWNs

Broadcast algorithms in general MWNs can be classified into stateful and stateless ones according to (Heissenbuttel *et al.* n.d.). Stateful algorithms utilize network information, like neighborhood state. In stateless algorithms, nodes make fully localized decisions only using information from themselves and the received packet.

10.1.1 Stateful broadcast

Stateful broadcast algorithms aim at minimizing redundancy from the global or local point of view. They require neighbor information and need extra overheads for neighbor information exchange.

10.1.1.1 Neighbor-designating methods

In neighbor-designating methods, the forwarder is the one who makes the decision about which of its neighbors should relay the packet. Basically, research in this category aims to find approximate algorithms for the minimum connected dominant set (MCDS), which is NP-hard. Typical algorithms include multipoint relaying (MPR) (Qayyum *et al.* 2002), and the dominant pruning (Lim and Kim 2000; Lou and Wu 2002).

10.1.1.2 Self-pruning based methods

In self-pruning based methods, each node decides its own forwarding status independently and locally. Dai and Wu (2004) proposed a general framework for self-pruning based methods, and pointed out that basic elements for these methods include: 1. Identifying the existence of a *replacement path*; 2. Identifying the existence of an alternative *cover set*; 3. The assignment of node priority.

Stateful broadcast algorithms are mainly suitable for static MWNs. In mobile MWNs with highly dynamic topology, neighbor information is often outdated and stateful algorithms always incur a large overhead in acquiring neighbor information, which in turn causes performance degradation.

10.1.2 Stateless broadcast

The stateless broadcast protocols are especially suitable for resource-limited networks with frequent topology changing. In Heissenbuttel *et al.* (n.d.), a stateless broadcast algorithm (DDB) is proposed, which allows nodes to make locally optimized forwarding decisions without neighborhood information. The core idea is to let nodes that can cover larger additional areas broadcast first and then suppress other nodes that do not cover an additional area. The only information needed is node location, and the previous hop's location which are embedded in the received packet. The rebroadcast delay of a node is dynamically adjusted according to received packets from neighbors during the timer period. The number of retransmissions and the broadcast delays is are reduced, while the delivery rate under node mobility and traffic congestion remains high.

The DDB algorithm is tolerant of link failure and packet loss as it utilizes the concept of dynamical forwarding delay which is similar to opportunistic forwarding. However, DDB still assumes the UDG model with a fixed transmission range, so it cannot take advantage of the nodes that received a packet but are located outside of the transmission radius. In this sense, DDB is not fully opportunistic. Also, it does not consider the reliability issue in the realistic physical layer model.

10.2   Related works on broadcasts in VANETs

Earlier works on broadcast protocols in VANETs mostly assumed the ideal propagation model, i.e., whether a packet reception is successful merely depends on a fixed transmission range. Examples include the UMB (Korkmaz and Ekici 2004) and SB (Fasolo *et al.* 2006). They designate the furthest node that receives a packet as the relay node to maximize the one-hop progress (Korkmaz and Ekici 2004) or minimize the one-hop delay (Fasolo *et al.* 2006).

However, as shown by empirical studies, the wireless channel in VANETs is far from perfect (Taliwal *et al.* 2004). Instead, channel fading is the primary challenge and has a major impact on broadcast reception rates (Torrent-Moreno *et al.* 2004, 2006). The suggested realistic propagation model on the highway is the *Nakagami model* (Torrent-Moreno *et al.* 28–25 Sept., 2005), where the packet reception probability (PRP) of single broadcast decreases with the distance, which aggravates the "broadcast storm" problem. The *probability-based methods* (Ni *et al.* 1999; Wisitpongphan *et al.* 2007b) simply let each node rebroadcast a packet with some probability. However, the probabilistic forwarding decision still results in redundant rebroadcasts and does not solve the broadcast storm problem.

10.2.1   Opportunistic forwarding in VANETs

As we have seen in the previous chapters, *opportunistic forwarding* (Biswas and Morris 2005; Zorzi and Rao 2003) is a promising way to deal with lossy links in multihop wireless networks. It exploits the "spacial diversity" enabled by the broadcast nature of the wireless medium, so that the probability of at least one node forwarding a transmitted packet is greatly increased. In this way, each transmission is useful with high probability, and the forwarding delay and the number of transmissions can be reduced.

The concept of opportunistic forwarding has recently been applied to event-driven emergency message broadcast in VANETs. Existing works include contention-based dissemination (CBD) (T-Moreno 2007), emergency message dissemination for vehicular environments (EMDV) (Torrent-Moreno *et al.* 2009), contention-based forwarding (Torrent-Moreno *et al.* 2006), OB-VAN (Blaszczyszyn *et al.* 2008) and location-based flooding (Oh *et al.* 2006). In these works, in order to maximize the "hop progress" of the message in its each rebroadcast, the common idea is to employ a contention process where the furthest node in the message dissemination direction is opportunistically selected. This is often achieved by letting each node that receives the same rebroadcast packet set up a contention timer, in which the backoff time is inversely proportional to the distance of that node to the sender. As a result, the nodes located nearer to the sender backoff longer times and quit contention whenever they hear rebroadcasts (or ACK signals) from a node that has larger hop progress.

However, under the presence of lossy wireless links, the previous schemes could incur unnecessary duplicate rebroadcasts (especially under congested channel) (Torrent-Moreno *et al.* 2009), which takes up the previous bandwidth of the VANET and may adversely affect the performance of other messaging services such as periodical safety beacon messages. This is because the rebroadcast of a vehicle located furthest in the WM dissemination direction may not be heard by other contending nodes due to channel fading. Although these duplicate rebroadcasts can enhance the reliability of WM reception, currently the "duplicate level" is often controlled by adjusting the threshold count of received duplicate messages for each contending node to decide whether to quit the contention. However, it is very difficult to select the appropriate parameters so that a desired WM reception reliability level is achieved, without causing more redundant rebroadcasts.

In addition, the contention processes could result in undesirably large end-to-end WM propagation delay. For nodes to decide whether to participate in contention, there is always some "contention range" chosen by each forwarder upon its rebroadcast. If this range is too large, even the furthest node would wait for a long time before rebroadcast since the contention count-down time tends to be prolonged. Towards solving this problem, in EMDV (Torrent-Moreno *et al.* 2009) Torrent-Moreno *et. al.* proposed to first designate the furthest node (who would rebroadcast without delay) within a "forwarding range" that is shorter than the communication range, and if that node does not receive the rebroadcast then other nodes continue participating in normal contention processes like that in CBD (T-Moreno 2007). However, the issue of how to select an optimal forwarding range has not been addressed.

In summary, the inefficiency of the existing schemes results from two aspects. 1. Nodes make forwarding decisions based on heuristic guesses of whether neighbors have received the same packet or not. The reliability requirement (PRR) of the WM propagation is not considered, and no optimizations have been made so far. 2. In the coordination mechanisms, rebroadcast messages are employed as "implicit acknowledgements", which always subject to channel fading and collisions. Thus it is hard to suppress unnecessary redundant rebroadcasts effectively. In this chapter, we solve these two problems accordingly, by 1. explicitly considering the reliability (PRR) as one of the relay node selection criteria, in that we minimize the number of rebroadcasts to satisfy a given PRR requirement; 2. designing a more reliable and efficient broadcast coordination mechanism, where a short, explicit "broadcast acknowledgement" (BACK) is send out (at the base rate) before each WM's rebroadcast, which has larger communication range than the normal WMs. The BACK not only effectively suppresses redundant rebroadcasts but also clears the channel for the rebroadcast. With BACK, the contention (relay selection) processes can be optimized for higher reliability and lower end-to-end delay.

10.2.2   The reliability issue in VANET broadcast

ElBatt *et al.* 2006 studied one-hop periodic broadcast in cooperative collision warning applications. They characterized the tradeoffs between the packet inter-reception latency, application broadcast rate and transmission range. For multihop WM broadcasts, Resta *et al.* (2007) analyzed theoretically the tradeoff between vehicles' probability to receive a WM within time t and the link-level reliability. But their channel model is oversimplified, and no distributed protocol has been proposed. On the other hand, in this chapter we cast insight on the application-level tradeoff between packet reception performances and the overall transmission overhead, under a realistic channel model.

10.2.3   Broadcast in partitioned VANETs

VANETs turn out to be partitioned (or disconnected) frequently, especially under sparse vehicular traffic, which falls into the delay-tolerant network (DTN) paradigm. Ilias and Cecilia (2007) proposed an opportunistic event dissemination protocol that employs cache and periodic replay mechanisms to keep a message alive in an area. In Abuelela *et al.* (2008), the authors proposed a routing protocol that uses local routing in connected clusters and store-carry-forward at cluster boundaries in order to reduce latency and overheads. In this chapter, we also extended OppCast into a DTN-compatible broadcast scheme, where the protocol adaptively switches between normal dissemination and store-carry-and-forward modes based on vehicles' local traffic densities. Recently, Tonguz *et al.* (2010) proposed DV-CAST, which extended the work in Wisitpongphan *et al.* (2007b) to handle network disconnections. The rebroadcast decisions in DV-CAST are also made based on local topology information; however, there are no reliability (or broadcast success rate) guarantees in DV-CAST. Unlike all the previous works, in our extension we focus on finding the optimal switching traffic density beyond which unnecessary redundant transmissions will be incurred, and below which desired PRR requirements cannot be fulfilled.

10.3   Problem statement

10.3.1   Model and assumptions

In this chapter, we present our event-driven warning message (WM) broadcast protocol using a highway scenario. Figure 10.1 shows the system model, which is a line-topology highway that may have multiple lanes. The VANET consists of vehicles equipped with on board units (OBUs) that can communicate with each other. Suppose a safety-related event (e.g. an accident) happens somewhere, where the source vehicle's OBU begins to disseminate WMs towards the interested region (IR) via broadcast. The IR is defined as the road segment of length $\mathscr{L}$ along the message dissemination direction, which can either include only the co-directional lanes or lanes in both directions. The source is called the *origin of IR*, and the

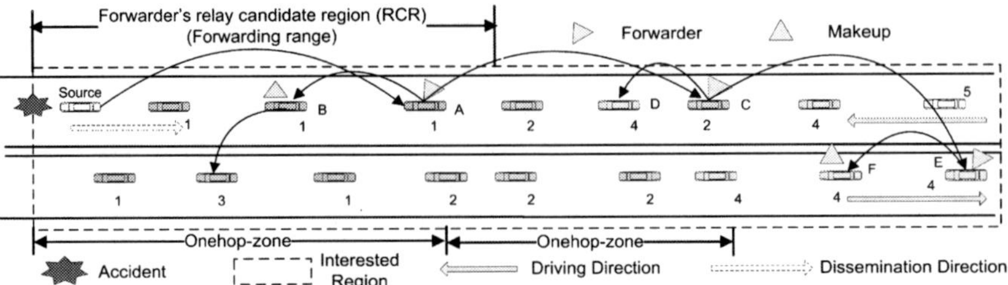

Figure 10.1   VANET model and overview of the broadcast scheme. The number i near each vehicle indicates that the vehicle receives the WM upon the ith (re)broadcast. © 2011 IEEE. Reprinted, with permission, from Z. Yang, M. Li and W. Lou, "CodePlay: Live Multimedia Streaming in VANETs using Symbol-Level Network Coding", IEEE Transactions on Mobile Computing (TMC), 2011.

other boundary of IR is called the *end of IR*. As the width of the highway is far less than the length of IR, for simplicity we model the vehicles to be located in one dimension.

We assume vehicles are GPS-capable and each vehicle obtains its location in real time. This is also assumed by many other works in the literature (C-Varghese *et al.* 2006; Korkmaz and Ekici 2004; Nadeem *et al.* 2006; T-Moreno 2007; Tonguz *et al.* 2010; Torrent-Moreno *et al.* 2009; Wisitpongphan *et al.* December 2007b). When GPS is not available (e.g. in tunnels), vehicles use complementary methods to estimate their locations. For example, this can be done by combining the vehicle speed measured by the speedometer with the GPS map. Vehicles are also are aware of the existence and locations of all neighboring vehicles, as they broadcast one-hop beacon messages every 100 ms (http://www.standards.its.dot.gov/Documents/advisories/dsrc advisory.htm). These beacons are routine safety messages, and warning messages are event-driven. They share the control channel (Jiang *et al.* 2006).

The *network of interest* can be modeled as an undirected graph $G(V, E)$, where V is the set of nodes within the IR. We adopt the probabilistic radio propagation model, where the probability that a node v at a distance x from a node u receives a broadcast packet directly from u is expressed as a (decreasing) function with x, which is denoted as $P_r(x)$. This function accounts for channel fading; it can either be derived from a propagation model Killat and Hartenstein (n.d.) or measured from practice. Assuming identical node transmission power, we can associate each bidirectional link $l = (u, v) \in E$ with a packet reception probability $P_r(d(u, v))$, where $d(u, v)$ is the distance between u and v. In addition, the packet reception at each vehicle is assumed to be independent.

10.3.2   Objectives

In this chapter, we aim to provide a reliable broadcast service in terms of ensuring the *packet reception ratio* (PRR) of the network of interest to be larger than a threshold P_{th} ($PRR \geq P_{th}$, $P_{th} \in (0, 1)$). This is called the *network PRR requirement*.

Definition 10.1 *(Packet reception ratio PRR)* Given a network G and a source s, *PRR* is defined as the percentage of nodes in IR that receive a WM originated from s.

In the meantime, it is also very important for vehicles to be warned in a timely manner. Since the broadcast reception delay $(t_{v,m})$ of WM m at each vehicle v is related to v's distance to the source (d_v), we define the individual dissemination rate as $d_v/t_{v,m}$. The *dissemination rate* is then defined as the individual dissemination rates averaged among all WMs sent and vehicles in the IR. Therefore, the second goal is to reach high dissemination rate. The PRR and dissemination rate capture the application level performances.

Finally, in WM broadcast it is desirable to minimize the transmission overhead, which is defined as the expected total number of transmissions incurred for each WM. Unnecessary transmissions take up bandwidth, increase the channel access delay and the chance of packet collision. This, in turn, degrades the broadcast performance.

10.4   Overview of OppCast

OppCast consists of two types of broadcast phases: *fast-forward-dissemination* (FFD) and *makeup-for-reliability* (MFR). Intuitively, the FFD phase uses relatively long hops to advance the WM towards the end of IR for the purpose of fast propagation. The FFD phase is realized via relaying the WM by a series of *forwarder nodes* that lie successively along the message dissemination direction, where each next-hop forwarder node's distance to the previous one is relatively large. These forwarder nodes thus divide the IR into several *one-hop zones*. Due to lossy links and the independent reception assumption, however, vehicles within these one-hop zones may not all receive the packet upon one relay node's transmission. Thus we use additional make up transmissions that constitute MFR phases to ensure the PRR of the network. In particular, in the MFR phase of each one-hop zone, a minimal set of *makeup nodes* are successively selected until the expected packet reception probability of each node within that one-hop zone is larger than a pre defined P_{th}. In order to satisfy the PRR requirement of the whole network, the reliability of each hop's forwarder node selection is ensured by a retransmission mechanism. By deriving the optimal parameter that controls the length of the one-hop zones, the PRR requirement is satisfied with the least transmission overhead. Note that, we refer to both forwarder and makeup nodes as relay nodes.

The concept of opportunistic forwarding is exploited by OppCast in every transmission to enhance the WM reception reliability and minimize the broadcast latency. Each relay node's (re)broadcast triggers the selection of next forwarder or/and makeup nodes, where the selection of each type of relay is associated with a different relay candidate region (RCR). In the RCR, each node is a potential candidate of the next relay node; due to the broadcast nature of the wireless medium, this greatly enhances the probability that at least one relay candidate receives and rebroadcasts the WM, especially when the network is dense. In the FFD phase, each node (forwarder candidate) that received a (re)broadcast from a previous forwarder contends to become a forwarder in a distributed manner. To maximize the hop progress, each forwarder candidate computes a backoff delay that is inversely proportional to the distance from it to the previous forwarder. The one with the largest hop progress will rebroadcast first and become the forwarder. In order for the forwarder to reliably and efficiently suppress other forwarder candidates, we propose the use of an *explicit broadcast acknowledgement* (BACK) message before each (re)broadcast. The BACK is a short message with longer range than ordinary event-driven WMs, because it is sent at the base rate while WMs are sent at a higher rate. In this way, the previous (re)broadcast is acknowledged, and redundant rebroadcasts are avoided. For each MFR phase, the above contending mechanism is also used to select the makeup nodes, where the selection priority is set as how much additional reception reliability each makeup candidate can bring to its neighbors.

The BACK is a key component in OppCast, and in order to realize the above concepts, we design an opportunistic broadcast coordination function (OBCF) as the underlying MAC protocol used in each broadcast transmission. Specifically, we use the BACK as a way to clear the channel for the subsequent rebroadcast (similar to the function of clear-to-send (CTS) in IEEE 802.11 unicast), which can suppress most of the hidden terminals. Furthermore, we enhance the backoff delay function in previous works, to reduce the hop delay and the possibility of packet collisions. As a result, it is ensured with high probability that in each RCR of each transmission, only one relay is selected.

Due to the use of BACKs, we are able to carry out optimizations on relay selection. To minimize the total number of incurred transmissions for each event-driven WM, we carefully consider the tradeoff between WM dissemination rate and the transmission count. Central to this tradeoff is the length of the RCR for selecting forwarders, namely *forwarding range* (FR). We found the optimal FR, given the vehicle traffic density and the PRR requirement.

In addition, we extend OppCast to handle the partitioned, sparse VANETs. The vehicles in the opposite road of the source are employed as data mules only if a forwarder indicates that its local traffic density is smaller than a certain threshold. The protocol therefore adaptively switches between the normal (fast propagation) mode and store-carry-and-forward mode according to the local traffic densities.

Next we use a simple example to illustrate the broadcast process of OppCast in a well-connected VANET (the normal mode). Figure 10.2 shows a bidirectional highway with the WM source in the upper lane. After the first broadcast by the source, the vehicles with number 1 receive the WM. Those vehicles inside the forwarding range of the source start the forwarder contention process, and node A, which is furthest from the source, becomes the forwarder and sends a BACK before it actually rebroadcasts. After A's second rebroadcast, node B contends and becomes the makeup for the first one-hop zone between source and A (as B derives that the minimum packet reception probability for nodes in that zone is smaller than P_{th}), and node C is selected as the next hop forwarder. Node B's third rebroadcast actually covers the rest of nodes in the first one-hop zone that had not received the WM, while node C's fourth rebroadcast forms the second one-hop zone between

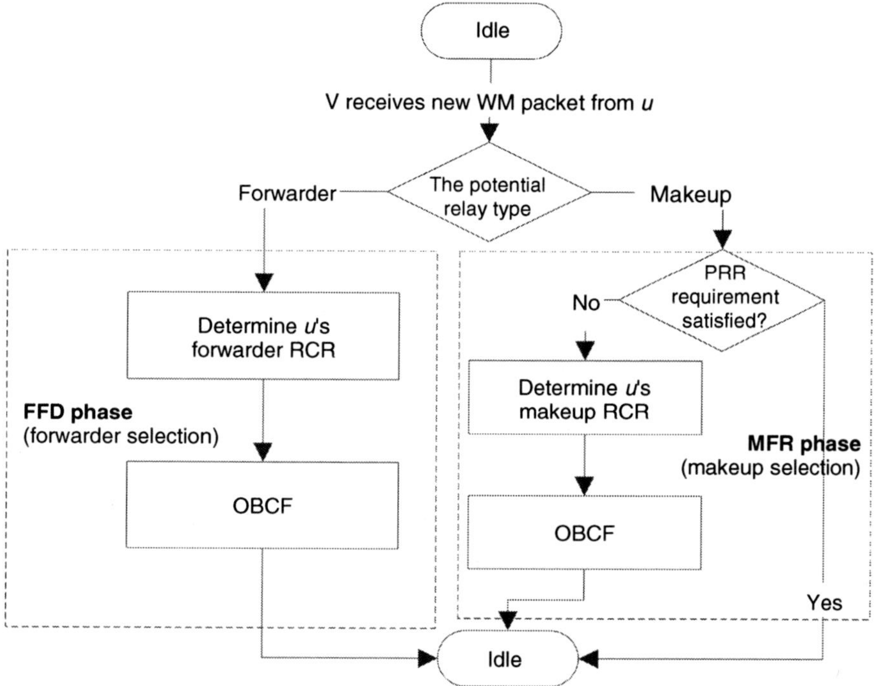

Figure 10.2   The high-level flow chart of OppCast (at node v when receiving a WM from node u). Reproduced by permission of © 2009 IEEE.

A and C. Now all the nodes in the first and second one-hop zones compute the packet reception probability for others, and found that to be larger than P_{th}. No makeup nodes are therefore further selected. In the mean time, the WM is being propagated further in the dissemination direction. A high-level protocol flow chart of OppCast is given in Figure 10.2.

10.5   OppCast: main design

In this section, we describe the components of OppCast. We begin by introducing the FFD and MFR phases from the high level, and then present the underlying OBCF MAC protocol used in relay selection for each transmission. The main notation in this chapter is summarized in Table 10.1.

10.5.1   Fast-forward dissemination

The goal of the FFD phase is to achieve fast WM propagation by using relatively long forwarding hops. Immediately after each forwarder F rebroadcasts, all nodes within the *forwarder RCR* that receive the WM are *candidate forwarders* and participate in the contention process to become the next hop forwarder. For F, its

Table 10.1   Summary of main notation

RCR	Relay candidate region
FR	The forwarding range (length of forwarder's RCR)
IR	The interested region
PRR	Packet reception ratio
F	A forwarder
$G(V, E)$	The modeled graph of the VANET in IR
P_{th}	The required PRR threshold
$d(u, v)$	Distance between two nodes u and v
$P_r(x)$	Packet reception probability at a distance x
$\xi(v)$	Accumulated packet reception probability (APRP) at node v
$E[NT]$	Expected total number of transmissions
$M_{\ell,\lambda}$	A makeup node at the ℓth level and λth branch
$Z_{\ell,\mu}$	A subzone of ℓth level and number μ
$W_{\ell,\mu}$	The middle point used to select makeup node $M_{\ell,\mu}$
Δt_v	The backoff delay of node v for relay contention
ρ	Vehicle density

forwarder RCR is a road segment from *F* towards the end of IR, whose length is called the *forwarding range* (FR). The priority of the candidate forwarders increases with the hop progress (their distance to *F*), in order to maximize the dissemination rate. Ideally, the candidate forwarder that is furthest to the previous forwarder should become the next hop forwarder. Each candidate forwarder sets a backoff timer that is inversely proportional to the hop progress, and a BACK is sent out immediately after this timer expires to suppress others. To ensure that only *one* forwarder is selected for each hop, the BACK itself should be reliable and not prone to collision. Consequently, an one-hop zone is formed between two successive forwarders, whose length is upper bounded by FR; and the index of the one-hop zone increases one by one. Note that, in order to propagate the WM all through the IR, retransmissions are adopted by each forwarder to ensure that a next hop forwarder is selected. The details of the forwarder selection mechanism are presented in the OBCF in Section 10.5.3.

Apart from the OBCF, the other key design issue here is how to choose an appropriate value for the parameter FR. Intuitively, the larger the FR is, the faster a WM can be disseminated. However, this may result in a larger transmission count. Recall that we adopt a probabilistic propagation model, if the actual hop progress is too large the expected percentage of nodes within the one-hop zone that receives the WM will be lower than the desired PRR. Thus more makeup nodes within the one-hop zone will be needed to help rebroadcast the WM and more transmissions could, in turn, slow down the overall dissemination rate. On the other hand, if the FR is too small, there will be many redundant rebroadcasts since the transmissions of two successive forwarders overlap with each other. We will therefore focus on minimizing the expected total number of transmissions $E[NT]$. This involves knowledge of the MFR phase, so we defer the derivations to Section 10.6.

10.5.2 Makeup for reliability

If a node u receives a WM from the kth forwarder and u is located in the one-hop zone created by the $k - 1$th and kth forwarder, u will run through the distributed makeup selection process. A key concept in the MFR phase is the accumulated packet reception probability (APRP) of a node, which captures the idea that for any node u located in an one-hop zone, the probability of receiving a WM packet m increases as m is consecutively rebroadcasted multiple times by the relay nodes near u (the concrete definition of APRP will become clear in the following). The objective of the MFR phase is to ensure that the APRPs of all the nodes in each one-hop zone are larger than the given threshold P_{th}. If all the nodes' APRPs are larger than P_{th} in each one-hop zone, the PRR requirement of the network can be ensured. To minimize the number of makeups in each one-hop zone, the idea is to give the highest priority to a node whose rebroadcast can maximize the minimum APRP of all the nodes in the one-hop zone.

We first illustrate the intuition of the makeup selection algorithm using the scenario in Figure 10.3. For the time being, assume for simplicity that the left forwarder F_L is the source of WM packet m. After the one-hop zone $Z_{0,0}$ is formed, we already have the left and right forwarders rebroadcasted. Since packet reception probability decreases with distance, the middle of $Z_{0,0}$ is covered with the least APRP. Intuitively, selecting a node $M_{1,0}$ in the middle (or nearest to the middle) is most helpful for increasing the minimum APRP for other nodes in $Z_{0,0}$. After $M_{1,0}$ broadcasts, it divides $Z_{0,0}$ into two sub zones. Similarly, the middle points of these subzones have the least APRP, and again new makeups closest to the middle points are selected. This process is continued until the minimum APRP of all nodes in all sub zones are larger than P_{th}.

Before we give a more rigorous treatment, we introduce some notation. The makeups form a binary tree, which is indexed by level ℓ and branch λ. A makeup is denoted as $M_{\ell,\lambda}$, $\lambda \in [0, \ldots, 2^{\ell-1} - 1]$. The depth of the tree is bounded by a maximum level.[2] At level ℓ, the makeups split the one-hop zone into 2^ℓ sub zones, denoted as $Z_{\ell,\mu}$ (its number $\mu \in [0, \ldots, 2^\ell - 1]$). Each ℓth level sub zone $Z_{\ell,\mu}$ is defined by scanning the one-hop zone from left to right and assigning two consecutive relay nodes at level $l \le \ell$ as its left and right boundaries $(x_L^{\ell,\mu}, x_R^{\ell,\mu})$. The one-hop zone is regarded as $Z_{0,0}$, which is bounded by $x_L^{0,0}$ and $x_R^{0,0}$, coordinates of the left and right forwarders (F_L and F_R). The right forwarder is regarded as the 0th level makeup.

Next we show how the APRP is defined and evaluated. Since a node u can hardly receive a rebroadcast packet from relays far away from it, we only consider contributions of rebroadcasts from specific nearby relays that are in the same one-hop zone with u. For a particular WM packet m and node u, we define a set of

---

[2] This will not cause a broadcast storm because the maximum level needed is small and bounded, and OBCF greatly reduces packet collisions.

locally visited nodes by m, which consists of relay nodes on the tree branch leading to u : $\{F_L, F_R, M_{1,\lambda_1}, \ldots, M_{\ell,\lambda_\ell}\}$, where ℓ is the level of the newest sub zone containing u. For example, if u is located in $Z_{2,1}$, which happens to be the newest sub zone containing u, the locally visited nodes are $\{F_L, F_R, M_{1,0}, M_{2,0}\}$. This is a conservative estimation of APRP as we have neglected the contributions of rebroadcasts from (possible) relays on sibling branches and those in other one-hop zones.

Upon receiving a WM m, each node u locally estimates the APRP of each neighbor node v within $Z_{\ell,\mu}$ iteratively, based on the locally visited nodes:

$$\xi_0(v) = 1 - (1 - P_r(d(F_L, v)))(1 - P_r(d(F_R, v)))$$
$$\xi_\ell(v) = 1 - (1 - P_r(d(M_{\ell,\lambda_\ell}, v)))(1 - \xi_{\ell-1}(v)) \tag{10.1}$$

where $\xi_\ell()$ denotes the ℓth iteration. If the minimum APRP: $\min_{v \in Z_{\ell,\mu}} \xi_\ell(v) \geq P_{th}$, then u knows the APRP requirement is satisfied. Otherwise, if u is in the makeup relay candidate region of M_{ℓ,λ_ℓ}, it becomes a makeup candidate and starts the OBCF according to its priority. The makeup RCR of M_{ℓ,λ_ℓ} is simply the sub zones created by M_{ℓ,λ_ℓ}. For example, in Figure 10.3, makeup RCR of F_R is $Z_{0,0}$, and that of $M_{1,0}$ consists of two subzones $Z_{1,0}$ and $Z_{1,1}$.

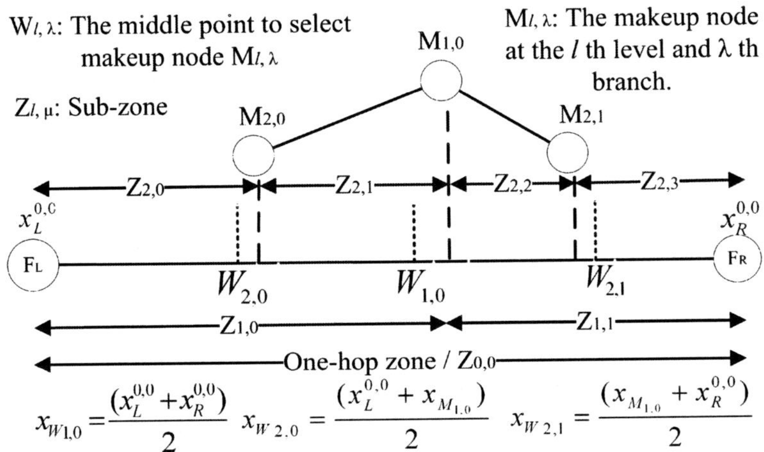

$$x_{W1,0} = \frac{(x_L^{0,0} + x_R^{0,0})}{2} \qquad x_{W2,0} = \frac{(x_L^{0,0} + x_{M_{1,0}})}{2} \qquad x_{W2,1} = \frac{(x_{M_{1,0}} + x_R^{0,0})}{2}$$

Figure 10.3   A makeup selection tree, maximum level = 2. Reproduced by permission of © 2009 IEEE.

The priority of u reflects the minimum APRP of nodes in $Z_{\ell,\mu}$ after u rebroadcasts m, which is denoted as $\xi^*|u$. This can be calculated by doing another iteration on Equation (10.1). For mathematical convenience, let us define the *APRP function*: $\Phi_{\ell,\mu}(x)$, $x \in [x_L^{\ell,\mu}, x_R^{\ell,\mu}]$ over each sub zone as a function of location coordinate x, which can be regarded as the APRP at location x given the rebroadcasts of the locally visited nodes $\{F_L, F_R, M_{1,\lambda_1}, \ldots, M_{\ell,\lambda_\ell}\}$. It is easy to see that for each node v, $\Phi_{\ell,\mu}(x_v) = \xi_\ell(v)$. Next, we claim that the priority of u decreases as the distance of u to the middle of the sub zone that u is located in increases.

Theorem 10.1 *Function* $\Phi_{\ell,\mu}(x)$ *is concave. If it is symmetrical with respect to the middle point* $W_{\ell+1,\mu}$ *of sub zone* $Z_{\ell,\mu}$, *then for any sequence of nodes* $\{i_0, i_1, \ldots, i_n\}$ *within* $Z_{\ell,\mu}$ *such that* $d(i_0, W_{\ell+1,\mu}) < d(i_1, W_{\ell+1,\mu}) < \ldots < d(i_n, W_{\ell+1,\mu})$, *we have*

$$\xi^* | W_{\ell+1,\mu} > \xi^* | i_0 > \ldots > \xi^* | i_n.$$

Proof. The concavity of $\Phi_{\ell,\mu}(x)$ is straightforward. Let $\Phi_{\ell+1}(x, x_M)$, $x, x_M \in Z_{\ell,\mu}$ denote the $\ell + 1$th level APRP given a node at x_M broadcasts, where $Z_{\ell,\mu}$ consists of $Z_{\ell+1,2\mu}$ and $Z_{\ell+1,2\mu+1}$. We use $W_{\ell+1,\mu}$ to represent a middle point and its coordinate interchangeably. It can be seen from the properties of concave and symmetrical functions that $W_{\ell+1,\mu}$ is the minimum point of $\Phi_{\ell,\mu}(x)$. Then $\Phi_{\ell+1}(x, W_{\ell+1,\mu})$, $x \in Z_{\ell,\mu}$ is also symmetric w.r.t $W_{\ell+1,\mu}$:

$$\begin{aligned}
&\Phi_{\ell+1}(2W_{\ell+1,\mu} - x, W_{\ell+1,\mu}) \\
&= 1 - (1 - P_r(|W_{\ell+1,\mu} - (2W_{\ell+1,\mu} - x)|))(1 - \Phi_{\ell,\mu}(2W_{\ell+1,\mu} - x)) \\
&= 1 - (1 - P_r(|x - W_{\ell+1,\mu}|))(1 - \Phi_{\ell,\mu}(x)) \\
&= \Phi_{\ell+1}(x, W_{\ell+1,\mu})
\end{aligned} \tag{10.2}$$

So there are two minimal points, x_L^* and x_R^* in $[x_L^{\ell+1,2\mu}, x_R^{\ell+1,2\mu}]$ and $[x_L^{\ell+1,2\mu+1}, x_R^{\ell+1,2\mu+1}]$ respectively, which are both equal to the minimum value of $\Phi_{\ell+1}(x, W_{\ell+1,\mu})$ in $Z_{\ell,\mu}$. In the following, we pick a point $x_{i_0} > W_{\ell+1,\mu}$ from the sequence of nodes within $Z_{\ell,\mu}$. First, we show that the minimum value of $\Phi_{\ell+1}(x, x_{i_0})$ is smaller than that of $\Phi_{\ell+1}(x, W_{\ell+1,\mu})$. At point x_L^*, we have $\Phi_{\ell+1}(x_L^*, x_{i_0}) < \Phi_{\ell+1}(x_L^*, W_{\ell+1,\mu})$:

$$\begin{aligned}
&\Phi_{\ell+1}(x_L^*, x_{i_0}) - \Phi_{\ell+1}(x_L^*, W_{\ell+1,\mu}) \\
&= (P_r(|x_{i_0} - x_L^*|) - P_r(|W_{\ell+1,\mu} - x_L^*|))(1 - \Phi_{\ell,\mu}(x_L^*)) < 0
\end{aligned} \tag{10.3}$$

as $P_r(x)$ is monotonically decreasing and $x_{i_0} > W_{\ell+1,\mu}$. Therefore, $\xi^* | W_{\ell+1,\mu} > \xi^* | i_0$. Similarly, for $x_{i_0} < W_{\ell+1,\mu}$, $\Phi_{\ell+1}(x_R^*, x_{i_0}) < \Phi_{\ell+1}(x_R^*, W_{\ell+1,\mu})$. Immediately, for any two nodes $\{i_0, i_1, \ldots, i_n\}$ within $Z_{\ell,\mu}$ such that $d(i_0, W_{\ell+1,\mu}) < d(i_1, W_{\ell+1,\mu}) < \ldots < d(i_n, W_{\ell+1,\mu})$, $\xi^* | W_{\ell+1,\mu} > \xi^* | i_0 > \ldots > \xi^* | i_n$.

Note that the above optimality is derived under the assumption that $\Phi_{\ell,\mu}$ is a symmetrical function. In reality, $\Phi_{0,0}$ is strictly symmetrical; as the level of broadcast increases, $\Phi_{\ell,\mu}(x)$ deviates from being symmetrical gradually because of the impact of the broadcasts of other lower-level relay nodes. However, the deviation degree is small if the level is small (Li and Lou 2008). In practice, to satisfy 99% PRR, it is usually enough for the maximum level to be 2.

Remark. Although the makeup selection algorithm in the MFR phase may seem complex, it should not be interpreted as an overkill for the whole scheme. Rather, it is a necessary component in OppCast to achieve the desired PRR using a minimal number of rebroadcasts. We present it in the above way in order to include the general case. Indeed, the only parameter in the algorithm is the maximum makeup level, and the algorithm at each node is simple enough. For the reception probability $P_r(x)$, one can use an empirical model suitable for the VANET such as the one in Killat and Hartenstein (n.d.). It also incurs a lower transmission count than the straightforward strategy where each node contends to become a makeup node using a random priority. Moreover, nodes can make transmission decisions in an on-demand fashion to satisfy a desired PRR requirement, which is a feature not possessed by existing schemes in the literature.

10.5.3   Broadcast coordination in OppCast

In the following, we present OBCF, which is the underlying mechanism for selection of both forwarder and makeups. A broadcast coordination process (or contention process) is started immediately after each rebroadcast by a relay node. The primary goal is to let the relay candidates agree on the actual relay nodes in a distributed way, and for the selected relays to perform a collision-free broadcast. The OBCF consists of the following components: 1. a process for the relay candidates to contend for the relaying opportunity; 2. a resource reservation mechanism to avoid collision and suppress hidden terminals; 3. a retransmission mechanism to prevent the WM from dying out. Its process is generally described as follows.

1. When a node v receives a WM m for the first time (from node u), if v is in the RCR of u, v becomes a relay candidate. Then v sets a broadcast backoff timer (BBT) for m and calculates m's backoff delay. Also, v sets a *self allocation vector* (SAV) at MAC layer. The SAV suspends the transmission of other types of packets from v itself until there are no ongoing OBCF processes. This design provides packet-level priority access for WMs, since the WMs are safety critical and have the highest priority in VANETs.

2. If v senses a busy signal from the physical layer, v will pause all its BBTs that are still counting down in order to prevent collision and to keep its BBTs on the same page with that of other nodes. When the physical layer indicates idle again, v will resume all the paused BBTs.

3. If the BBT for m expires without receiving a broadcast acknowledgement (BACK), node v becomes a relay node for packet m and sends a short MAC-layer BACK **at the base rate** to suppress other candidates. After the BACK has been transmitted, and after a short inter frame space (SIFS), the WM is sent immediately at the data rate (higher than the base rate). The BBTs for other WMs are also paused during transmission.

4. If v receives a BACK for packet m from another node w before its own BBT expires, and if w is a relay node that contends for the same relaying opportunity with v, v will cancel the BBT. After that, v clears the SAV, and sets a *network allocation vector* (NAV) to reserve the time period for the WM that follows the BACK to suppress hidden terminals. Also, v pauses all of its own BBTs.

5. The OBCF process for m finishes when the NAV for m expires, or v finishes broadcast of m as a relay node. The SAV of v is cleared only if there are no OBCF processes going on, or when a NAV is set.

6. Each source or forwarder F sets an additional, recurring *retransmission timer* (different from BBT) after transmitting m for the first time. This timer expires after every period of *MAX_WAIT_TIME* (the maximum delay of receiving a BACK from a forwarder, which is adaptively set; see below). This timer is only canceled when F receives a BACK that acknowledges the reception of m from a forwarder or makeup belonging to an one-hop zone with a higher index than F. Otherwise, whenever this timer expires, F retransmits m until the maximum allowed number of retransmissions, *MAX_RETX*, is reached.

The timeline of events is illustrated in Figure 10.4.

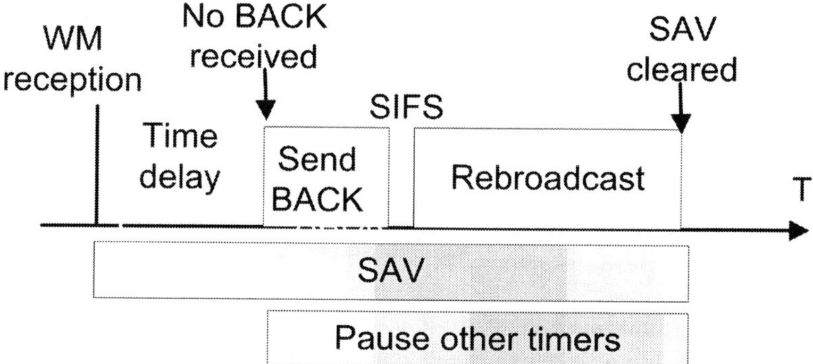

(a) Actual relay: time delay is the shortest, and becomes the first to send a BACK

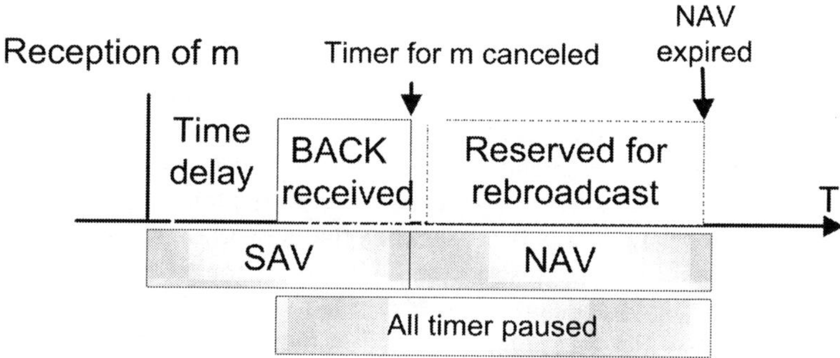

(b) Potential relay: received BACK before timer expires; timer paused and then canceled

Figure 10.4   Time domain illustration of OBCF. Reproduced by permission of © 2009 IEEE.

A key element of OBCF is the delay function in BBT. A higher priority implies a smaller backoff delay. Observe that, for both types of relays, the RCR is a road segment, and the priority of nodes in a RCR increases/decreases monotonically from one end to the other. In the FFD phase, a node with a larger hop progress should have a higher priority in rebroadcast. In the MFR phase, a node closer to the center of a sub-zone (which is the boundary of RCR) should rebroadcast earlier. So, in both cases, the delay can be expressed by a function of the distance.

A straightforward way to set the timer is to let the delay be inversely proportional to the distance from the sender. However, two or more nodes that happen to be very close in space cannot be distinguished by this method. Thus, we propose an *enhanced slotted delay function*, where the RCR is divided into multiple equal-length spacial segments. The length of a segment adapts to the local vehicle density, which results in one node per segment on average. Each spacial segment corresponds to a time slot, where the central time of each slot increases linearly with the segment number, while a random jitter is used to separate potentially multiple nodes in the same segment to prevent collision. In this way, even if two nodes were very close spatially and were in the same segment, when one of them transmits first, the other node can hear it. A guard time is also placed between adjacent time slots, which provides a minimum difference between backoff times for nodes in adjacent segments, in order to enforce the priorities of nodes in different segments. Although the nodes' priorities are not strictly followed within the same segment, the impact of this reduces with the increase of traffic density as the segment length decreases.

Let x_I denote the boundary of RCR towards which the delay should increase, and x_D denote the boundary towards which the delay should decrease. For a node v located in u's RCR, v's backoff delay Δt_v used for its BBT is calculated as:

$$S_v = \lfloor \frac{|x_v - x_D|}{L} \rfloor, L = 1000/\rho, x_v \in [x_I, x_D] \tag{10.4}$$

$$\Delta t_v = \begin{cases} [S_v \cdot (T + \delta) + T \cdot Rand(0, 1)], & x_v \in [x_I, x_D] \\ \infty & \text{otherwise,} \end{cases} \tag{10.5}$$

where S_v is the segment number of v, L is the segment length (ρ is the vehicle density in # of vehicles/km which can be estimated distributively), T is the maximum delay range of nodes in a segment, δ is a guard time which is used to separate two neighboring segments. Note that, by the above construction, Δt_v is always small for an actual forwarder. This is because when the network is sparse, although the actual forwarder may not be close to the boundary of the RCR, each segment is long and the forwarder's segment number is small. When the network is dense, it is more probable that the forwarder is located in the first few segments.

In the above, the segment length is $L = 1000/\rho$. On average, in distance L, there will be 1 vehicle. The actual number of vehicles in each segment depends on the vehicle distribution; but as ρ will be locally estimated by each relay node (see Section 6.1.2) our method can effectively ensure a small number of vehicles in each segment.

To set the parameter *MAX_WAIT_TIME* in the retransmission timer, we estimate the upper bound on the time delay that a forwarder receives a BACK. Choose a d_{max} to be larger than the maximum possible RCR length, then according to Equation (10.4), $S_v = \lfloor \frac{d_{max}}{L} \rfloor$. When no BACKs are received from nodes in the segments with numbers smaller than S_v except the one equal to S_v, we have $MAX\_WAIT\_TIME \approx (T + \delta) \cdot S_v$. In this chapter, we set $d_{max} = 1000$, thus $MAX\_WAIT\_TIME = (T + \delta) \cdot \rho$.

The OBCF has several advantages. First, redundant transmissions are eliminated more effectively. Because the BACK is transmitted at the base rate, for which the threshold of received signal-to-interference-and-noise ratio (SINR) is lower at a receiver than using the data rate, it can be received by most of the relay candidates. Second, the one-hop delay is small. This is because 1. in OBCF a node pauses its timers when its detects a busy channel which prevents a collision, and 2. in the relatively rare case that a node is in a nearby segment with a relay but does not hear the latter's BACK, as the BACK is very short (its transmission takes around $80\,\mu s$ when the payload length is 14 bytes), choosing $T + \delta$ to be larger than the BACK transmission time can prevent most of the BACK collisions from happening. In our simulations, we found that $T = 80\,\mu s$ and $\delta = 20\,\mu s$ are good values. Third, BACK is also used to suppress the hidden terminals. As illustrated in Figure 10.5, the transmission range of BACK is larger than twice that of the WM, which means most of the hidden terminals to the WMs are avoided.

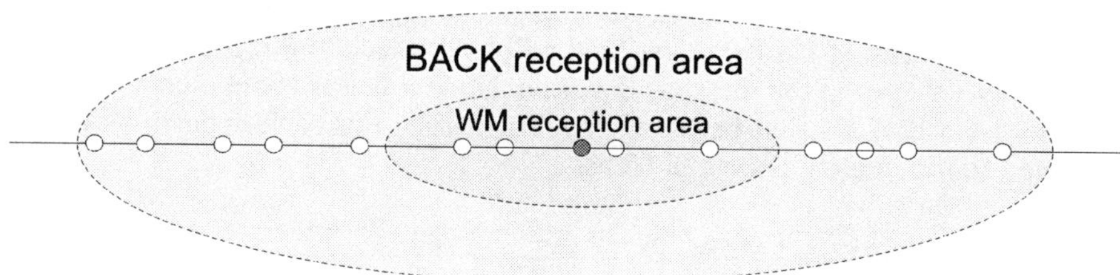

Figure 10.5   BACK suppresses hidden terminals. Reproduced by permission of © 2009 IEEE.

10.5.4   Extension to disconnected VANET

When the VANET is sparse, for example at nights or when the penetration rate is low, the VANET tends to be disconnected (Wisitpongphan *et al.* 2007a) and the WM cannot be propagated through the whole IR. Thus we need additional mechanisms to ensure reliable reception of the WM by vehicles in the interested region. To this end, we extend OppCast to handle this situation, and still assume the WM is disseminating from west to east.

Like many routing schemes for disconnected VANETs (Nadeem *et al.* 2006; Wisitpongphan *et al.* 2007a), for WM broadcast in this chapter we also take advantage of bidirectional mobility, i.e., we utilize the vehicles driving in opposite direction to store-carry-and-forward the WM packets. While the idea is simple, for OppCast, we need to design the protocol carefully so that the three previously proposed objectives are still attained. Thus, several questions arise. 1. Which vehicles should be chosen to perform store-carry-and-forward and when should they do it? 2. How to ensure the WM reception reliability in the network when store-carry-and-forward is adopted? 3. How should the parameters in the protocol be adjusted under sparse VANETs? 4. Can we still preserve the advantages of using BACKs?

To answer these questions, we allow the protocol to switch adaptively between the normal dissemination mode (as previous described) and the store-carry-and-forward mode, according to local network topology. The straightforward solution is that, in "connected" parts such as vehicle clusters a message should propagate fast until it reaches the end of a cluster; while the end vehicle in a cluster performs store-carry-forward to enhance reliability (PRR). This is also the idea in existing schemes (Abuelela *et al.* 2008; Tonguz *et al.* 2010). However, it can hardly guarantee the desired level of reliability because this strategy neglects the differences in the local traffic densities. The normal mode is adopted even if there are only a few (> 0) vehicles around a relay node, where the relay's rebroadcast may not be heard by any vehicles in the message direction, which results in message propagation stopping.

Thus in this chapter, in addition to using the straightforward store-carry-forward condition, we propose to employ local traffic density as a decision factor, i.e., a relay node u switches to store-carry-forward mode whenever its local traffic density ($\widehat{\rho}(u)$) is smaller than a threshold. To implement this function, a *carry_flag* is used, which is set to 1 if $\widehat{\rho}(u) < \rho_{th}$, and will be set back to 0 if u's rebroadcast is received by vehicles in the message direction. Node u learns about others' message receptions via BACKs sent before their rebroadcasts. For each node v, its locally estimated traffic density is computed as:

$$\widehat{\rho(v)} = \frac{\text{Number of neighboring vehicles within range } CR \text{ in the message direction}}{CR},$$

$$(10.6)$$

where CR is the *communication range* of the WM, defined equivalently (for the same transmission power) under the two-ray ground propagation model. We only count the neighbors in the message direction because the FFD of WM is mainly impacted by the density of those vehicles. The detailed forwarding rules are as follows:

1. We assume the source vehicle is west-bound in the following. For any forwarder (or source) F driving towards west, upon receiving a WM m for the first time, if its locally estimated traffic density ($\widehat{\rho(F)}$) is smaller than a *threshold density* ρ_{th} or there are no vehicles in the message direction (east), it will carry the packet and set a *carry_flag* $= 1$. (For example, the node A in Figure 10.6(a)). Only if F receives a BACK from a relay in the east will it set *carry_flag* $= 0$. If F has not received such a BACK after it retransmits m *MAX_RETX* times, it will store the packet. Later when F receives a beacon message from a vehicle driving opposite and also located in the east of F, if *carry_flag* $= 1$, a rebroadcast of m will be triggered and exactly the same OBCF is invoked (Figure 10.6(b)). Note that, the other node is very close to F because the beacon rate is usually high (for example, ten beacons/s), thus the single-hop reliability is very high in this case.

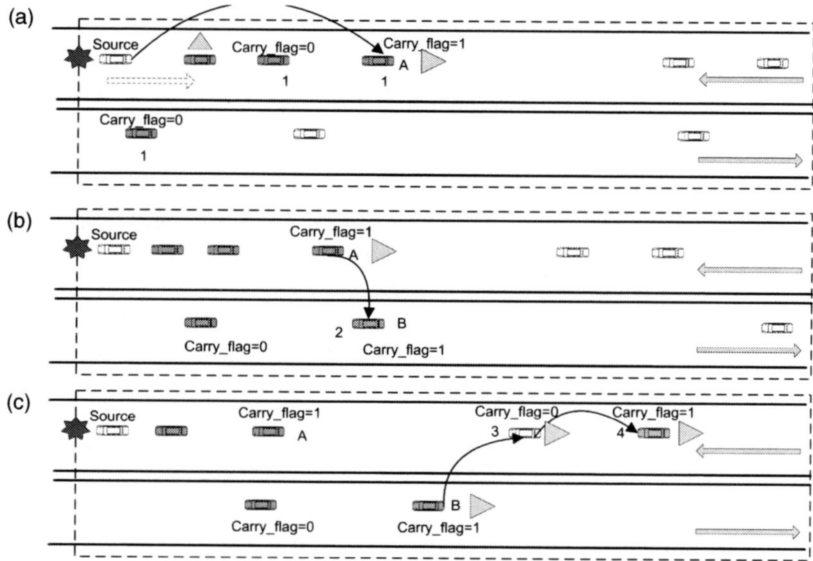

Figure 10.6 Dissemination process of a WM in the OppCast-extension for the sparse VANET. Legends are the same as for Figure 10.2. © 2011 IEEE. Reprinted, with permission, from Z. Yang, M. Li and W. Lou, "CodePlay: Live Multimedia Streaming in VANETs using Symbol-Level Network Coding", IEEE Transactions on Mobile Computing (TMC), 2011.

2. For any vehicle V driving towards east, upon receiving WM m from a forwarder F, if F's local traffic density $\widehat{\rho(F)} < \rho_{th}$ (contained in m), V will temporarily set $carry\_flag = 1$. It does not matter whether $carry\_flag = 1$ or 0, if the number of neighbors in the east is larger than 0, V participates in relay selection as usual. This is to let the WM propagate to the end of a "cluster" as fast as it can and to reduce the total transmission count. If $carry\_flag = 1$, only when V later receives a BACK from another relay in the east and also driving towards east will V set $carry\_flag = 0$. Otherwise, if sometime later V receives a beacon message from a node in the east but driving towards the west, FFD will be triggered at V according to OBCF (Figure 10.6(c)). If V is the head of a cluster driving east, V will always carry the WM, and thus reliability is ensured with high probability.

In the above way, the WM can be still disseminated with high speed in connected clusters, where "connected" should be interpreted in probabilistic sense, under opportunistic forwarding and retransmissions. For "disconnected" parts, the opposite directional vehicles act as data mules. Here we need to determine a suitable ρ_{th}. If ρ_{th} is too small, a WM may not be able to get through the network which affects the PRR; if ρ_{th} is too large, there will be redundant rebroadcasts by data mules. We will therefore discover the optimal ρ_{th} in Section 10.6.

10.5.5   Implementation issues

We first give the definition of "neighbors" adopted in the implementation. For a node v to become a neighbor of node u, u needs to receive at least one beacon message from v recently within a reasonable amount of time, such as five times the beacon broadcast interval.

In addition, we need to ensure that the receivers of each (re)broadcast agree on a unified RCR to determine whether to participate in contention processes. We therefore include the locally calculated (optimal) forwarding range of the sender in every broadcast packet from a forwarder/source and each node that receives the packet uses the received value in the packet instead of its own. For the additional makeups to decide their RCRs, they simply rely on the locations of previous visited nodes piggybacked in the rebroadcast packets.

Next, we present the format of the WM packet header in Figure 10.7. The first $2...19$ bytes are used for FFD phases, and $20...4k + 29$ bytes are used for MFR phases, where k is the maximum level of makeups in a one-hop zone. The fields x_{left} and x_{right} stand for the boundary locations of the current one-hop zone or sub zone. The list of locations of visited nodes in current one-hop zone include the left and right forwarders, and visited makeups; it is used in MFR to calculate the APRP. Although when $k = 2$ the header length is 37B, it is relatively small compared to the message length, which is usually more than 200B.

									Left forwarder	Right forwarder	... kth makeup
UID	Type of previous hop	Location of the source	X_left	X_right	One-hop zone index	Local optimal boundary range	Level of makeup node	Subzone index	List of locations of visited nodes in current one-hop zone		
Bytes 1	2	6	10	14	15	19	20	21			4k+29

Figure 10.7   The WM packet header format. © 2011 IEEE. *Reprinted, with permission, from Z. Yang, M. Li and W. Lou, "CodePlay: Live Multimedia Streaming in VANETs using Symbol-Level Network Coding", IEEE Transactions on Mobile Computing (TMC), 2011.*

In the BACK messages, headers include little information: the one-hop zone index, makeup level, and the sub zone index. The one-hop zone index increases by one when the WM traverses another forwarder, and upon receiving a BACK, a node can determine whether to cancel its own rebroadcast. In practice, in order to enhance the reliability of BACK to suppress redundant transmissions, an additional rule is used. The BACK from a forwarder with a one-hop zone index $k + i, i \geq 0$ suppresses the forwarder candidates in the kth one-hop zone, and a higher level makeup suppresses lower level ones in the same one-hop zone.

The pseudo-code of the OppCast protocol at a running at node u is given in Algorithm 10.1.

10.6   Parameter optimization

In this section, we optimize two parameters in OppCast: the forwarding range *FR* and threshold density ρ_{th}.

10.6.1   Optimize the forwarding range

As mentioned in Section 10.5.1, the goal here is to minimize the expected total number of transmissions $E[NT]$. Stated formally, we want to find the *FR*

$$Min \qquad E[NT] \tag{10.7}$$

$$s.t. \ \forall v \in G, PRR \geq P_{th} \tag{10.8}$$

Algorithm 10.1 Distributed OppCast algorithm running at node u

---

Input: Node u receives a WM m;
If u is in the IR of m, and m is a new packet
  If $type_{prevhop} = $ `forwarder` and $x_u > x_{right}$ //Enter FFD;
    If $(x_u - x_{right}) <$ the local optimal boundary range (LBR)
      Set $x_D = x_{right} + LBR$, $x_I = x_{right}$; //the RCR
      Compute Δt_u and run OBCF; //potential forwarder
      If u becomes a forwarder
        Use $\widehat{\rho_u}$ to compute the new local optimal boundary range;
        Set $x_{left} = x_{right}$, $x_{right} = x_u$;
        Increase the one-hop zone index by 1 and broadcast m.
  If $x_{left} < x_u < x_{right}$ //potential makeup
    If $type_{prevhop} = $ `forwarder` //Initialize MFR
      Makeup node level $\ell = 0$; sub-zone index $\mu = 0$;
      If $\min_{v \in Z_{0,0}} \xi_\ell(v) \geq P_{th}$, then **exit**;
      Set the visited node location list $V_L : \{x_0 = x_{left}, x_1 = x_{right}\}$;
    If $type_{prevhop} = $ `makeup` //Enter MFR;
    $x_{prev} \leftarrow$ last element in V_L;
    If $x_u > x_{prev}$
      $x_{left} = x_{prev}$, $\mu = 2\mu + 1$;// boundary of current sub-zone;
    Else, $x_{right} = x_{prev}$, $\mu = 2\mu$;
    If $\min_{v \in Z_{\ell,\mu}} \xi_\ell(v) \geq P_{th}$, then **exit**;
    $x_W = (x_{left} + x_{right})/2$;
  If $\ell \leq maxlevels$
    $\ell + +$, append x_u to V_L;
    If $x_u \geq x_W$
      $x_D = x_W$, $x_I = x_{right}$; //The RCR;
    Else if $x_u < x_W$
      $x_D = x_W$, $x_I = x_{left}$; //The RCR;
    Calculate Δt_u and run OBCF;

The optimization is targeted at the well connected case (vehicle density is larger than ρ_{th}), whereas it is straightforward to show that under the strategies of the FFD and MFR phases, the constraint is satisfied with high probability.

Thus, we first compute the expected number of transmissions ($E[NT]$), based on which the optimal *FR* is derived. We introduce the centralized solution to find the optimal *FR* and then propose a distributed, locally optimized version. The centralized solution takes as input the average vehicle density ρ of IR, and approximates the $E[NT]$. Since $E[NT]$ has no closed form expression, the optimal *FR* that minimizes $E[NT]$ is sought out by sampling and searching.

Let us consider a one-dimensional VANET where the IR length $\mathscr{L}$ is sufficiently large. Assume there are no redundant transmissions and no packet collisions. Further, we assume there are enough relay candidates so that the PRP requirement can always be satisfied. Finally, the Rayleigh fading model is used for pairwise PRP function: $P_r(d) = exp(-\frac{P_{rxth}}{P_{ref}}d^2)$, where P_{rxth} is the reception threshold power, P_{ref} is the reference receive power at distance 1 m by free space propagation model.

The total number of transmissions NT can be expressed as:

$$NT = \sum_{i=1}^{X} (M_i + \omega_i) \tag{10.9}$$

$$\sum_{i=1}^{X} Y_i = \mathscr{L} \tag{10.10}$$

where X is the number of one-hop zones, and M_i is the number of makeups in the ith one-hop zone. Y_i is the length of the ith one-hop zone, and $M = M(Y)$ is a single-variable function of Y, while Y is related to both *FR* and ρ. ω is the number of retransmissions made by the ith forwarder. Since Y_i are i.i.d. random variables, X is also a random variable.

Therefore, we have

$$\begin{aligned} \mathscr{L} &= E[\textstyle\sum_{i=1}^{X} Y_i] = E_X[E_{Y|X}[\textstyle\sum_{i=1}^{X} Y_i|X]] \\ &= E[Y_1|X = 1]P(X = 1) + E[Y_1 + Y_2|X = 2]P(X = 2) + \dots. \end{aligned} \tag{10.11}$$

For an approximation, we neglect the dependence between X and Y_i (i.e. $E\left[\sum_{i=1}^{X} Y_i|X = i\right] \approx E\left[\sum_{i=1}^{X} Y_i\right] = i \cdot E[Y])$, then

$$\mathscr{L} = E[X] \cdot E[Y] \tag{10.12}$$

and thus

$$E[NT] = E[X](E[M] + E[\omega]) \tag{10.13}$$

where $E[X]$ is the average number of one-hop zones, $E[M]$ is the average number of makeups in each one-hop zone, $E[Y]$ is the average one-hop zone length, $E[\omega]$ is the expected retransmission count of each forwarder.

We then approximate $E[Y]$ and $E[M]$ by fixing the inter space between successive vehicles to[3] $L = 1000/\rho$.

$$E[Y] = \sum_{k=1}^{N} kL \cdot P_F(k, L), \quad N = \lfloor\frac{FR}{L}\rfloor \tag{10.14}$$

---

[3] The uniform distribution of vehicle positions is adopted in performance evaluation.

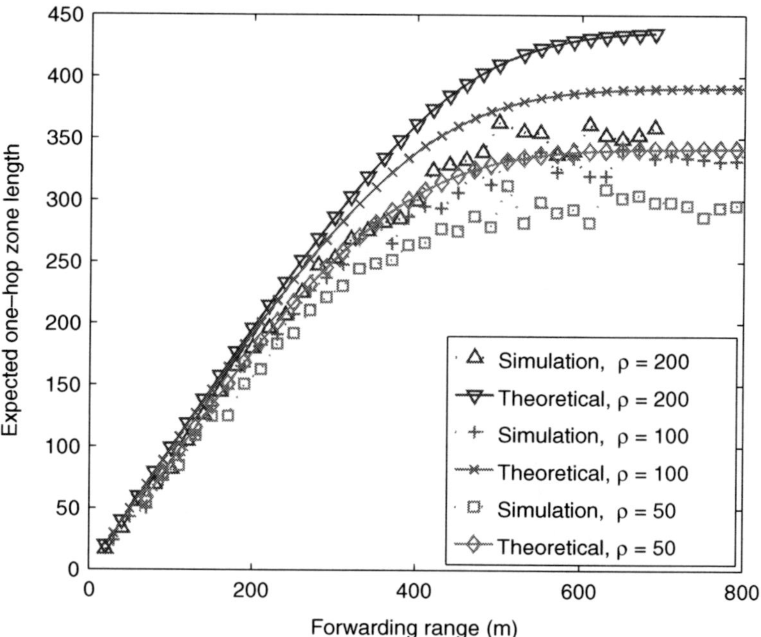

Figure 10.8   Numerical validation of E[Y]. Reproduced by permission of © 2009 IEEE.

where $P_F(k, L) = P_r(kL) \prod_{j=k+1}^{N} (1 - P_r(jL))$. From Figure 10.8, we can see that the above equation yields a good approximation to the average one-hop zone length. Similarly,

$$E[M] = \sum_{k=1}^{N} M(kL) \cdot P_F(k, L) \tag{10.15}$$

where $M(kL)$ is the number of makeups needed in an one-hop zone of length kL under the ideal case where each makeup locates in the middle of its parent's subzone.

For each forwarder, the expected number of retransmissions to be made is:

$$E[\omega]' = \frac{1}{1 - \prod_{i=1}^{\lfloor \frac{FR}{L} \rfloor} [1 - P_r(iL)]}, \tag{10.16}$$

and $E[\omega] = min\{E[\omega]', MAX\_RETX\}$.

Finally, the expected total number of transmissions is obtained by Equation (10.13). $E[NT]$ is a function of both FR and ρ; however, it has no closed form solution. Under a fixed ρ, the optimal FR that minimizes $E[NT]$ can be obtained by searching FR from L to R_c (e.g. 500 m), by setting the sampling interval to a small enough value, e.g., 10 m.

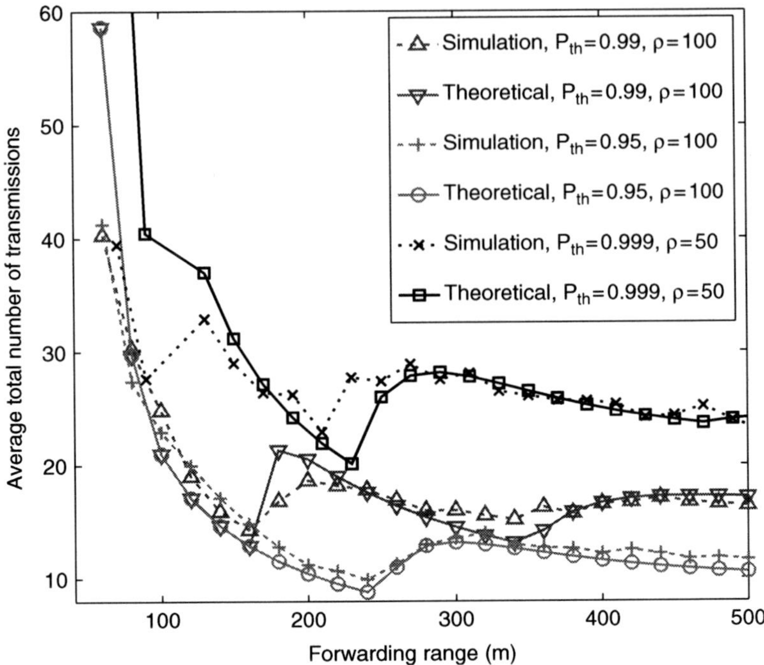

Figure 10.9   Numerical validation of $E[NT]$, $\mathscr{L} = 2km$. Reproduced by permission of © 2009 IEEE.

10.6.1.1   Theoretical insights

First, we carry out simulations to verify the above results. An idealized version of the protocol (referred to as IDEAL) is implemented in NS2, where the BACK can be reliably received by all nodes in the network. The averaged vehicle density in IR is adopted as an input in IDEAL.

Figure 10.9 compares the theoretical value of $E[NT]$ to the average number of transmissions in IDEAL. The theoretical values are close to the simulated values for all the vehicle densities and P_{th} shown, and the same for the optimal points of *FR*.

Interestingly, the optimal *FR* also exhibits opportunistic behavior, depending on the required PRR and vehicle density. In Figure 10.10, the optimal *FR* increases and decreases recurringly as the P_{th} increases. The reason is twofold. 1. On the one hand, using some particular "longer hops" reduces the number of transmissions. Note that the $E[NT] \sim E[Y]$ function has multiple local minimas[4]. Using a more distant minimal point not only reduces the number of hops but also contributes to the APRP of the other nodes. 2. On the other hand, the longer a hop is, the less possible it is for a WM to reach that far. The $E[Y]$ is upper-bounded when ρ is fixed.

---

[4] Because as $E[Y]$ increases beyond these local minima points, the number of makeups per hop will first increase and then remain fixed.

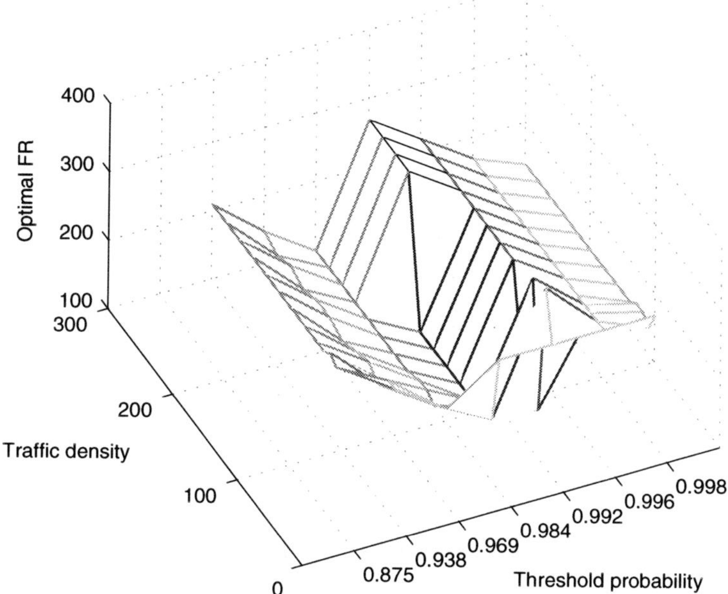

Figure 10.10   Optimal forwarding range, $\mathscr{L} = 5km$. Reproduced by permission of © 2009 IEEE.

For example, when $CR = 250\,\mathrm{m}$, $E[Y]$ is always less than $350\,\mathrm{m}$. For $P_{th} = 0.95$, the first two local minimal points are $E[Y] = 220$ and 450, which implies the *FR* corresponding to the first one is optimal. Therefore, the above results indicate that when the network is well connected, *the best strategy is try to use long hops opportunistically but only when it is feasible to reach that long hop statistically.*

Finally, in Figure 10.11 the average total number of retransmissions increase linearly when the threshold probability increases inverse exponentially towards 1. This reveals the intrinsic tradeoff between the desired WM reception ratio and transmission count, which is a helpful result for WM broadcast applications in VANETs.

10.6.1.2   Distributed algorithm

In OppCast, a distributed algorithm is used to set the *FR* because global vehicle density information is not available. Each node calculates its own optimal *FR* based on the local vehicle density estimated from its direct neighbor nodes' locations. A node is considered to be a neighbor as long as a beacon is heard from it within 1 s. Each forwarder estimates its local traffic density $\widehat{\rho(v)}$, based on which a local optimal *FR* is derived and included in every rebroadcast packet, and every vehicle that receives it use the same *FR* included in the packet. In addition, an upper limit (for example, $1000\,\mathrm{m}$) is imposed on the *FR* to prevent the hop delay from being too large.

Note that, in the OppCast extension to the sparse VANET, due to a short optimal *FR* at some traffic densities, while the vehicles that received the WM may all be

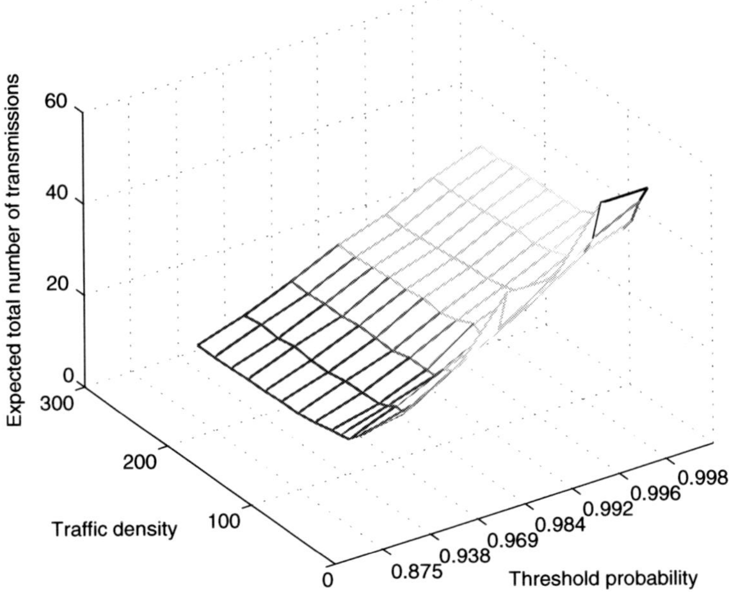

Figure 10.11   Expected total number of transmissions, $\mathscr{L} = 5km$. $CR = 250m$ in all figures. © 2011 IEEE. Reprinted, with permission, from Z. Yang, M. Li and W. Lou, "CodePlay: Live Multimedia Streaming in VANETs using Symbol-Level Network Coding", IEEE Transactions on Mobile Computing (TMC), 2011.

located outside the *FR*, there is a small chance of forwarder shortage. To deal with this situation, we let each forwarder increase its *FR* by $2\times$ after it retransmits the same packet another time, until reaching the limit.

10.6.1.3   Discussion

Our optimization is carried out using the equal inter distance vehicle distribution model (regularly distributed). However, through Figure 10.8 and Figure 10.9, it can be seen that the results are not so sensitive to the uniform vehicle distribution, which is a common mobility model adopted in most previous works. In fact, the variations in vehicle densities in the regular and uniform models are both small. In reality, some vehicles may travel in platoons; one may wonder if such mobility pattern would affect the effectiveness of optimization. Let us imagine a well-connected platoon of length 1 km; the vehicles in it can often be regarded as nearly regularly distributed. Since the distributed algorithm is based on local traffic density estimation, and the range of "locality" is really restricted to $CR = 250\,m$ (Equation (10.6)), the algorithm is expected to work well for vehicles within the platoon. Near the boundaries of the platoon, if the local density falls below ρ_{th}, the vehicle density experiences large variation; store-carry-forward will be used where the optimization does not play an important role. If the local density are larger than ρ_{th}, the variations in density are relatively small and our algorithm still applies well.

10.6.2   Optimal threshold density

Above a certain threshold density, the network is connected and the PRR requirement can be satisfied, but a higher threshold is unnecessary, which incurs redundant
transmissions. Thus, we first calculate the probability that the VANET is connected,
which means successive forwarders can be selected to propagate the WM towards
the end of IR. Still using the simplified model in the above, for a given ρ, the
probability that a forwarder is selected for one hop equals

$$P_F = \sum_{k=1}^{N} P_F(k, L), \; N = \lfloor \frac{FR_{opt}}{L} \rfloor, \tag{10.17}$$

where FR_{opt} is the optimal FR. The expected number of hops equals $\frac{\mathscr{L}}{E[Y]}$, where
$E[Y]$ is computed from Equation (10.14), by substituting FR with FR_{opt}. Thus

$$P_{connect} = P_F^{\frac{\mathscr{L}}{E[Y]}}. \tag{10.18}$$

The result is shown in Figure 10.12. It can be seen that when $P_{th} = 0.95$, the
optimal ρ_{th} is between 15-20.

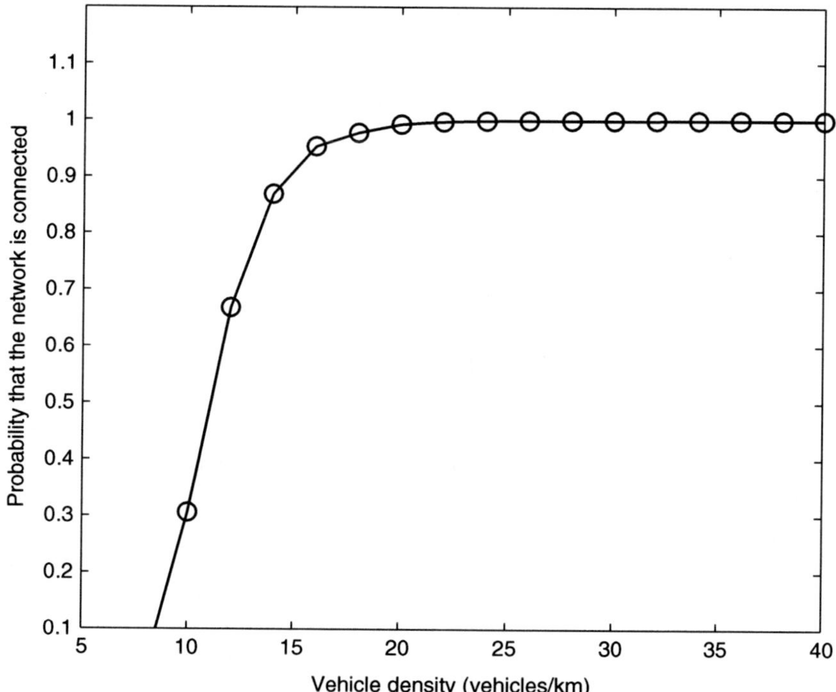

Figure 10.12   Probability that the VANET is connected, CR = 250m, $\mathscr{L}$ = 5km. ©
2011 IEEE. Reprinted, with permission, from Z. Yang, M. Li and W. Lou, "CodePlay:
Live Multimedia Streaming in VANETs using Symbol-Level Network Coding", IEEE
Transactions on Mobile Computing (TMC), 2011.

10.7   Performance evaluation

In this section, we evaluate the performance of OppCast. The compared protocols are as follows.

- Slotted p-persistence broadcast (Wisitpongphan *et al.* December 2007b) (*Slotted-p*). Upon receiving a packet from j, a node i rebroadcasts the packet with a fixed probability q after the backoff delay T_{ij}, if it receives the WM packet for the first time and has not received any duplicates during the delay. Otherwise, it drops the packet. The delay-distance function is slotted and linear. *Slotted-p* is shown to be the best among the probability-based protocols (Wisitpongphan *et al.* 2007b). We set $\tau = 5$ms, $N_S = 5$, $q = 0.5$ (the settings used in Wisitpongphan *et al.* 2007b) and the forwarding range $R = CR = 250$ m in the simulations.

- Contention-based dissemination (CBD) (T-Moreno 2007), a typical broadcast protocol also based on opportunistic forwarding. It does not differentiate between relay nodes, and uses WM as implicit ACKs. A node in the forwarding range will set a backoff timer upon receiving a WM for the first time; it cancels the timer only if it receives duplicate WMs during the backoff process, otherwise it rebroadcasts. The delay-distance function is continuous and linear. We set the maximum backoff delay to be 10 ms, which is below the value (50 ms) adopted by (T-Moreno 2007) (since our $CR = 250$ which is smaller than the one used in (T-Moreno 2007), and the channel tends to be less congested). Also, we set $R = CR = 250$ m.

Meanwhile, the IDEAL protocol is also compared, which can be regarded as a lower-bound to the transmission overhead as it has no collisions and redundant transmissions. The proposed protocols are named by appending the threshold PRR to the protocol type, e.g., for OppCast95, $P_{th} = 95\%$.

10.7.1   Simulation setup

OppCast and its extension (OppCast-Ext) is implemented in NS-2.33 (http://www.isi.edu/nsnam/ns), which supports probabilistic propagation models. The parameters are summarized in Table 10.2. The other PHY and MAC layer parameters follow the default settings of IEEE 802.11p. The Rayleigh fading model is used, which is a special case of the Nakagami model.

For the data traffic model, every vehicle periodically generates ten beacon messages every second for routine safety applications. The beacons have the same CR and transmission rates as WMs. Also, each (west bound) vehicle located between 1 km and 2 km randomly generates urgent event-driven WMs at an average messaging rate of r packets/s, according to the Poisson arrival model.

Each WM's IR is chosen to be the road segment between its source and the east end of the highway. Three sets of experiments are conducted. In the first one, the traffic density ρ ranges from 20–200 vehicles/km while the messaging rate r is fixed to 0.1 packets/s. For the second, ρ is fixed to 80 while r is varied

Table 10.2 Parameter settings

Maximum time slot length, guard time	80 μs, 20 μs
CR for WM and BACK	250 m, 628 m
Transmission rates for WM and BACK	12 mbps, 3 mbps
Tx power, CSThresh, Noise floor	10, −96, −98 dBm
WM, Beacon and BACK length	292, 72, 14 bytes
MAX_RETX	3
Vehicle distribution	Uniformly random
Range of global vehicle density	5–200 cars/km
Vehicle speed	Randomly sampled from 72–108 km/h
Road length, IR length	6 km, 4–5 km (two lanes/direction)
Maximum makeup level	2

from 0.01 to 10 packets/s, and for the third, ρ changes from 5 to 60. In the third case, some vehicles in the opposite driving direction will definitely not receive the WM due to disconnection, so we set the IR to include only the two co-directional lanes. Each simulation run lasts for 10–200 s, and a random scenario is generated where vehicles are uniformly distributed. Figures 10.13–10.15 show the results where each point is averaged from five repetitive runs with different topologies generated using different seeds, and the error bars indicate a 95% confidence interval.

10.7.2 Results for OppCast without extension

10.7.2.1 WM reception ratio

We first fix $r = 0.1$, and change ρ. In Figure 10.13(a), when $\rho = 60 \sim 200$, Opp-Cast99 maintains an average PRR of above 99%, and that of OppCast95 is higher than 98%. This shows that OppCast indeed satisfies the PRR requirement when the network is well connected. The average PRR turns out to be higher than the thresholds because the PRR requirement is taken as a minimum requirement in each MFR phase. When the network is sparse, i.e., $\rho = 20 \sim 50$, the PRR of OppCast protocols is still higher than 90%, which is much higher than Slotted-p and CBD. The advantage is primarily because of the FFD phase trying to guarantee that the forwarders span the whole network. The PRR in this case is lower than required because there may not be enough relay nodes due to network partition.

Figure 10.13(b) shows the PRR results of the second experiment. It can be seen that the PRR requirement in OppCast is always satisfied when r is small to moderate. The decrease of PRR only happens when message generation is very dense, i.e., $r > 1$. However, the PRR of OppCast is still much higher than Slotted-p and CBD in this case, while OppCast introduces much lower overheads. Similar results can be observed for the dissemination rate. This shows OppCast is more scalable, i.e., more capable of handling saturated message traffic situations than other protocols.

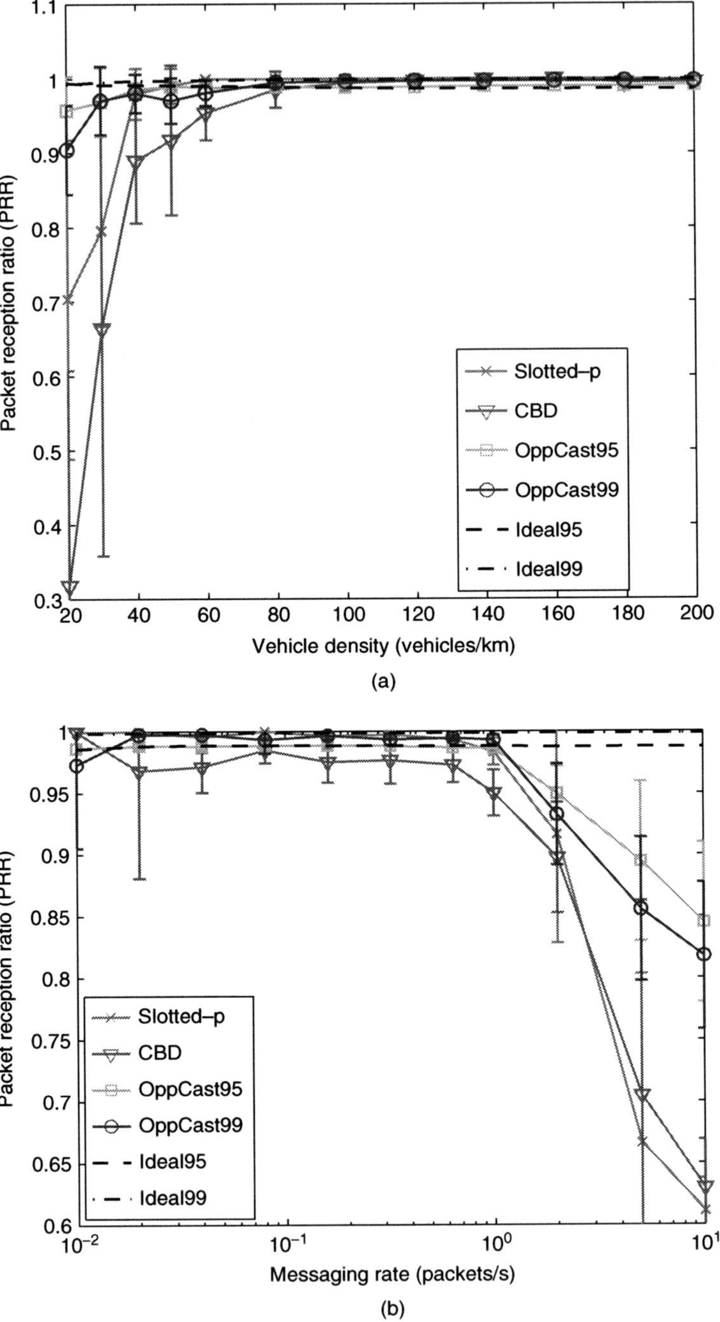

*Figure 10.13   Simulation results: WM reception ratio (OppCast without extension).
(a): fix $r = 0.1$, change ρ. (b): fix $\rho = 80$, change r. © 2011 IEEE. Reprinted,
with permission, from Z. Yang, M. Li and W. Lou, "CodePlay: Live Multimedia
Streaming in VANETs using Symbol-Level Network Coding", IEEE Transactions on
Mobile Computing (TMC), 2011.*

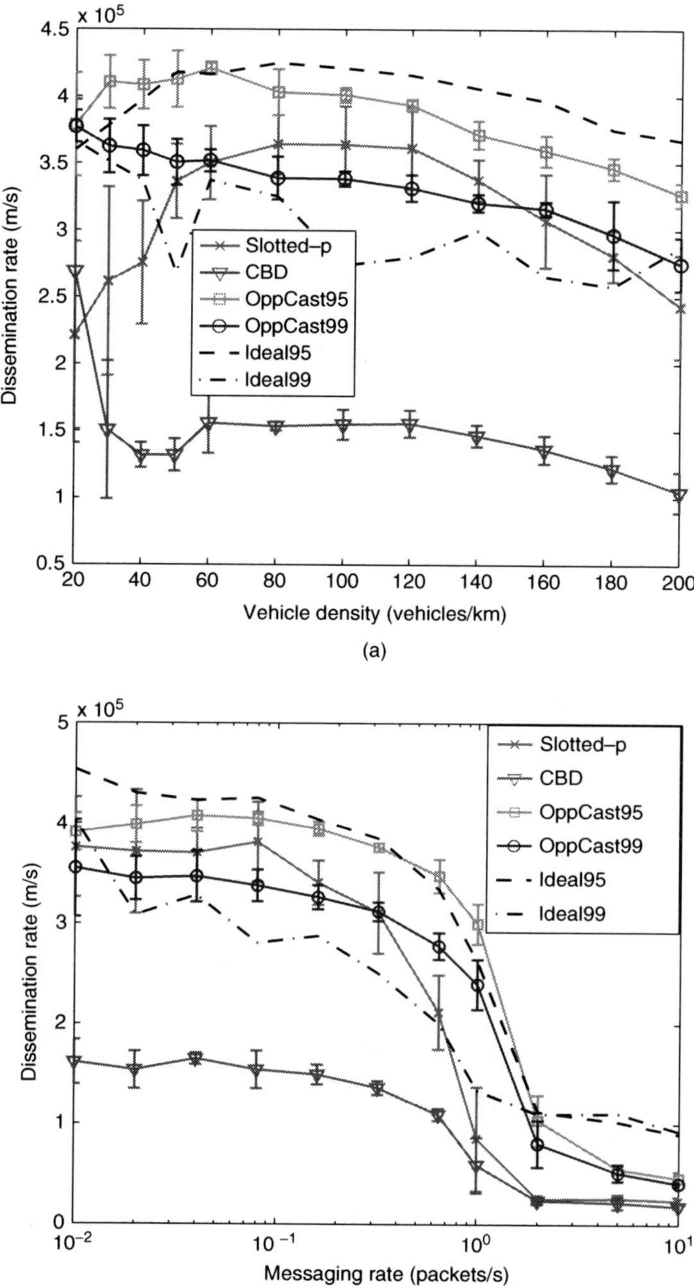

Figure 10.14 Simulation results: dissemination rate (OppCast without extension). (a): fix $r = 0.1$, change ρ. (b): fix $\rho = 80$, change r. © 2011 IEEE. Reprinted, with permission, from Z. Yang, M. Li and W. Lou, "CodePlay: Live Multimedia Streaming in VANETs using Symbol-Level Network Coding", IEEE Transactions on Mobile Computing (TMC), 2011.

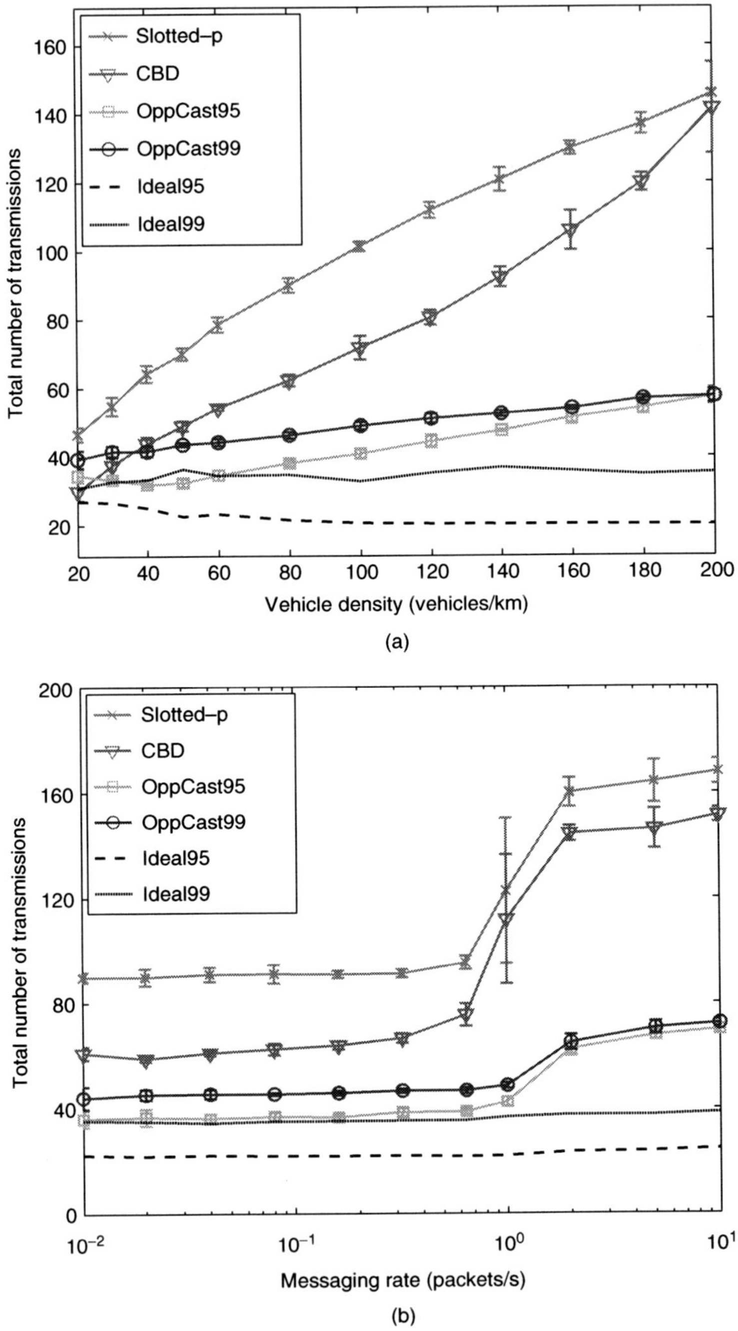

Figure 10.15 Simulation results: transmission count (OppCast without extension).
(a): fix r = 0.1, change ρ. (b): fix ρ = 80, change r. © 2011 IEEE. Reprinted,
with permission, from Z. Yang, M. Li and W. Lou, "CodePlay: Live Multimedia
Streaming in VANETs using Symbol-Level Network Coding", IEEE Transactions on
Mobile Computing (TMC), 2011.

10.7.2.2   Dissemination rate and delay

From Figure 10.14(a), it can be seen that the dissemination rate of OppCast95 is the highest except for IDEAL95, for all the vehicle densities shown. Similar results are shown in Figure 10.14(b), where OppCast95's dissemination rate is still among the highest for all the messaging rates. This can be attributed mainly to the opportunistic forwarding concept adopted in OBCF, which always utilizes the most distant forwarder candidate so that the one-hop delay is minimized. On the other hand, for OppCast99, although the achieved reliability is a little higher than OppCast95, the dissemination rate is smaller. It turns out that the reduced dissemination rate is a cost to enhance the WM reception ratio in OppCast.

To further investigate the dynamics of WM dissemination in OppCast and see why it performs better, we show in Figure 10.16 the end-to-end delay results of each vehicle in the IR correlated with its distance to the source, for a typical WM disseminated in VANET with traffic density equal to 80 vehicles/km (well connected). Remarkably, the last vehicle in the IR receives the WM within about

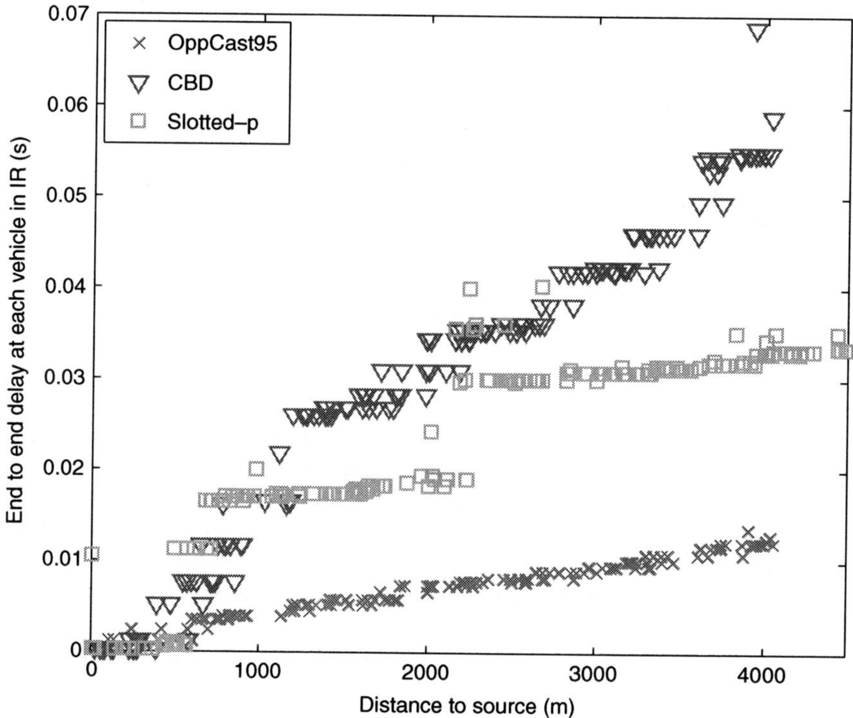

Figure 10.16   The end-to-end delay of each vehicle in the IR versus distance to the source, $\rho = 80$, $r = 0.1$. © 2011 IEEE. Reprinted, with permission, from Z. Yang, M. Li and W. Lou, "CodePlay: Live Multimedia Streaming in VANETs using Symbol-Level Network Coding", IEEE Transactions on Mobile Computing (TMC), 2011.

12 ms, which is much less than that of the CBD and slotted-p. The delay-distance curve also increases smoothly, showing that there is little gap between reception times of successive rebroadcasts (read from the *y*-axis). This shows the effectiveness of the carefully designed coordination mechanism for relay selection (OBCF), where average hop-delays in the order of $10-100\,\mu s$ can indeed be achieved. Moreover, although we used makeup nodes so that nodes that missed the WM in the FFD phase can receive it later, the introduced delay variance is negligible.

In contrast, other protocols do not enjoy the same level of fast propagation. For CBD, there are obvious time gaps between consecutive rebroadcasts, which are partially due to its large backoff delay in relay contention processes. However, the maximum backoff delay (10 ms) is already much less than the adopted value in (T-Moreno 2007). By studying Figure 10.16 in more detail, one can observe that some time gaps are relatively long, and many next-hop relays are located near their previous hops. This suggests that during relay contention process in CBD, due to channel fading and poor coordination, packet collisions happen more frequently, resulting in sub optimal relays being selected, while in OppCast, the FFD phase is employed to propagate the WM towards the end of IR in the first place. To guarantee this we use the BACK, by which the channel is cleared before each rebroadcast, and candidate relays in each newly traversed one-hop zone suspend their counting down timers during BACK to give priority to the forwarding WM. Slotted-p is somewhat different, in that the continuous propagation periods are longer than that of CBD, however the gaps are even larger. The former is naturally due to zero delay for relays outside of the contention region; but the latter indicates that packet collisions are even worse. This is mainly because Slotted-p is still controlled flooding; although using a coarse-grained slotted timer function, it cannot completely eliminate the broadcast storm. The above results show that redundant transmissions indeed undermine broadcast performance to a large extent and the explicit BACK mechanism in OBCF is effective and necessary.

10.7.2.3 Transmission overhead

The transmission overhead is evaluated by the total number of WM packet rebroadcasts incurred per WM sent by the source. As the length of a BACK is quite small compared with a WM packet, we neglect the overhead caused by BACKs. In Figure 10.15(a), as vehicle density increases to 200, the total number of transmissions incurred by OppCast95 and OppCast99 is about 40% of that of CBD. More importantly, the overhead increases more slowly with respect to vehicle density than in Slotted-p and CBD, because the relay selection mechanisms are optimized, and the OBCF is effective in reducing redundant transmissions and packet collisions in the presence of lossy links. In CBD, because of channel fading, the rebroadcast of relays cannot be heard by many other relay candidates, which leads to a large number of redundant transmissions. On the other hand, in OppCast, using BACK, we can exert more fine-control over the selection of makeups, which turns out to be less than three per one-hop zone. The above indicates that the high reliability and fast dissemination are achieved in a resource-efficient way in OppCast.

10.7.2.4   The tradeoffs

The OppCast95 achieves competitively high PRR and the highest dissemination rate using the smallest number of transmissions. The OppCast99 achieves higher PRR than OppCast95 in most scenarios, but uses more transmissions and leads to slower dissemination. As we aim to achieve multiple goals at the same time in the design of OppCast, this reflects the fundamental tradeoff between them: to achieve higher reliability, more transmissions are needed, which in turn causes greater broadcast latency. When the PRR is already close to 1, a marginal gain in PRR would also demand noticeably more transmissions, and would result in a big reduction in the dissemination rate, as in the case of OppCast99. *Thus, using a lower PRR goal, such as 95%, is better than 99% in this sense.*

On the other hand, the Slotted-p exploits a different tradeoff: it uses aggressive rebroadcasts to achieve high reliability and a relatively high dissemination rate. However, this is not very resource efficient because it consumes a much larger portion of the VANET bandwidth. Moreover, too many transmissions adversely affect the dissemination rate, as one can see from Figures 10.13–10.15 and Figure 10.16.

Note that, in our comparisons, we have not extended the Slotted-p and CBD to allow a forwarder to perform multiple retransmissions as is the case in OppCast. This could be done to enhance their PRRs under the disconnected case; however, the gain is very small when the network is well connected and it results in even more transmission overheads.

10.7.2.5   How reliable is the BACK?

Next, we investigate deeper about the reliability of BACK in OppCast, and discuss how the broadcast performance will be affected by BACK. Ideally, BACK should achieve three goals: 1. acknowledge the transfer of relaying opportunity and suppress all redundant relay candidates; 2. inform the previous forwarder to cancel retransmission; 3. suppress hidden terminals to reserve the channel for WM broadcast. In IDEAL protocol, all these goals are achieved perfectly. But in reality, BACK is still subject to losses. This comes from either fading, or collisions between BACK and its hidden terminals. Consequently, there may exist redundant relay nodes, redundant rebroadcasts or WM collisions.

In Figures 10.13–10.15, the performance degradation of OppCast w.r.t. IDEAL is also shown. When message traffic is dense ($r > 1$), the PRR in OppCast is smaller than that of IDEAL. As PRR is the primary goal in OppCast, when a BACK is not heard by a relay node, it tends to use more rebroadcasts to guarantee PRR. We then show the reliability of BACK by showing the total number of (re)broadcasts for each WM, which is broken down into number of relays and retransmissions by forwarders in Table 10.3.

For both OppCast95 and OppCast99, when $\rho = 120$ the number of relays is one-third more than their IDEAL counterparts (optimal), which make up the major part of the redundant (re)broadcasts, but this is acceptable. The number of retransmissions in OppCast95 increases faster with ρ, because OppCast99 uses more makeups that send BACKs to cancel forwarders' retransmissions. The redundant transmissions lead to PRR over-provisioning in OppCast.

Table 10.3   Average number of relays and retransmissions

ρ	IDEAL95		OppCast95		IDEAL99		OppCast99	
60	24.0	1.5	27.5	6.8	39	1.5	41.5	3.8
120	22.0	1.3	31.6	13.0	36	1.3	49.5	4.0

© 2011 IEEE. Reprinted, with permission, from Z. Yang, M. Li and W. Lou, "CodePlay: Live Multimedia Streaming in VANETs using Symbol-Level Network Coding", IEEE Transactions on Mobile Computing (TMC), 2011.

In order to reduce the redundant (re)broadcasts, the BACK has to be reliably received by more nodes in the network. However, using a longer communication range for BACK is not necessarily better, which will cause the exposed terminal problem. This can be seen from the lower dissemination rate of IDEAL when the channel load (ρr) is low to moderate (Figures 10.14(a),(b)). We believe that, to balance the goals of suppressing hidden terminals and avoiding exposed terminals, it is a good choice to set BACK's range to be around twice of WM's CR. The

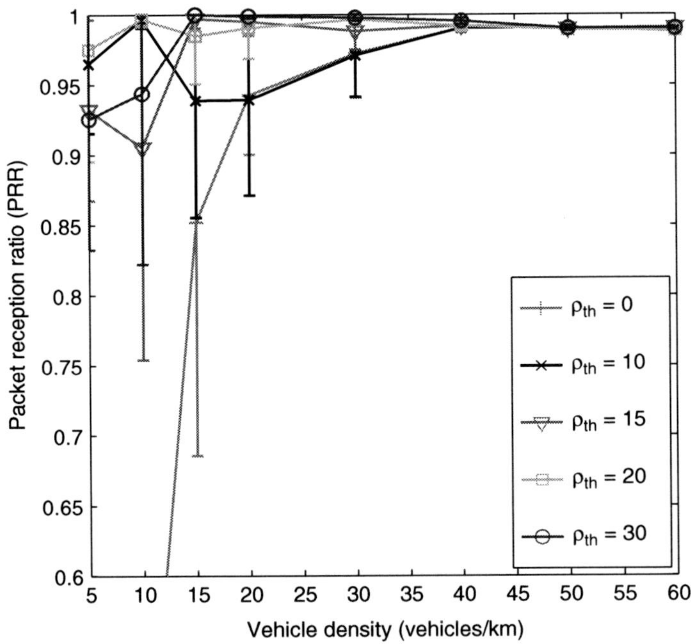

Figure 10.17   Simulation results of PRR for OppCast95-Ext under sparse VANET, $P_{th} = 0.95$, $r = 0.1$. The case $\rho_{th} = 0$ corresponds to OppCast without extension. (Each point is averaged from 10 repetitive runs with different topologies.) © 2011 IEEE. Reprinted, with permission, from Z. Yang, M. Li and W. Lou, "CodePlay: Live Multimedia Streaming in VANETs using Symbol-Level Network Coding", IEEE Transactions on Mobile Computing (TMC), 2011.

intuitive explanation is that if the BACK's range is smaller than twice of WM's CR there will be hidden terminals that cannot be suppressed; on the other hand, if the BACK's range is more than twice WM's CR, the nodes within the area from twice of WM's CR to the BACK's range will become exposed terminals.

10.7.3   Results for OppCast with extension

In Figures 10.17–10.19 we show the results of OppCast extension under sparse VANETs. The effects of using different threshold densities ρ_{th} are clear from the figures. When the ρ_{th} increases from 0 to 30, the achieved PRR tends to increase for all the densities, the dissemination rate decreases and the transmission overheads increase. These results are in line with intuition because the higher the ρ_{th}, the more often data mules are used and more redundant transmissions are incurred, which makes the channel more saturated and thus increases the channel access delay. The redundant transmissions increase dramatically with the ρ_{th} as we do not allow a data mule to cancel its carrying status for WMs, in order to ensure the PRR.

Moreover, jointly considering the PRR requirements ($PRR \geq P_{th} = 0.95$) and the transmission overhead, one can find that the optimal threshold density equals 20 from Figures 10.17–10.19. This matches well with our theoretical result derived in the last section, where the optimal ρ_{th} is between 15 and 20.

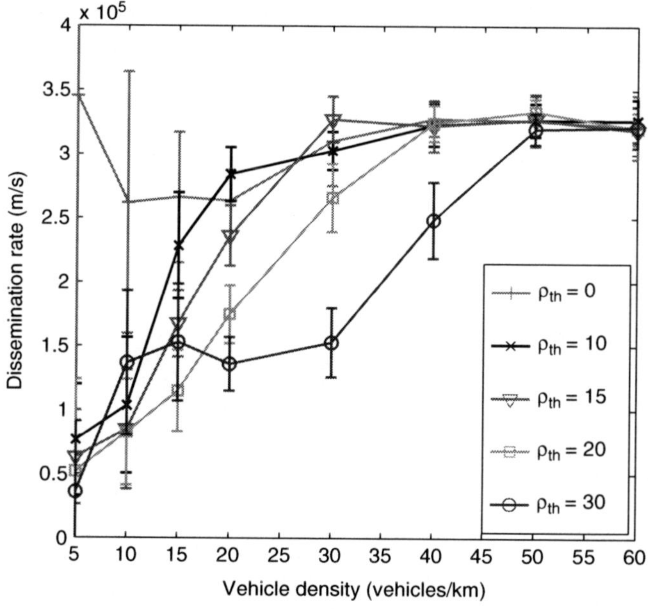

Figure 10.18   Simulation results of dissemination rate for OppCast95-Ext under sparse VANET, $P_{th} = 0.95$, $r = 0.1$. © 2011 IEEE. Reprinted, with permission, from Z. Yang, M. Li and W. Lou, "CodePlay: Live Multimedia Streaming in VANETs using Symbol-Level Network Coding", IEEE Transactions on Mobile Computing (TMC), 2011.

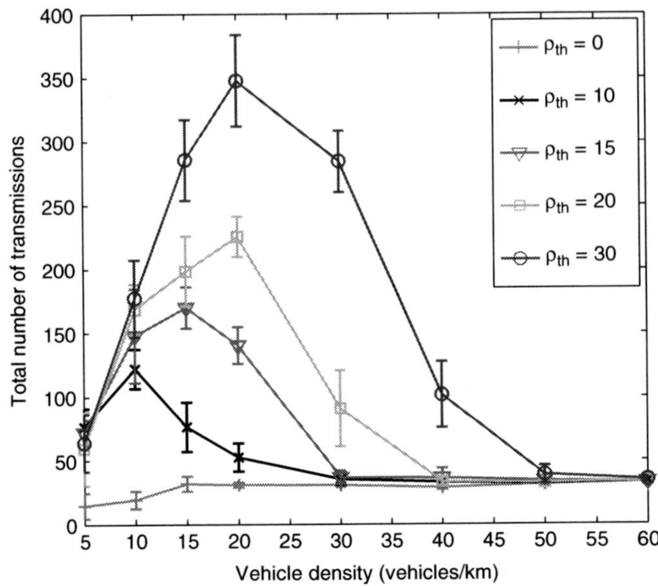

Figure 10.19 *Simulation results of transmission count for OppCast95-Ext under sparse VANET, $P_{th} = 0.95$, $r = 0.1$. © 2011 IEEE. Reprinted, with permission, from Z. Yang, M. Li and W. Lou, "CodePlay: Live Multimedia Streaming in VANETs using Symbol-Level Network Coding", IEEE Transactions on Mobile Computing (TMC), 2011.*

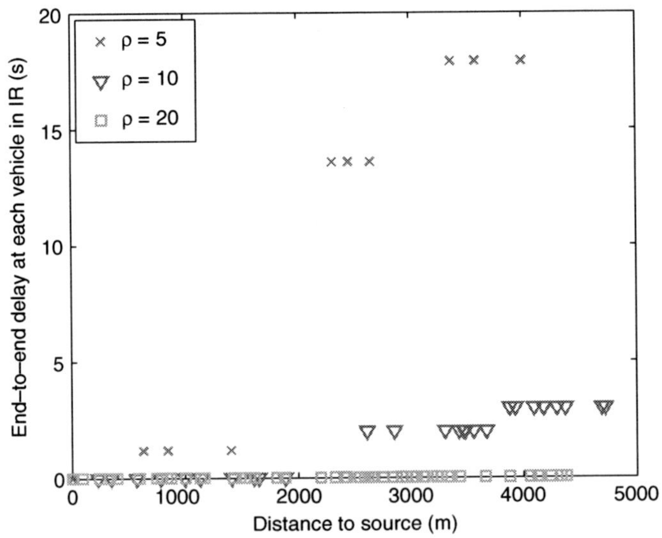

Figure 10.20 *The end-to-end delay of vehicles under sparse VANET, OppCast95-Ext, $\rho_{th} = 20$. © 2011 IEEE. Reprinted, with permission, from Z. Yang, M. Li and W. Lou, "CodePlay: Live Multimedia Streaming in VANETs using Symbol-Level Network Coding", IEEE Transactions on Mobile Computing (TMC), 2011.*

To reveal the WM dissemination dynamics under sparse VANET, we show the delay-distance graph in Figure 10.20. The plateaux indicate the connected parts in the network. We can see that OppCast still allows a WM to propagate very fast within connected platoons, while always relaying the WM to the next platoon successfully. This is ensured by the retransmission mechanism by the forwarders.

10.8 Conclusion

In this chapter, we studied the opportunistic broadcast of event-driven WMs in VANETs. The multihop broadcast of WMs needs to satisfy multiple stringent performance requirements such as high reliability (packet reception ratio), low end-to-end latency and low transmission overheads, which is especially difficult to achieve due to the lossy wireless environment and fast-changing topology in VANETs. Thus, a fully distributed opportunistic broadcast protocol (OppCast) for the multihop dissemination of event-driven warning messages in VANETs was proposed. Aiming at achieving high WM reception reliability and fast dissemination in a resource-efficient way, the concept of opportunistic routing was exploited in the link layer at each hop to enhance reception reliability and provide a small hop delay. Meanwhile, at the network layer, we propose a double-phase broadcast method, in which fast propagation is ensured by one phase, and the desired reliability level is ensured by the other. As a key idea in OppCast, we use explicit broadcast acknowledgements (BACK) in rebroadcast contention so that the optimal relays can always be selected with a high probability and the undesired redundant rebroadcasts are dramatically reduced. Through extensive simulations we have shown that, compared with state-of-the-art protocols, OppCast achieves a higher WM packet reception ratio and a higher dissemination rate using a smaller amount of transmissions. More importantly, the BACK has been shown to be a more reliable and effective approach than implicit acknowledgements adopted in previous works. In addition, OppCast was extended to handle disconnected VANET scenarios, in which the optimal threshold density to switch between normal dissemination and store-carry-and-forward scheme is characterized. Our results revealed the intrinsic tradeoff and intricate interplay between WM reception reliability, dissemination rate and overheads. We believe it will provide valuable guidelines to protocol design in VANETs and mobile WMNs.

References

Abuelela M, Olariu S and Stojmenovic I 2008 Opera: Opportunistic packet relaying in disconnected vehicular ad hoc networks *IEEE MASS '08*, pp. 285–294.

Alshaer H and Horlait E 2005 Optimized adaptive broadcast scheme for inter-vehicle communication *IEEE VTC*, pp. 2840–2844.

Biswas S and Morris R 2005 Exor: opportunistic multi-hop routing for wireless networks. *SIGCOMM Comput. Commun. Rev.* **35**(4), 133–144.

Blaszczyszyn B, Laouiti A, Muhlethaler P and Toor Y 2008 Opportunistic broadcast in vanets (ob-van) using active signaling for relays selection *ITS Telecommunications, 2008*, pp. 384–389.

C-Varghese J, Chen W, Altintas O and Cai S 2006 Survey of routing protocols for inter-vehicle communications *MobiQuitous*, pp. 1–5 IEEE.

Dai F and Wu J 2004 Performance analysis of broadcast protocols in ad hoc networks based on self-pruning. *IEEE Trans. Parallel Distrib. Syst.* **15**(11), 1027–1040.

Dutta A, Saha D, Grunwald D and Sicker D 2009 Smack: a smart acknowledgment scheme for broadcast messages in wireless networks. *SIGCOMM Computing Communication Review* **39**(4), 15–26.

ElBatt T, Goel SK, Holland G, Krishnan H and Parikh J 2006 Cooperative collision warning using dedicated short range wireless communications *VANET '06*, pp. 1–9. ACM.

Fasolo E, Zanella A and Zorzi M 2006 An effective broadcast scheme for alert message propagation in vehicular ad hoc networks. *IEEE ICC* **9**, 3960–3965.

Heissenbuttel M, Braun T, Walchli M and Bernoulli T n.d. Optimized stateless broadcasting in wireless multi-hop networks *INFOCOM 2006*.

Ilias L and Cecilia M 2007 Opportunistic spatio-temporal dissemination system for vehicular networks *MobiOpp '07*, pp. 39–46. ACM.

Jiang D, Taliwal V, Meier A, Holfelder W and Herrtwich R 2006 Design of 5.9ghz dsrc-based vehicular safety communication. *IEEE Wireless Communications* **13**(5), 36–43.

Killat M and Hartenstein H n.d. An empirical model for probability of packet reception in vehicular ad hoc networks. *EURASIP J. Wirel. Commun. Netw.* 2009, 1–12.

Korkmaz G and Ekici E 2004 Urban multi-hop broadcast protocol for inter-vehicle communication systems *Proceedings of VANET '04*, pp. 76–85 ACM.

Li M and Lou W 2008 Opportunistic broadcast of emergency messages in vehicular ad hoc networks with unreliable links *QShine '08*, pp. 1–7.

Li M, Yang Z and Lou W 2011 Codeon: Cooperative popular content distribution for vehicular networks using symbol level network coding. *IEEE Journal on Selected Areas in Communication* **29**(1), 223–235.

Lim H and Kim C 2000 Multicast tree construction and flooding in wireless ad hoc networks *MSWIM '00: Proceedings of the Third ACM International Workshop on Modeling, Analysis and Simulation of Wireless and Mobile Systems*, pp. 61–68. ACM.

Lou W and Wu J 2002 On reducing broadcast redundancy in ad hoc wireless networks. *IEEE Transactions on Mobile Computing* **1**(2), 111–123.

Mangharam R, Rajkumar R, Hamilton M, Mudalige P and Bai F 2007 Bounded-latency alerts in vehicular networks *MoVE*, pp. 55–60.

Nadeem T, Shankar P and Iftode L 2006 A comparative study of data dissemination models for vanets *Third Annual International Conference on Mobile and Ubiquitous Systems: Networking Services*, pp. 1–10.

Ni SY, Tseng YC, Chen YS and Sheu JP 1999 The broadcast storm problem in a mobile ad hoc network *IEEE/ACM MobiCom*, pp. 151–162.

Oh S, Kang J and Gruteser M 2006 Location-based flooding techniques for vehicular emergency messaging *IEEE MobiQuitous*, pp. 1–9.

Qayyum A, Viennot L and Laouiti A 2002 Multipoint relaying for flooding broadcast messages in mobile wireless networks *HICSS '02: Proceedings of the 35th Annual Hawaii International Conference on System Sciences* **9**, 28.

Resta G, Santi P and Simon J 2007 Analysis of multi-hop emergency message propagation in vehicular ad hoc networks *ACM MobiHoc*, pp. 140–149, New York, NY.

Taliwal V, Jiang D, Mangold H, Chen C and Sengupta R 2004 Empirical determination of channel characteristics for dsrc vehicle-to-vehicle communication *ACM VANET '04*. ACM.

Tang K and Gerla M 2000 Mac layer broadcast support in 802.11 wireless networks *MILCOM 2000*, vol. 1, pp. 544–548.

T-Moreno M 2007 Inter-vehicle communications: Assessing information dissemination under safety constraints *WONS*, pp. 59–64 IEEE.

Tonguz O, Wisitpongphan N and Bai F 2010 Dv-cast: A distributed vehicular broadcast protocol for vehicular ad hoc networks. *IEEE Wireless Communications* **17**(2), 47–57.

Torrent-Moreno M, Jiang D and Hartenstein H 2004 Broadcast reception rates and effects of priority access in 802.11-based vehicular ad-hoc networks *ACM VANET '04*. ACM.

Torrent-Moreno M, Killat M and Hartenstein H 28-25 Sept., 2005 The challenges of robust inter-vehicle communications. *IEEE VTC 2005* **1**, 319–323.

Torrent-Moreno M, Mittag J, Santi P and Hartenstein H 2009 Vehicle-to-vehicle communication: Fair transmit power control for safety-critical information. *IEEE Transactions on Vehicular Technology* **58**(7), 3684–3703.

Torrent-Moreno M, Schmidt-Eisenlohr F, Fussler H and Hartenstein H 2006 Effects of a realistic channel model on packet forwarding in vehicular ad hoc networks *IEEE WCNC*, pp. 385–391.

Wisitpongphan N, Bai F, Mudalige P, Sadekar V and Tonguz O 2007a Routing in sparse vehicular ad hoc wireless networks. *IEEE JSAC* **25**(8), 1538–1556.

Wisitpongphan N, Tonguz O, Parikh J, Mudalige P, Bai F and Sadekar V December 2007b Broadcast storm mitigation techniques in vehicular ad hoc networks. *IEEE Wireless Communications* **14**(6), 84–94.

Xu Q, Mak T, Ko J and Sengupta R 2004 Vehicle-to-vehicle safety messaging in dsrc *ACM VANET '04*, pp. 19–28.

Zorzi M and Rao RR 2003 Geographic random forwarding (geraf) for ad hoc and sensor networks: Multihop performance. *IEEE Transactions on Mobile Computing* **2**(4), 337–348.

11

Conclusions and future research

11.1 Summary

The essential idea of opportunistic routing is to exploit the broadcast nature and space diversity provided by the wireless medium. By having multiple forwarding candidates, the successful rate of each transmission can be much improved. However, a good OR protocol is to decide which set of nodes (in contrast to which single node) are good to form the forwarding candidate set and how they should be prioritized. Although we are taking opportunities, we want the packets to be routed to the destination through a set of paths that are statistically optimal. In this book, we presented principles of the local behavior of OR, we analyzed the capacity, throughput and energy efficiency of OR, we developed a new candidate coordination scheme for OR, and we designed secure link quality measurement mechanism.

In Chapter 2, we found and proved properties of the local behavior of OR and the associated candidate selection and prioritization issues. In Chapter 3 we proposed an energy-efficient geographic opportunistic routing framework and corresponding local candidate selection and prioritization algorithms. The contributions of Chapters 4 and 5 present an analytical model to compute the end-to-end throughput bound and capacity of OR in multiradio, multichannel and multirate wireless networks. In Chapter 6, we presented a new, efficient candidate coordination scheme, which takes advantage of physical layer information. Chapter 7 demonstrated how network coding can be integrated with opportunistic routing. We studied the performance of multirate GOR under a contention-based medium-access

Multihop Wireless Networks: Opportunistic Routing, First Edition.
Kai Zeng, Wenjing Lou and Ming Li.
© 2011 John Wiley & Sons, Ltd. Published 2011 by John Wiley & Sons, Ltd.

scenario in Chapter 8. In Chapter 9, we presented a secure link quality measurement mechanism, which is able to prevent a malicious attacker from reporting a fake link quality. We further investigated the security issues about opportunistic routing. Opportunistic broadcast in VANETs is studied in Chapter 10. We summarize our results by chapter below.

- Chapter 2. In this Chapter, we generalized the definition of EPA for an arbitrary number of forwarding candidates in GOR. Through theoretical analysis, we showed that the maximum EPA can only be achieved by giving the forwarding candidates closer to the destination higher relay priorities when a forwarding candidate set is given. We give the analytical result of the upper bound of the EPA that any GOR can achieve. We also showed that giving an available next-hop neighbor set with M nodes, the maximum EPA achieved by selecting r ($1 \leq r \leq M$) nodes is a strictly increasing and concave function of r. We proved that a feasible subset of the available next-hop neighbor set that achieves that maximum EPA is contained in at least one feasible subset with more nodes. We also showed that increasing the maximum EPA is consistent with increasing the one-hop reliability. Least-cost opportunistic routing is introduced and important properties about it are presented. Then two polynomial time algorithms that find the least-cost opportunistic path are introduced.

- Chapter 3. In this chapter, we studied the geographic opportunistic routing strategy with both routing and energy efficiencies as the major concerns. We proposed a new routing metric, which evaluates EPA per unit of energy consumption so that energy efficiency can be taken into consideration in routing. By leveraging the proved findings in Chapter 2, we proposed two localized candidate-selection algorithms with $O(M^3)$ and $O(M^2)$ running time in the worst case, respectively, and $\Omega(M)$ in the best case, where M is the number of available next-hop neighbors. The algorithms efficiently determine the forwarding candidate set that maximizes the proposed new metric for energy efficiency, namely, the EPA per unit of energy consumption. We further proposed an EGOR framework applying the node selection algorithms to achieve the energy efficiency. Simulation results show that EGOR achieves better energy efficiency than geographic routing and blind opportunistic protocols in all the cases while maintaining a very good routing performance. Our simulation results also show that the number of forwarding candidates necessary to achieve maximum energy efficiency is mainly affected by the reception to transmission energy ratio but not by the node density under a uniform node distribution. Although the EPA can be maximized by involving the greatest number of nodes in GOR, in terms of energy efficiency, only a very small number of forwarding candidates (around two) are needed on average. This is true even when the energy consumption of reception is far less than that of transmission.

- Chapter 4. Taking into consideration wireless interference and the unique properties of OR, we proposed a new method of constructing transmission conflict graphs and presented a methodology for computing the end-to-end throughput bounds (capacity) of OR, giving forwarding strategies. We formulate the maximum end-to-end throughput problem of OR as a maximum-flow linear programming subject to the transmission conflict constraints and effective forwarding rate constraints on each link in different concurrent transmission sets. We also proposed two metrics for OR under multirate scenarios. One is *expected medium time* (EMT) and the other is *expected advancement rate* (EAR). Based on these metrics, we proposed the distributed and local rate and candidate selection schemes: LMTOR and MGOR, respectively. We validated the analysis results by simulation, and compared the throughput capacity of multirate OR with single-rate ones under different settings, such as different topologies, source–destination distances, the number of forwarding candidates and node densities. We showed that OR has great potential to improve end-to-end throughput under different settings, and our proposed multirate OR schemes achieved higher throughput bound than any single-rate GOR. We observed some features of OR: 1. the gain in end-to-end capacity decreases when the number of forwarding candidates is increased. When the number of forwarding candidates is larger than 3, the throughput almost remains unchanged; 2. there exists a node density threshold, higher than which 24 mbps GOR performs better than 12 mbps GOR, and lower than which the opposite is the case. The threshold is about 5.5 and 10.9 neighbors per node at 12 mbps for line and square topologies, respectively.

- Chapter 5. We proposed a unified framework to compute the capacity of opportunistic routing between two end nodes in single/multiradio/channel multihop wireless networks by allowing dynamic forwarding strategies. Our model accurately captures the unique property of OR that multiple outgoing links sharing the same transmitter can be virtually scheduled at the same time under particular rate constraints. We also studied the necessary and sufficient conditions for the schedulability of a flow demand vector associated with a transmitter to its forwarding candidates in a concurrent transmission set. We further proposed an LP approach and a heuristic algorithm to obtain an opportunistic forwarding strategy scheduling that satisfies a flow demand vector. Our methodology can not only be used to calculate the end-to-end throughput bound of OR and TR in multiradio/channel multihop wireless networks, but can also be used to study OR behaviors (such as candidate selection and prioritization) in multiradio multichannel systems. Leveraging our analytical model, we found that OR can achieve comparable or even better performance than TR using less radio resources.

- Chapter 6. We analyzed the candidate coordination problem in opportunistic routing and, based on these analyses, we proposed a new coordination scheme "fast slotted acknowledgment" (FSA), which takes full advantage of the channel-detection approach to reach an agreement among multiple candidates. We compared FSA with state-of-the-art schemes and simulation results show that it achieves better performance in all the metrics, especially in time delay. The simulation also validated that FSA can achieve a similar level of performance as ideal coordination where relay priority can be ensured and duplicate packet forwarding is avoided.

- Chapter 7. We further investigated the transmission coordination problem in opportunistic routing and studied how the strict coordination requirement in ExOR can be greatly eased using network coding. First, we reviewed MORE, a well-known state-of-the-art MAC-independent opportunistic routing protocol. After identifying the inadequacy of MORE in dealing with multicast and broadcast, we demonstrated how network coding can be integrated with opportunistic listening in wireless broadcast. In particular, we looked into a class of challenging problem – mobile content distribution (MCD) in VANET, where large files are broadcasted proactively from a few APs to vehicles inside an interested area. To combat the lossy wireless transmissions in VANETs, we leveraged symbol-level network coding (SLNC), which exploits symbol-level diversity to achieve better error tolerance compared with traditional packet-level network coding (PLNC). We then quantitatively characterized the advantages of SLNC compared with PLNC from two aspects, namely higher throughput and spacial-reusability. Using two typical MCD applications as examples – popular content distribution and live multimedia streaming, we presented two novel push-based broadcast schemes – CodeOn and CodePlay, respectively. The common ideas underlying the two schemes included a prioritized relay selection algorithm that opportunistically maximizes the usefulness of transmitted content and a simple transmission coordination mechanism which exploits the higher spacial reusability brought by SLNC. Compared with state-of-the-art pull-based contend distribution protocols, CodeOn and CodePlay both achieve significant performance gains. Through simulation study, the gain can be partly attributed to the use of SLNC and partly attributed to the new push-based protocol design. Thanks to the use of opportunistic listening and network coding (especially SLNC) the challenging problem of designing a MCD protocol in VANETs is solved elegantly.

- Chapter 8. We studied multirate geographic opportunistic routing (MGOR) in a contention-based scenario, and examined the factors that affect its throughput, which includes multirate capability, candidate selection, prioritization, and coordination. Based on our analysis, we proposed the local metric, the *Opportunistic Effective One-hop Throughput* (OEOT), to characterize the tradeoff between the packet advancement and medium time cost under different data rates. We further proposed a rate and candidate selection algorithm to approach the local optimum of this metric. We also presented a multirate link quality-measurement mechanism. Simulation results show that MGOR incorporating our algorithm achieves better throughput and delay performance than the corresponding opportunistic routing and geographic routing operating at any single rate. It indicates that OEOT is a good local metric to achieve high end-to-end throughput and low delay for MGOR.

- Chapter 9. We identified attacks on opportunistic routing and proposed countermeasures. We investigated the existing link quality measurement mechanisms, and analyzed the security vulnerabilities in them. A common fact in all the existing LQM mechanisms is that a node's knowledge about the forward PRR from itself to its neighbors is informed by its neighbors. We then proposed a broadcast-based secure LQM mechanism that prevents a neighboring node from maliciously claiming a higher measurement result. Our mechanism has very low computation, storage, and communication overheads and thus can be implemented in resource-constraint sensor networks as well as mesh networks. Our SLQM mechanism can be applied easily to unicast-based and cooperative LQM with slight modifications. We further discussed attacks on existing opportunistic routing protocols and studied the resilience of opportunistic routing on packet dropping attacks.

- Chapter 10. We studied the opportunistic broadcast of event-driven warning messages (WMs) in VANETs. The multi-hop broadcast of WMs needs to satisfy multiple stringent performance requirements such as high reliability (packet-reception ratio), low end-to-end latency and low transmission overheads, which is especially challenging to achieve due to the lossy wireless environment and fast-changing topology in VANETs. Thus, we proposed a fully distributed opportunistic broadcast protocol (OppCast) for multihop dissemination of WMs in VANETs. With the aim of achieving high WM reception reliability and fast dissemination in a resource-efficient way, the concept of opportunistic routing is exploited in the link layer at each hop to enhance reception reliability and provide small hop delay. In particular, we used explicit broadcast acknowledgements (BACK) in rebroadcast contention so that the optimal relays can always be selected with a high probability. Meanwhile, at the network layer, we proposed a double-phase broadcast method, in which fast propagation is ensured by one phase and the desired reliability level is ensured by the other. Through extensive simulations we showed that, compared with state-of-the-art protocols, OppCast achieved higher WM packet reception ratio, higher dissemination rate using smaller amount of transmissions. Our results reveal the intrinsic tradeoff and intricate interplay between WM reception reliability, dissemination rate and overhead and we believe it will provide valuable guidelines to protocol design in VANETs and mobile WMNs.

11.2   Future research directions

The frameworks proposed in Chapters 4 and 5, which compute the throughput bound and capacity of OR need to find all the feasible concurrent transmitter sets, which is a NP-complete problem. How to find, efficiently, a good subset of all the CTSs to approach the optimal solution within a controllable gap could be an interesting topic. Some heuristic algorithms similar to that in Tang *et al.* (2007), or a column generation technique (Zhang *et al.* 2005) may be adopted to serve this purpose.

In multiradio multichannel networks, it is interesting to investigate distributed algorithms to solve the channel assignment and candidate selection issues. As we discussed in Chapter 5, there is a tradeoff between spatial diversity and multiplexing when applying OR in multiradio multichannel networks. Designing an efficient algorithm that leads to near-optimal channel assignment and candidate selection is desirable.

Routing metrics with various performance objectives, such as maximizing throughput, minimizing delay, and maximizing energy efficiency, can be studied and the tradeoff between conflicting goals can be analyzed and considered for OR.

Another potential direction for research is to investigate further the error of link quality (PRR) estimation and its impact on the OR performance in different types of networks, and design protocols accordingly that are robust to estimation error. We plan to break down this task into three subtasks.

First, geographical routing has been studied extensively in the literature in networks where location information is available to the nodes, which is true in many applications of multihop wireless networks. Geographic opportunistic routing has been proposed as an efficient routing scheme in such networks. In GOR, the Euclidean distance between nodes is known and can be used as the cost function in routing. We can start with GOR in wireless sensor networks where the distance is a fixed value and not affected by the link estimation error. We can design cost functions that are less affected by PRRs but represent the spatial diversity along the path.

Second, a local OR decision depends on OETT and EMT in Chapter 4. Opportunistic ETT only depends on the local PRRs whereas EMT depends on remote PRRs through D_is. Thus the impact of link estimation error is propagated through the network by D_is. For very dynamic networks, such as mobile ad hoc networks and vehicular networks, the link condition may change very fast, which may diminish the benefit from our optimization based on link error estimation. In our second step of understanding the impact of estimation error on routing performance, we will study D_is that are less sensitive to such changes. One option is to use the cost based on the traditional routing, which is less affected by the link estimation error than OR. The goal is to mitigate such impact from remote nodes and to focus on the impact on local estimations. Another option is to develop on-demand protocols (Johnson *et al.* 2001; Perkins and Royer 2001). Like multipath on-demand routing (Marina and Das 2001; Ye *et al.* 2003; Zeng *et al.* 2005), multiple replies can be enabled from the destination and nodes learn its local spacial diversity opportunity and report it in the reply messages. Spacial diversity along the paths can then be taken into consideration in routing decisions.

Third, after understanding the impact on each local decision, we will then extend the investigation to the whole paths. This study will be in a relatively stable setting such as sensor networks and mesh networks. We may adopt ideas similar to fisheye state routing (FSR) (Pei *et al.* 2000), which allows multilevel routing information exchange depending on the distance to the destination. The focus is to control the routing overhead while trying to take advantage of OR and path diversity on a larger scale. This study will help us to gain a deeper understanding of the OR and the capability of gaining performance benefits in the face of inaccurate link-quality estimation. We believe that the theoretical results and insights from this research will be valuable to the research community and crucial to the design of practical and efficient OR protocols that approach optimal performance.

Other than performance, security is another major concern in multihop wireless networks. Opportunistic routing, due to its indeterministic nature, should be more robust to many attacks aiming to disrupt routing and data forwarding functions. Generally, OR is more resilient than single-path traditional routing under packet-dropping attacks. However, in some cases, single-path routing may be better – for example, where the path length is short. It is worthwhile studying in depth the theoretical foundations of opportunistic routing under different networking configurations and attacking models and quantifying the damage or performance degradation that would result. It will be valuable to design secure OR protocols and integrate them into existing security framework to provide more robust and more secure information delivery service.

With recent advanced wireless communication technology, such as MIMO (multiple-input multiple-output), OFDM (orthogonal frequency-division multiplexing) and cognitive radio, it will be interesting to study how opportunistic routing can be integrated with these physical layer technologies to optimize network performance.

References

Johnson DB, Maltz DA and Broch J 2001 Dsr: The dynamic source routing protocol for multi-hop wireless ad hoc networks In *Ad Hoc Networking* (ed. Perkins CE) Addison-Wesley chapter 5, pp. 139–172.

Marina MK and Das SR 2001 On-demand multipath distance vector routing in ad hoc networks *9th International Conference on Network Protocols*, Riverside, CA.

Pei G, Gerla M and Chen TW 2000 Fisheye state routing: A routing scheme for ad hocwireless networks *International Conference on Communications (ICC)*.

Perkins CE and Royer EM 2001 The ad hoc on-demand distance-vector protocol In *Ad Hoc Networking* (ed. Perkins CE) Addison-Wesley, New York, NY, chapter 6, pp. 173–219.

Tang J, Xue G and Zhang W 2007 Cross-layer design for end-to-end throughput and fairness enhancement in multi-channel wireless mesh networks. *IEEE Transactions on Wireless Communications* **6**(10), 3482–3486.

Ye Z, Krishnamurthy SV and Tripathi SK 2003 A framework for reliable routing in mobile ad hoc networks *IEEE INFOCOM*, San Francisco CA.

Zeng K, Ren K and Lou W 2005 Geographic on-demand disjoint multipath routing in wireless ad hoc networks *IEEE MILCOM*, Atlantic City, NJ.

Zhang J, Wu H, Zhang Q and Li B 2005 Joint routing and scheduling in multi-radio multi-channel multi-hop wireless networks *IEEE Broadnets*.

Index

Access Point, xiv
Accumulated packet reception
probability, 242
acknowledgement, 112, 115, 135, 198,
233, 246
area of interest, 138

broadcast acknowledgement, 236, 240,
271, 279
broadcast backoff timer, 246

Carrier Sensing Multiple Access with
Collision Avoidance, 122
clear-to-forward, 117, 224
clear-to-send, 112, 117, 240
concurrent transmission set, 18, 61–3,
66, 68, 93, 277
connected dominant set, xv, 232, 234
conservative CTS, 63, 64, 67
constant bit rate, 124, 203, 218
contention-based dissemination, 260
contention-based forwarding, xiv, 8,
112, 223, 235

DDB, 234, 235
dedicated short range communications,
137
Delay/Disruption Tolerant Network, 13
Distributed Inter Frame Space, 118

DSDV, 4
Dynamic Source Routing, 6

emergency message dissemination for
vehicular environments, 235
energy detection, 148, 150, 180
energy-efficient geographic
opportunistic routing, 17, 39
error correction coding, 146
Expected Advancement Rate, 75
expected medium time, 17, 62, 71, 73,
74, 87, 277
expected one-hop throughput,
76, 123
Expected Opportunistic Transmission
Count, 33
expected packet advancement, xiv, 27,
28, 57, 75, 198
expected transmission count, 72, 213
expected transmission time, 14, 71
exponentially weighted moving
average, 202

fast-forward-dissemination, 239
fast slotted acknowledgment, xv, 18,
132, 277
fisheye state routing, 280
forwarding priority scheduling
problem, 98

Multihop Wireless Networks: Opportunistic Routing, First Edition.
Kai Zeng, Wenjing Lou and Ming Li.
© 2011 John Wiley & Sons, Ltd. Published 2011 by John Wiley & Sons, Ltd.

forwarding range, 236, 240, 242, 252, 260

geographic opportunistic routing, xiv, 7, 17, 39, 59, 71, 276, 278
geographic routing, 14, 17, 19, 28, 40, 41, 49, 51–4, 59, 75, 115, 193, 194, 202, 205, 214, 276, 278
GeRaF, xiv, 7, 39, 112, 223–4
Global Positioning System, 21, 140
greedy CTS, 64, 67, 68

Intelligent Transportation Systems, 3
interested region, 237, 242, 249

LDPC, 10
least cost opportunistic routing, xiv, 33, 34, 38
linear programming, xiv, 17, 61, 65, 67, 85, 89, 91, 96, 99, 277
link quality measurement, xiv, xv, 1, 9, 19, 20, 193, 202, 203, 213, 214, 219, 224, 225, 227, 275, 278
live multimedia streaming, 133, 137, 138, 172, 190, 278
LRR, 179–81, 188

makeup-for-reliability, 239
Management Information Base, 216
mobile ad hoc networks, xiv, 1, 8, 13
mobile content distribution, 18, 133, 136, 137, 188, 231, 278
multiple-input multiple-output, 281

NCDD, 143, 182, 184

OB-VAN, 235
OFDM, 281
on board units, 237
opportunistic broadcast, xv, 20, 233, 240, 271, 279
opportunistic broadcast coordination function, 233, 240
Opportunistic Effective One-hop Throughput, 19, 199, 278

opportunistic ETT, 71
opportunistic forwarding, xiv, 14, 18, 20, 40, 64, 89, 91, 93, 96, 99–101, 107, 108, 198, 223, 225, 232, 235, 239, 251, 260, 265, 277
Opportunistic routing, xiii, 10, 39, 49, 51, 206
Optimized Link State Routing, 6

Packet Delivery Ratio, 48
packet level network coding, 137, 141, 188, 278
packet reception probability, 235, 238–43
Preamble detection, 122

quality of service, 137–9, 142

random linear network coding, 12, 135
ready-to-forward, 224
ready-to-send, 112
reception to transmission energy ratio, 17, 52, 59, 276
relay candidate region, 239, 244
road side units, 137, 138
route error, 5
route reply, 5
route request, 5, 7

self-allocation vector, 246
short inter frame space, 195, 246
signal to interference and noise ratio, 148, 249
Signal to Noise Ratio, 48
slotted acknowledgment, xv, 18, 111, 118, 132, 277
symbol level network coding, 18, 133, 188, 278

Topology Control, 6
traditional routing, xiv, xv, 1, 5, 7–9, 14–16, 33, 39, 61, 65–8, 71, 76, 89–90, 115, 121, 134, 136, 205,

213, 215, 217–19, 226–8, 232,
    280–1
Transport Control Protocol, 140

unit disk graph, 232

vehicular ad hoc networks, 1, 133, 271

warning messages, 231, 271, 279
Wireless mesh networks, 2
Wireless sensor networks, 2, 39
WLAN, 2, 11, 122